Statistical and T

Statistical and Thermal Physics
An Introduction

Second Edition

Michael J.R. Hoch
School of Physics, University of the Witwatersrand
Johannesburg, South Africa

CRC Press
Taylor & Francis Group
Boca Raton London New York

CRC Press is an imprint of the
Taylor & Francis Group, an **informa** business

Second edition published 2021
by CRC Press
6000 Broken Sound Parkway NW, Suite 300, Boca Raton, FL 33487-2742

and by CRC Press
2 Park Square, Milton Park, Abingdon, Oxon, OX14 4RN

© 2021 Taylor & Francis Group, LLC

First edition published by CRC Press 2011

CRC Press is an imprint of Taylor & Francis Group, LLC

Library of Congress Cataloging-in-Publication Data
Names: Hoch, M. J. R. (Michael J. R.), 1936- author.
Title: Statistical and thermal physics : an introduction / Michael J.R.
Hoch, School of Physics, University of the Witwatersrand, Johannesburg, South Africa.
Description: Second edition. | Boca Raton : CRC Press, 2021. | Includes
bibliographical references and index.
Identifiers: LCCN 2021005852 | ISBN 9780367461348 (paperback) |
ISBN 9780367464103 (hardback) | ISBN 9781003028604 (ebook)
Subjects: LCSH: Matter—Properties—Mathematical models. | Statistical physics. | Thermodynamics.
Classification: LCC QC171.2 .H63 2021 | DDC 530.13—dc23
LC record available at https://lccn.loc.gov/2021005852

ISBN: 978-0-367-46410-3 (hbk)
ISBN: 978-0-367-46134-8 (pbk)
ISBN: 978-1-003-02860-4 (ebk)

Typeset in Times
by codeMantra

To my wife Renée

Contents

PART I Classical Thermal Physics: The Microcanonical Ensemble

SECTION IA Introduction to Classical Thermal Physics Concepts: The First and Second Laws of Thermodynamics

SECTION IB   Microstates and the Statistical Interpretation of Entropy

SECTION IC Applications of Thermodynamics to Gases and Condensed Matter, Phase Transitions, and Critical Phenomena

PART II Quantum Statistical Physics and Thermal Physics Applications

SECTION IIA The Canonical and Grand Canonical Ensembles and Distributions

SECTION IIB   Quantum Distribution Functions, Fermi–Dirac and Bose–Einstein Statistics, Photons, and Phonons

SECTION IIC   The Classical Ideal Gas, Maxwell–Boltzmann Statistics, Nonideal Systems

SECTION IID   The Density Matrix, Reactions and Related Processes, and Introduction to Irreversible Thermodynamics

Preface

Thermal and statistical physics concepts and relationships are of fundamental importance in the description of systems that consist of macroscopically large numbers of particles. This book provides an introduction to the subject at the advanced undergraduate level for students interested in careers in basic or applied physics. The subject can be developed in different ways by taking either macroscopic classical thermodynamics or microscopic statistical physics as topics for initial detailed study. Considerable insight into the fundamental concepts, in particular temperature and entropy, can be gained in a combined approach in which the macroscopic and microscopic descriptions are developed in tandem. This is the approach adopted here.

This book consists of two major parts, within each of which there are several sections, as detailed below. A flowchart that shows the chapter sequence and the interconnection of major topics covered is given at the end of this introduction. Part I is divided into three sections, each made up of three chapters. The basics of equilibrium thermodynamics and the first and second laws are covered in Section IA. These three chapters introduce the reader to the concepts of temperature, internal energy, and entropy. Two systems, ideal gases and ideal noninteracting localized spins, are extensively used as models in developing the subject.

Chapters 4–6 in Section IB provide a complementary microscopic statistical approach to the macroscopic approach of Section IA. Considerable insight into both the entropy and temperature concepts is gained, and the Boltzmann expression for the entropy is given in terms of the number of accessible microstates in the fixed energy, microcanonical ensemble approach. This relationship is of central importance in the development of the subject. Explicit expressions for the entropy are obtained for both a monatomic ideal gas and an ideal spin system. The entropy expressions lead to results for the other macroscopic properties of these two systems. It is made clear that for ideal gases in the high-temperature, low-density limit, quantum effects can be neglected. The need to allow for the indistinguishable nature of identical particles in nonlocalized systems is emphasized. Chapter 6 introduces the third law of thermodynamics through the use of expressions for the entropy and the temperature parameter obtained in Chapter 5. The four laws of thermodynamics are given in a concise form. Following Section IB, the reader who is interested in pursuing statistical physics can proceed directly to Chapter 10 in the second half of this book.

In the final section of Part I, Section IC, the Helmholtz and Gibbs thermodynamic potentials are introduced, and the Gibbs potential is used to discuss equilibrium conditions in multicomponent systems. The power of thermodynamics is demonstrated in the description of processes for both gases in Chapter 7 and condensed matter in Chapter 8. The Maxwell relations are obtained and used in a number of situations that involve adiabatic and isothermal processes. Chapter 9 concludes this section with a discussion of phase transitions and critical phenomena.

Chapter 10 in Section IIA gives a brief introduction to probability theory, mean values, and three statistical ensembles that are used in statistical physics. The partition function is defined in terms of a sum over energy states, and the ideal localized spin system is used to illustrate the canonical ensemble approach. The grand canonical ensemble and the grand sum are discussed in Chapter 11. It is shown that for large systems in contact with thermal and particle reservoirs, the fluctuations in the system's energy and particle number are extremely small. It follows that the different ensembles are equivalent.

Section IIB is concerned with quantum statistics. Chapter 12 reviews the quantum mechanical description of systems of identical particles and distinguishes fermions and bosons. Chapters 13 and 14, respectively, deal with the ideal Fermi and Bose gases. Expressions for the heat capacity and magnetic susceptibility are obtained for the Fermi gas, while the Bose–Einstein condensation at low temperatures is discussed for the Bose gas. These chapters are illustrated with applications to a variety of systems. For example, the Fermi–Dirac statistics is used to treat white dwarf stars and neutron stars. The radiation laws and the heat capacity of solids are discussed in Chapter 15, which deals with photons and phonons. The cosmic microwave background radiation is considered as an illustration of the Planck distribution.

In Section IIC, Chapter 16 returns to the ideal gas treated in the classical limit of the quantum distributions, which automatically allows for the indistinguishable nature of identical nonlocalized particles. The internal energy of molecules is included in the partition function for the classical gas. The equipartition of energy theorem for classical systems is discussed in some detail. Nonideal systems are dealt with in Chapter 17 in terms of the cluster model for gases and the mean field approximation for spins. The Ising model for interacting spins is introduced, and the one-dimensional solution of the Ising model is given for the zero applied field case. An introduction to Fermi liquid theory is followed by a discussion of the properties of liquid helium-3 at low temperatures. This chapter concludes with a phenomenological treatment of the Bose liquids and the properties of liquid helium-4.

Section IID deals with special topics that include the density matrix, chemical reactions, and an introduction to irreversible thermodynamics. Chapter 18 introduces the density matrix formulation with applications to spin systems and makes a connection to the classical phase space approach. Topics covered in Chapter 19 are the law of mass action, adsorption on surfaces, and carrier concentrations in semiconductors. Chapter 20 deals with irreversible processes in systems not far from equilibrium, such as thermo-osmosis and thermoelectric effects.

For a one-semester course, the important sections that should be covered are Sections IA, IB, IIA, and IIB. If students have had prior exposure to elementary thermodynamics, much of Section IA may be treated as a self-study topic. Problems given at the end of each chapter provide opportunities for students to test and develop their knowledge of the subject. Depending on the nature of the course and student interest, material from Sections IC, IIC, and IID can be added.

In preparing the second edition of this book, a number of new topics have been included to complement and extend the material covered in the first edition. New chapter sections are *Energy Sources* (Chapter 1), *Statistical Ensembles* (Chapter 5), *Thermodynamics of Black Holes* (Chapter 6), *Boltzmann and Gibbs Entropy Equations, Shannon Entropy and Information Theory* (Chapter 10), and *Mean Field Theory of Magnetism* and *Antiferromagnetic Heisenberg Chains* (Chapter 17). A number of minor corrections and additions have been made to the first addition text.

A flowchart for statistical and thermal physics topics covered in Chapters 1–16 is provided below:

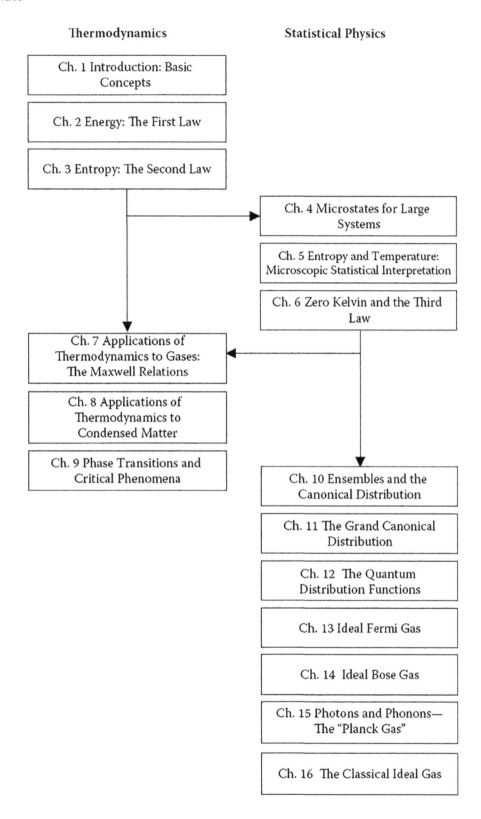

Thermodynamics   Statistical Physics

Ch. 1 Introduction: Basic Concepts

Ch. 2 Energy: The First Law

Ch. 3 Entropy: The Second Law

Ch. 4 Microstates for Large Systems

Ch. 5 Entropy and Temperature: Microscopic Statistical Interpretation

Ch. 6 Zero Kelvin and the Third Law

Ch. 7 Applications of Thermodynamics to Gases: The Maxwell Relations

Ch. 8 Applications of Thermodynamics to Condensed Matter

Ch. 9 Phase Transitions and Critical Phenomena

Ch. 10 Ensembles and the Canonical Distribution

Ch. 11 The Grand Canonical Distribution

Ch. 12 The Quantum Distribution Functions

Ch. 13 Ideal Fermi Gas

Ch. 14 Ideal Bose Gas

Ch. 15 Photons and Phonons— The "Planck Gas"

Ch. 16 The Classical Ideal Gas

Acknowledgments

My thanks go to numerous colleagues at the University of the Witwatersrand, Johannesburg, and at Florida State University, Tallahassee, together with the National High Magnetic Field Laboratory, for helpful discussions on the concepts described in this book. In teaching the material, I have learnt a great deal from the interactions I have had with many students. Their comments and responses to questions have often been enlightening.

The editors at Taylor and Francis/CRC Press have provided much guidance and encouragement which is greatly appreciated.

Finally, I wish to thank my family for their continuing support. In particular, I owe a great deal to my wife Renée who prepared most of the figures and maintained an ongoing interest in the book project.

Author

Michael J.R. Hoch completed his Honours and Master of Science degrees in physics at the University of Natal (now KwaZulu-Natal) in his home town Pietermaritzburg, South Africa. He was awarded a Commonwealth Scholarship to study at the University of St Andrews, Scotland, where he completed his PhD degree in 1963. He returned to a senior lectureship at the University of Natal and established a small research group of graduate students. Following a sabbatical year at Bell Laboratories in Murray Hill, NJ, he accepted a position at the University of the Witwatersrand, Johannesburg in 1970. After becoming a professor of physics, he served as head of department from 1982 until 1994. Together with his colleagues, he established the Condensed Matter Physics Research Unit and became its first director. The unit was upgraded in 2000 and became the Institute for Materials Physics. In 1991, he and Professor Richard Lemmer organized an international summer school on Low-Temperature Physics, with the edited proceedings published in the Springer Lecture Notes in the Physics series. He served as Dean of the Faculty of Science in 1997–1998.

Following his retirement at the end of 2001, he accepted the position of Visiting Scientist at the National High Magnetic Field Laboratory at Florida State University in Tallahassee, FL, where he carried out research and, for several years, also served as an adjunct professor, giving undergraduate lectures in the Department of Physics. He returned to South Africa in 2017 and is an emeritus professor in the School of Physics at the University of the Witwatersrand. He is an elected fellow of the Royal Society of South Africa and a member of the Academy of Science of South Africa.

PHYSICAL CONSTANTS

Avogadro's number, N_A	6.023×10^{23} mol^{-1}
Bohr magneton, μ_B	9.27×10^{-24} J T^{-1}
Boltzmann's constant, k_B	1.38×10^{-23} J K^{-1}
Electron charge, e	1.60×10^{-19} C
Electron mass, m_e	9.11×10^{-31} kg
Gas constant, R	8.314 J mol^{-1} K^{-1}
$\hbar$	1.055×10^{-34} J s
Nuclear magneton, μ_n	5.05×10^{-27} J T^{-1}
Permeability constant, μ_o	$4\pi \times 10^{-7}$ Hm^{-1}
Permittivity constant, ε_o	8.554×10^{-12} F m^{-1}
Planck's constant, h	6.624×10^{-34} J s
Proton mass, m_p	1.67×10^{-27} kg

CONVERSION FACTORS

1 atmosphere (atm)	1.01×10^5 Pa
1 electron volt (eV)	1.60×10^{-19} J
1 Joule (J)	107 erg
1 liter (L)	10^{-3} m^3
1 mass unit (u)	1.66×10^{-27} kg

Part I

Classical Thermal Physics

The Microcanonical Ensemble

Section IA

Introduction to Classical
Thermal Physics Concepts
The First and Second Laws of Thermodynamics

1 Introduction
Basic Concepts

1.1 STATISTICAL AND THERMAL PHYSICS

The subject of statistical and thermal physics is concerned with the description of macroscopic systems made up of large numbers of particles of the order of Avogadro's number $N_A = 6.02 \times 10^{23} \text{mol}^{-1}$. The particles may be atoms or molecules in gases, liquids, and solids or systems of subatomic particles such as electrons in metals and neutrons in neutron stars. A rich variety of phenomena are exhibited by many particle systems of this sort. The concepts and relationships that are established in thermal physics provide the basis for discussion of the properties of these systems and the processes in which they are involved. Applications cover a wide range of situations from basic science, in many important fields that include condensed matter physics, astrophysics, and physical chemistry, to practical devices in energy technology.

The origins of modern thermal physics may be traced to the analysis of heat engines in the nineteenth century. Following this early work, a number of researchers contributed to the development of the subject of thermodynamics with its famous laws. By the end of the nineteenth century, thermodynamics, classical mechanics, and electrodynamics provided the foundation for all of classical physics. Today, thermodynamics is a well-developed subject, with modern research focused on special topics such as nonequilibrium thermodynamics. Application of the methods of thermodynamics to complex systems far from equilibrium, which include living organisms, presents a major challenge.

The microscopic classical statistical description of systems of large numbers of particles began its development in the late nineteenth century, particularly through the work of Ludwig Boltzmann. This approach was transformed by the development of quantum mechanics in the 1920s, which then led to quantum statistics that is of fundamental importance in a great deal of modern research on bulk matter. Statistical techniques are used to obtain average values for properties exhibited by macroscopic systems. The microscopic approach on the basis of classical or quantum mechanics together with statistical results has given rise to the subject known as statistical mechanics or statistical physics. Bridge relationships between statistical physics and thermodynamics have been established and provide a unified subject.

Under conditions of high temperature and low density, we shall find that it does not matter whether classical or quantum mechanical descriptions are used for a system of particles. At high densities and low temperatures, this is no longer true because of the overlap of the particles' wave functions, and quantum mechanics must be used, giving rise to quantum statistics. Under high-density, low-temperature conditions, the properties of a system depend in a crucial way on whether the particles that make up the system are fermions or bosons. Many fascinating phenomena occur in condensed matter as the temperature is lowered. Examples are ferromagnetism, superconductivity, and superfluidity. These important new properties appear fairly abruptly at phase transitions. Progress in the microscopic understanding and description of the behavior of these systems involves quantum mechanics and statistical physics ideas. Applications of quantum statistics are not confined to terrestrial systems and include astrophysical phenomena such as the microwave background radiation from the Big Bang and the mass–radius relationships for white dwarf stars and neutron stars. An important concept in thermal physics is that of entropy, which, as we shall see, increases with time as systems become more disordered. The increase of the entropy of the universe with the passage of time provides what is termed time's arrow. An interesting and unanswered question arises as to why the entropy of the universe was so low at the beginning of time.

There are a number of ways in which the subject of thermal physics may be approached. In this book, the elements of classical thermodynamics are presented in the first three chapters that comprise Section IA. The microscopic statistical approach is introduced in the three chapters of Section IB, which complement the thermodynamics in Section IA and provide additional insight into fundamental concepts, specifically entropy and temperature. Both classical and quantum mechanical descriptions are introduced in discussing the microstates of large systems, with the quantum state description preferred in the development of the subject that is presented in this book. This approach involves the application of the microcanonical ensemble methods to two model systems, the ideal gas system and the ideal spin system, for both of which the quantum states are readily specified with use of elementary quantum mechanics. The laws of thermodynamics are stated in a compact form, and their significance is heightened by the interweaving of macroscopic and microscopic approaches. Expressions for the entropy and chemical potential of the ideal gas are expressed in terms of the ratio of the quantum volume V_Q to the atomic volume V_A. V_Q is taken to be the cube of the thermal de Broglie wavelength, whereas V_A is the mean volume per particle in the gas. In the final three chapters in the first half of the book (Section IC), the thermodynamic approach is applied to the description of the properties of gases, and condensed matter.

In the second half of the book, emphasis is placed, initially, on statistical physics results and the canonical and grand canonical distributions presented in Section IIA. Following the introduction of quantum statistics ideas in Section IIB, the Fermi–Dirac and Bose–Einstein quantum distribution functions are derived. These functions are used in the discussion of the properties of ideal Fermi and Bose gases. Photon and phonon systems are then treated in terms of the Planck distribution. In Section IIC, canonical ensemble results are applied, first, to the ideal gas in the classical limit of the quantum distributions and then to the nonideal gases and spin systems. The final section (Section IID) of the book contains more advanced topics and includes an introduction to the density matrix and nonequilibrium thermodynamics.

The material in this book is concerned primarily with the description of systems at or close to equilibrium. This implies that, apart from fluctuations, which for macroscopic systems are generally small, the properties do not change with time or change only slightly with small changes in external conditions. Processes will often be considered to consist of a large number of infinitesimal changes, with the system of interest always close to equilibrium. As mentioned above, two special systems are used in the development of the subject: the ideal gas system and the ideal spin system. *Ideal* in this context implies that interactions between the particles are negligibly small and may be ignored. These two systems provide sufficient insight to permit generalization of the methods that are developed to many other systems. The ideal gas system will be viewed as consisting of N nonlocalized particles in a box or container of volume V. Because of their thermal motion, the particles exert a pressure, or force per unit area P, on the walls of the container. The ideal spin system consists of N localized spins, each with associated magnetic moment μ located in a magnetic field B.

In the development of the subject, use is made of concepts such as volume, pressure, work, and energy, which are familiar from classical mechanics. A further concept of fundamental importance is that of temperature, and this is discussed in the next section. Useful relations that involve temperature, such as equations of state and the equipartition of energy theorem, are given in later sections in this chapter. SI units are used throughout the book unless otherwise indicated.

1.2 TEMPERATURE

The concept of temperature has evolved from man's experience of hot and cold conditions with temperature scales devised on the basis of changes in the physical properties of substances that depend on temperature. Practical examples of thermometers for temperature measurement include the following:

- Constant volume gas thermometers, which make use of the pressure of a fixed quantity of gas maintained at a constant volume as an indicator

- Liquid in glass thermometers, which use the volume of a liquid, such as mercury or alcohol, contained in a reservoir attached to a capillary tube with a calibrated scale
- Electrical resistance thermometers, which use the variation of the resistance of a metal, such as platinum, or of a doped semiconductor, such as GaAs, to obtain temperature
- Vapor pressure and paramagnet thermometers for special purposes particularly at low temperatures

Most of the thermometers listed above are secondary thermometers that are calibrated against agreed standards. The constant volume gas thermometer has a more fundamental significance as explained in Section 1.3. For everyday purposes, various empirical scales have been established. Commonly used scales are the Fahrenheit and Celsius scales. For reasons that become clear below, we consider the Celsius scale, which chooses two fixed reference temperatures. The lower reference point is at 0°C, which corresponds to the triple point of water, the point at which water, ice, and water vapor coexist, and the higher reference point at 100°C, which corresponds to the steam point, where water and steam coexist at a pressure of 1 atm. Degrees Celsius are obtained by dividing the range between the triple point and the steam point into 100°. Figure 1.1 shows a schematic drawing of a triple-point cell.

Thermodynamics shows that it is possible to establish an *absolute* temperature scale called the *Kelvin scale* in honor of Lord Kelvin, who introduced it and first appreciated its significance. The absolute scale does not depend on the properties of a particular substance. Absolute zero on the Kelvin scale, designated as 0 K, corresponds to −273.16°C. For convenience, 1 K is chosen to correspond to 1°C. This gives $T(\text{K}) = t(°\text{C}) + 273.16$. We can gain insight into why the absolute zero of temperature is of fundamental importance by considering the equation of state for an ideal gas.

1.3 IDEAL GAS EQUATION OF STATE

An equation of state establishes a relationship among thermodynamic variables. For an ideal gas, the variables chosen are the pressure P, the volume V, and the absolute temperature T. Experiments carried out on real gases, such as helium, under conditions of low density have shown that the following equation describes the behavior of many gases:

$$PV = nRT, \tag{1.1}$$

where n is the number of moles of gas and R is a constant called the *gas constant* with a value of 8.314 J mol^{-1} K^{-1}. As mentioned above, the constant volume gas thermometer involves the measurement of the pressure of a constant volume of gas as a function of temperature. Figure 1.2 gives a sketch of a constant volume gas thermometer with a representative P versus T plot shown in Figure 1.3.

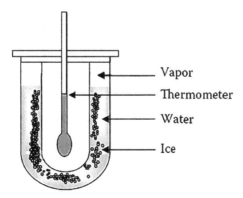

FIGURE 1.1   Schematic depiction of a triple-point cell in which water, ice, and water vapor coexist. The cell is used to fix 0°C on the Celsius scale.

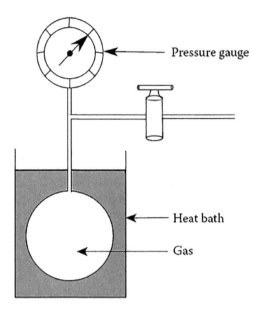

FIGURE 1.2   Sketch of a constant volume gas thermometer in which the pressure of a fixed volume of gas, held at various fixed temperatures by use of heat baths, is measured on a pressure gauge.

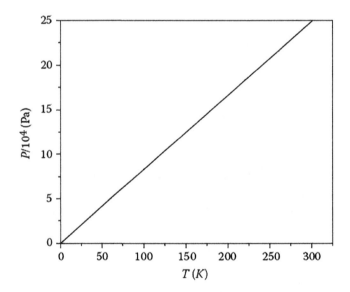

FIGURE 1.3   Calculated pressure variation with temperature for a constant mass of an ideal gas (10^{-1} mol) kept at constant volume (10^{-3} m³). The low-temperature portion of the graph, where gases in general liquefy as a result of intermolecular forces, is in practice obtained by extrapolation from higher temperatures.

From Equation 1.1, it follows that the temperature $T = 0$ K corresponds to a pressure $P = 0$ Pa. Zero pressure implies that the particles of the gas have zero kinetic energy at 0 K and do not exchange momentum with the walls of the container. We can therefore view the absolute zero of temperature as the temperature at which the energy of particles in the system is effectively zero. This is of fundamental significance.

As the temperature is lowered, gases normally liquefy, and most solidify at temperatures much higher than 0 K. This is because real gases have interactions between particles, which lead to departures from ideal gas behavior. Extrapolation from the high-temperature, low-density region, where gases obey the ideal gas equation of state, shows what would happen at much lower temperatures if the gas were to remain ideal. The ideal gas equation of state, expressed in Equation 1.1, is extremely useful in considering processes in which gases are involved. P, V, and T are called *state variables*, and because they are related by the equation of state, the specification of any two of the variables immediately fixes the value of the third variable. Examples of applications of the ideal gas law, as it is also called, are given in later chapters. The ideal gas equation provides a fairly good description of the behavior of many gases over a range of conditions. Under conditions of high density, however, the description may not be adequate, and empirical equations of state that work better under these conditions have been developed. Two of these equations are briefly discussed in the next section.

1.4 EQUATIONS OF STATE FOR REAL GASES

An important empirical equation of state that provides a fairly good description of the properties of real gases at high densities is the van der Waals equation:

$$\left(P+\frac{a}{V^2}\right)(V-b)=nRT. \tag{1.2}$$

Equation 1.2 is similar to the ideal gas equation in Equation 1.1 but with a pressure correction term a/V^2, which increases in importance with a decrease in volume, and a volume correction term b. The van der Waals constants a and b are determined experimentally for a given gas. The pressure correction term allows for interparticle interactions, and the volume correction term allows for the finite volume occupied by the particles themselves.

Another widely used empirical equation of state is the virial equation:

$$PV = nRT\left[1+\left(\frac{N}{V}\right)B(T)+\left(\frac{N}{V}\right)^2 C(T)+\cdots\right], \tag{1.3}$$

$B(T)$ and $C(T)$ are called the second and third virial coefficients, respectively, and are generally temperature-dependent. The correction terms become important as the volume decreases and the particle density N/V increases. Virial coefficients have been measured for a large number of gases and are available in tables. Further discussion of these two empirical real gas equations of state is given in later chapters.

1.5 EQUATION OF STATE FOR A PARAMAGNET

An ideal paramagnet consists of N particles, each of which possesses a spin and an associated magnetic moment μ proportional to the spin, with negligible interactions between spins. Real paramagnetic systems approximate ideal systems only under certain conditions, such as high temperature, and in magnetic fields that are not too large. A more detailed discussion of these conditions is given later in this book.

For an ideal paramagnet, experiment and theory show that the magnetic moment per unit volume, or magnetization M, is given by

$$M = \frac{CH}{T}, \tag{1.4}$$

with H an external applied magnetic field and C a constant called the *Curie constant*. In the SI system of units applied to ideal paramagnetic systems, we shall often, to a good approximation, take the field that the spins see as $H = B/\mu_0$, with $\mu_0 = 4\pi \times 10^{-7}$ Hm^{-1} the permeability of free space and B the magnetic induction in tesla. *M*, *H*, and *T* are state variables analogous to *P*, *V*, and *T*. Any two fix the value of the third variable. Like the ideal gas equation of state, the ideal paramagnet equation, called *Curie's law*, is very useful in calculations related to processes that involve changes in the state variables. Note that for $T \rightarrow 0$ K, Equation 1.4 predicts that *M* will diverge. This unphysical prediction shows that the equation breaks down at low temperatures, where the magnetization saturates after it reaches a maximum value with all spins aligned parallel to *H*. In many magnetic systems, the spins interact to some extent and order below a temperature called the *Curie point*. Examples are metals such as iron and nickel. The Curie–Weiss equation takes interactions into account and has the form

$$M = \frac{CH}{T - T_c}. \tag{1.5}$$

T_c is called the *Weiss constant* and has the dimensions of temperature. Equation 1.5 provides a satisfactory description of the magnetic properties of magnetic materials for $T > T_c$. For a given system at a particular temperature $T \simeq T_c$, spontaneous order among spins sets in and the system undergoes a phase transition. Values of T_c have been measured for many magnetic systems. Phase transitions and critical phenomena associated with these transitions are discussed in Chapter 9.

1.6 KINETIC THEORY OF GASES AND THE EQUIPARTITION OF ENERGY THEOREM

In Chapter 2, we deal with work and energy for thermodynamic systems, and this will lead to the first law of thermodynamics. It is helpful to have expressions for the total energy of a system in terms of thermodynamic variables. For ideal gases, kinetic theory provides an important result known as the equipartition of energy theorem, which we now consider along with other kinetic theory results. The kinetic theory of gases, which makes use of classical mechanics, is related to certain topics in statistical physics but is less general in scope and approach.

Although thermodynamics concerns itself with the macroscopic description of systems of large numbers of particles, kinetic theory considers the microscopic nature of fluid systems. In particular, for our purposes, kinetic theory provides a classical microscopic description of the properties of ideal gases in terms of the kinetic energy of the particles in the system. Particles in a gaseous system are in a constant motion and undergo collisions with each other and with the walls of the container. For a gas at a particular density and temperature, the collision processes may be characterized by a collision time τ, which is the mean time between collisions. As we shall see later in this section, τ is typically very short compared with the time scale on which measurements of any properties of the system are made. The rapid exchange of energy and momentum that the particles undergo makes it meaningful to consider average properties of the particles, such as the mean energy $\langle \varepsilon \rangle$ defined below. If there are *N* particles in the gas, the total energy is simply $E = N \langle \varepsilon \rangle$.

Classically, the mean kinetic energy may be written in terms of the mean square speed $\langle v^2 \rangle$ of the particles and each particle's mass *m* as $\langle \varepsilon \rangle = \frac{1}{2} m \langle v^2 \rangle$, where the mean square speed is given by

$$\langle v^2 \rangle = \int_0^\infty v^2 P(v) \mathrm{d}v. \tag{1.6}$$

Expressions for mean values are given in Chapter 10. We assume that the potential energy is negligibly small, and the total energy is simply the kinetic energy. The speed distribution $P(v) \, \mathrm{d}v$ in Equation 1.6 has the Maxwellian form:

$$P(v) \, \mathrm{d}v = 4\pi \left(\frac{m}{2\pi k_B T} \right)^{3/2} v^2 e^{-mv^2 / 2k_B T} \mathrm{d}v. \tag{1.7}$$

The Maxwellian speed distribution is derived in Chapter 16. In Equation 1.7, k_B is Boltzmann's constant, related to the gas constant R by means of Avogadro's number N_A and given by $k_B = R/N_A = 1.381 \times 10^{-23}\,\text{J K}^{-1}$. Elegant experiments using molecular beams have verified the speed distribution. Figure 1.4 shows the form of the Maxwellian distribution for helium gas at three temperatures. The distribution is not symmetrical about the value of the most probable speed at a particular temperature, which shows that some atoms or molecules in a given system have speeds much higher than the average value. For symmetry reasons, the average velocity is zero as discussed in Chapter 16.

Using Equation 1.7 in Equation 1.6 gives

$$\langle v^2 \rangle = 4\pi \left(\frac{m}{2\pi k_B T} \right)^{3/2} \int_0^\infty v^4 e^{-mv^2/2k_B T}\ \mathrm{d}v. \tag{1.8}$$

The integral in Equation 1.8 is of the form

$$I = \int_0^\infty e^{-\alpha x^2}\ \mathrm{d}x = \frac{3}{8}\sqrt{\pi}\alpha^{-5/2}. \tag{1.9}$$

Discussion of definite integrals of this type is given in Appendix A. The use of Equation 1.9 to evaluate Equation 1.8 results in $\langle v^2 \rangle = 3k_B T/m$, which may be rewritten as

$$\frac{1}{2}m\langle v^2 \rangle = \frac{3}{2}k_B T \ \text{ or } \ \langle \varepsilon \rangle = \frac{3}{2}k_B T. \tag{1.10}$$

This is an important and useful result for ideal gases. The mean square speed may be expressed in terms of Cartesian velocity components $\langle v^2 \rangle = \langle v_x^2 \rangle + \langle v_y^2 \rangle + \langle v_z^2 \rangle$. Because the choice of axes is arbitrary, symmetry requires $\langle v_x^2 \rangle = \langle v_y^2 \rangle = \langle v_z^2 \rangle$, and it follows that

$$\frac{1}{2}m\langle v_x^2 \rangle = \frac{1}{2}m\langle v_y^2 \rangle = \frac{1}{2}m\langle v_z^2 \rangle = \frac{1}{2}k_B T. \tag{1.11}$$

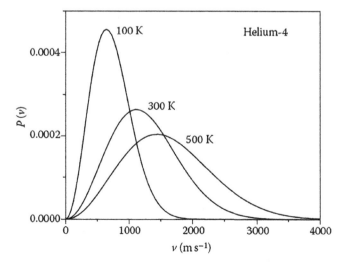

FIGURE 1.4 The Maxwell speed distribution for helium gas at three temperatures. The most probable speed and the RMS speed increase steadily with temperature.

In dealing with systems of large numbers of particles, it is useful to introduce the terminology *degrees of freedom*. Classical mechanics identifies the number of degrees of freedom of a system with the number of independent generalized coordinates that are used to describe the system. This topic is dealt with in some detail in Chapters 15 and 16. For N particles in a gas such as helium, which is monatomic, there are $3N$ degrees of freedom, which correspond to translational motion along three orthogonal directions for each of the N particles. The result in Equation 1.11 is an example of the equipartition of energy theorem, which is stated as follows: For a classical system of particles in equilibrium at a temperature T, each quadratic degree of freedom has mean energy $\frac{1}{2}k_{B}T$.

The term *quadratic degree of freedom* means that a given term in the energy may be written in the form $\varepsilon = \frac{1}{2}kq^{2}$, where q represents a generalized coordinate, such as a particle's velocity component or the displacement of an atom from its equilibrium position in a molecule. The number of degrees of freedom for a molecule in a gas will be designated as f. For a monatomic gas, we find $f = 3$ because there are three velocity components. Diatomic molecules have additional rotational and vibrational degrees of freedom. It turns out that, for diatomic molecules, the vibrational motions cannot be described by classical mechanics at normal laboratory temperatures of around 295 K, and these degrees of freedom only contribute significantly to the mean energy at much higher temperatures. Figure 1.5 shows the two rotational degrees of freedom that are important in our discussion of diatomic molecules.

Note that I_{y} is very small, and motions about the y axis are not described classically at any temperatures of interest. For a diatomic gas, we therefore take $f = 5$, giving for the total energy of the gas

$$E = \frac{5}{2}Nk_{B}T. \tag{1.12}$$

This discussion may be extended to triatomics and still larger molecules. In general,

$$\langle \varepsilon \rangle = f\left(\frac{1}{2}k_{B}T\right) \text{ and } E = Nf\left(\frac{1}{2}k_{B}T\right). \tag{1.13}$$

Equations 1.12 and 1.13 show that the internal energy of a fixed quantity of ideal gas with negligible interactions between the N particles is simply a function of the absolute temperature or $E = E(T)$. It follows that for processes in which the temperature does not change, called *isothermal processes*,

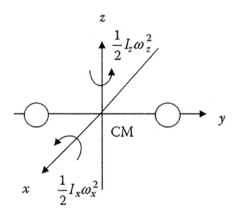

FIGURE 1.5 Rotational motions that contribute to the energy of a diatomic molecule. The moment of inertia of the molecule is much smaller about the y axis than about the x and the z axes. Quantum mechanics shows that we can ignore rotational motion about the y axis.

the internal energy of an ideal gas system remains constant. This is an important result that we make use of in Chapters 2 and 3 in the development of thermodynamics concepts and laws.

The equipartition of energy theorem may be justified with the aid of the ideal gas equation of state. Consider the mean pressure exerted on a surface by gas particles in a container. The force may be obtained as the rate of change of momentum of particles in collisions with a surface. Figure 1.6 represents a collision process in a container. As an idealization, assume that all collisions are elastic; the change in the momentum component p_x for a particle i in striking the smooth wall, as shown, is $\Delta p_{ix} = 2mv_{ix}$.

From Figure 1.6, the number of molecules n that strike area A in a time t is given by the ratio of the sum, for all particles i, of the volumes of cylinders of length $v_{ix}t$ and cross-sectional area A to the total volume or $n = \frac{1}{2}(A/V)\sum_{i=1}^{N} v_{ix}t$. The factor $\frac{1}{2}$ is introduced because on average, half the molecules in each small volume considered will have a velocity component in the $+x$ direction, whereas the other half will have a component in the $-x$ direction. Using Newton's second law, the rate of change of momentum divided by A gives the force per unit area, or the pressure, $P = (1/At)(mAt/V)\sum_i v_{ix}^2$.

Taking $\sum v_{ix}^2 = N\langle v_x^2 \rangle$, with $\langle v_x^2 \rangle = \frac{1}{2}\langle v^2 \rangle$ for symmetry reasons, we obtain $PV = \frac{2}{3}N\left(\frac{1}{2}m\langle v^2 \rangle\right)$. Comparison of this result with the ideal gas equation of state $PV = Nk_BT$ permits the identification

$$\langle \varepsilon \rangle = \frac{1}{2}m\langle v^2 \rangle = \frac{3}{2}k_BT. \tag{1.14}$$

This is the equipartition of energy theorem result for a monatomic gas. The root mean square (RMS) speed is

$$v_{RMS} = \sqrt{\langle v^2 \rangle} = \left(\frac{3k_BT}{m}\right)^{1/2} \tag{1.15}$$

Table 1.1 gives the calculated values of the RMS speed for molecules in a number of gases.

Although the RMS speeds are high, the molecules undergo frequent collisions with each other, and this limits how far they travel in a given direction in a particular time interval. The collision time τ for the molecules may be estimated as follows: We let the effective diameter of a molecule be a and consider two particles to have undergone a collision if their centers are within a distance a of each other. Consider a representative molecule that travels with mean speed $\langle v \rangle$, attach a disk of diameter $2a$ to this molecule, and shrink all the other molecules to points. In a time t, the representative molecule sweeps out a volume $\pi a^2 \langle v \rangle t$. The mean number of collisions per unit time v_c is

TABLE 1.1
Values of the RMS Speed at 295 K for a Number of Gases

Gas	Molecular Mass m (u)	$v_{RMS} = \frac{2713.4}{\sqrt{m}}$ (ms^{-1})
Helium	4.003	1357
Oxygen	15.999	678
Neon	20.179	604

Note: 1 u = 1.66 × 10^{-27}kg.

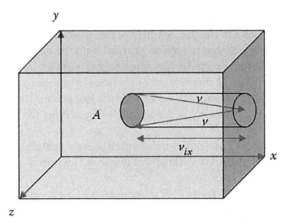

FIGURE 1.6 A representative particle in a container of volume V is shown making an elastic collision with a smooth wall. The distances traversed correspond to a unit time interval. The rate of change of momentum for all particles that strike the end wall gives the net force on the wall.

simply the volume swept out in that time multiplied by the number of molecules per unit volume or $v_c = (\pi a^2 \langle v \rangle)N/V$. The mean time between collisions is $\tau = 1/v_c = V/(N\pi a^2 \langle v \rangle)$. If allowance is made for the motion of the other molecules, a correction term $\dfrac{1}{\sqrt{2}}$ is introduced. With $V/N = k_B T/P$, from the ideal gas equation, we obtain

$$\tau = \frac{k_B T}{\sqrt{2}\left(P\pi a^2 \langle v \rangle\right)}. \qquad (1.16)$$

To a good approximation, we take $\langle v \rangle = v_{RMS}$. For helium gas at 1 atm and 295 K, with $a \sim 10^{-10}$ m, Equation 1.16 predicts $\tau \simeq 0.67$ ns. This collision time corresponds to a collision rate of $1.5 \times 10^9 \text{s}^{-1}$. The numerical estimate confirms the statement made earlier in this section that τ is much shorter than typical measurement times.

Kinetic theory is used to obtain expressions for the transport coefficients of gases, such as the viscosity coefficient η and the thermal conductivity coefficient κ. For completeness, expressions for these coefficients are given here without derivation. Details may be found in kinetic theory texts:

$$\text{Viscosity coefficient}: \eta = \frac{1}{3}\rho\tau\langle v^2 \rangle. \qquad (1.17)$$

$$\text{Themal conductivity coefficient}: \kappa = \frac{1}{3}\rho c\langle v^2 \rangle\tau. \qquad (1.18)$$

The collision time τ appears in both expressions as does ρ, the density of the gas. The specific heat c is discussed in Chapter 2.

1.7 THERMAL ENERGY TRANSFER PROCESSES: HEAT ENERGY

To describe many thermodynamic processes, it is necessary to introduce the concept of thermal energy transfer or heat flow. When two systems, such as two containers of an ideal gas at different temperatures, are brought into thermal contact, heat flow occurs. This corresponds to the transfer of energy at the microscopic level through numerous collision processes of particles with a partition separating the two systems. Figure 1.7 depicts a heat flow process of this kind.

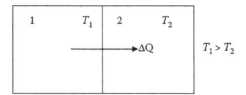

FIGURE 1.7 When two systems at different temperatures, with $T_1 > T_2$, are placed in thermal contact as shown, thermal energy, or heat, ΔQ is transferred from System 1 to System 2 at the microscopic level through collision processes.

As a consequence of energy transfer, the kinetic energy of molecules in Container 2 increases, whereas that of molecules in Container 1 decreases until equilibrium is reached with $T_1 = T_2$. The amount of heat is usually denoted ΔQ and is measured in energy units, for example, joules in the SI system. (Historically, the calorie, with 1 calorie = 4.18 J, was used as the heat unit, but this terminology is now less commonly used in physics.)

In the development of thermodynamics, the concepts of temperature, energy, and entropy play key roles. In Chapter 2, the first law of thermodynamics, which is based on the law of energy conservation, is introduced, and its use is illustrated in the processes for two model systems: the ideal gas and the ideal spin system. Chapter 3 deals with the second law of thermodynamics. The entropy concept is introduced following an analysis of heat engines and their efficiency in converting heat into work. The fundamental relation on the basis of the first and second laws of thermodynamics is given toward the end of Chapter 3. This relation encapsulates the important connections between internal energy, entropy, and work done in any given process and is a high point in the development of the subject.

1.8 ENERGY SOURCES

Following the development of steam engines in the eighteenth century, scientific interest in other types of heat engines grew in the early part of the nineteenth century. Heat engines operate by converting heat into useful work as discussed in Chapter 3. Early theoretical models made use of the ideal gas as the working substance, while practical internal combustion engines developed in the second half of the nineteenth century used air, which approximates an ideal gas under the conditions used. The efforts made to establish heat engine principles played a major role in the development of thermodynamics as a major scientific field.

In order to gain perspective on energy availability and future energy trends, it is instructive to examine the energy sources available on earth. These sources are first classified as either renewable or nonrenewable. Nonrenewable fossil fuels include natural gas, oil (petroleum, jet fuel), and coal. All of these fuels are hydrocarbons produced from the remains of plants and animals that accumulated millions of years ago. The heat produced by burning these fuels is thus originally derived from solar energy and is made available by oxidation of the carbon content.

Electricity is produced using high-pressure steam to drive steam turbines, which drive electricity generators. A modern multistage steam turbine is a form of heat engine generally powered by fossil fuels or nuclear reactors. It is interesting to compare the energy produced by the various fossil fuels. In the case of petroleum, the energy released as heat of combustion is 45 MJ kg^{-1}. Natural gas has a value around 50 MJ kg^{-1}, while for good quality coal, the value is lower at 25 MJ kg^{-1}. Natural gas and coal are largely used to generate electrical energy, which is sold to users in units of kW-h $\left(1\text{kW-h} = 3.6\,\text{MJ}\right)$. A representative figure of the daily consumption of electrical energy in developed countries is 44 MJ day^{-1} per individual with some variation from country to country.

The element uranium, and in particular uranium-235 isotope, which is used as the energy source in nuclear reactors, is another nonrenewable energy source. In a natural uranium light water reactor, 500 GJ kg^{-1} of energy is released during fission of uranium-235 into lighter nuclei. The release of energy is explained using the Einstein mass-energy relation $E = mc^2$ with m the mass loss that occurs in the uranium-235 fission process and c the speed of light in vacuum. Note that the energy per kg of unenriched uranium is a factor 10^4 larger than that of natural gas. Enriched uranium (3.5% uranium-235) provides a further factor 10 increase in the energy per kg. Allowing for the large atomic mass difference between ^{12}C and ^{235}U by considering the energy per *atom* increases the uranium energy production advantage by 20 to exceed 10^6. The fission energy is released as kinetic energy of fission fragments together with neutrons, which produce further fission processes, and gamma rays. Most of the fission energy ends up as heat, which is removed by a coolant passed through the reactor core. Steam from a nuclear reactor is used to drive a turbine-generator system to produce electricity.

Almost all renewable energy sources rely on radiant energy from the Sun reaching the Earth's surface. Geothermal energy from the Earth's interior is a notable exception. The Sun's radiation is distributed over a range of wavelengths from the infrared and the visible, which together amount to more than 90% of the radiation, into the ultraviolet (~7%). In order to quantify solar radiation, it is convenient to introduce the solar constant, which is the rate at which electromagnetic radiation from the Sun reaches Earth per unit area oriented perpendicular to the radiation direction. The solar constant is not a true constant but varies by a small amount due to the Earth's slightly elliptical orbit about the Sun and to variations in sunspot activity. Precise measurements of the solar constant made using instruments on Earth satellites give an average value of 1366 W m^{-2}. In traveling through the Earth's atmosphere, the solar radiation loses intensity through wavelength-dependent scattering and absorption processes. A number of other factors, including latitude, altitude, and cloud cover, further reduce the intensity of radiation reaching the Earth's surface. Allowing for a 50% reduction in intensity, it follows that an area of 20 m^2 in low- to mid-latitude regions of Earth receives about 49 MJ of solar radiation energy per hour, during peak hours of daylight. This is comparable to the heat energy produced by burning 1 kg of natural gas.

A variety of systems have been developed to convert the Sun's radiation directly into electrical energy. Solar panels consist of arrays of silicon photodiodes, which operate in the near-infrared and red-orange portion of the radiation spectrum and have conversion efficiencies approaching 30%. Large solar installations called heliostats use arrays of computer-controlled mirrors to focus the Sun's radiation onto a central tower facility where molten salt reaches to temperatures as high as 500°C. The stored heat n is used to produce superheated steam for electricity generation. These mirror arrays can produce hundreds of MW. Electricity generation using wind power involves assemblies of wind turbines each of which can provide several MW. The turbine blades operate when the wind speed exceeds 10–15 km h^{-1}.

Solar panels and heliostats need sunlight in order to operate, while wind turbines clearly need the wind to be blowing. As a consequence of variations in these conditions, it is necessary to use energy storage systems to provide energy on demand. Lithium-ion batteries and hydrogen fuel cells are examples of portable energy storage technology. Energy storage is an active research area.

Having identified the Sun's radiation as the primary source of available energy on Earth, it is of interest to examine the processes at work in the Sun's interior. The total energy radiated by the Sun per second is called the luminosity L. Taking the Earth's orbit about the Sun to be almost circular with radius $R_{ES} = 1.5 \times 10^{11}$ m and multiplying the solar constant by the area of the sphere of radius R_{ES} gives $L = 3.86 \times 10^{26}$ W. This enormous amount of radiant power is produced by nuclear fusion reactions in the core of the Sun. The energy released gradually diffuses to the Sun's surface. The luminosity has remained almost constant for billions of years.

The Sun is made up of the elements hydrogen (73%) and helium (25%) together with small amounts of heavier elements. As a result of the extremely high temperatures and pressures in the hot dense plasma at the Sun's core, fusion reactions occur continually. Four protons transform into helium-4 accompanied by the release of energy. The fusion reactions proceed through intermediate steps. Firstly, a small fraction of proton–proton collisions lead to the formation of a deuteron with one of the protons becoming a neutron via β – decay. Next, a deuteron and a proton combine to form helium-3, and finally, two helium-3 ions fuse to form helium-4 with the release of two protons into the plasma. The mass of a helium-4 ion is less than that of four protons by an amount $\Delta m = 0.029\, u = 4.81 \times 10^{-29}$ kg.. Einstein's mass–energy relation gives the energy released per fusion reaction as $\Delta E = 4.33 \times 10^{-12}$ J.

Using the Sun's luminosity, we can estimate the number of fusion reactions that occur per second in producing helium-4 as $N = L/\Delta E = 3.86 \times 10^{26} / 4.33 \times 10^{-12} = 8.9 \times 10^{37}\, s^{-1}$. Multiplying N by the mass of a heliun-4 ion gives the amount of heliun-4 produced per *second* as roughly 600 million metric tons.

Based on estimates of the proton concentration in the core, with radius 20% of the Sun's radius, the predicted lifetime of the Sun is around 10 billion years of which 4.5 billion years have elapsed. When there are insufficient protons in the hot dense core to maintain fusion processes, the Sun will become a red giant star large enough to include the orbits of Mercury and Venus. Eventually, the Sun will turn into a white dwarf star.

While uncontrolled fusion reactions are used in thermonuclear weapons on Earth, controlled fusion for large-scale energy generation has not yet been achieved and remains the subject of research. A major international research facility named ITER is being developed in France. The IIER research program will involve magnetic confinement of a plasma of deuterium and tritium ions. Fusion of deuterium and tritium ions will produce helium-4 plus energetic neutrons, which escape from the plasma. A surrounding blanket material converts the neutron energy into heat to power an electricity generator. Controlled fusion offers the prospect of abundant energy, but achieving this goal will require a major long-term commitment of resources.

PROBLEMS CHAPTER 1

1.1   A monatomic nonideal gas is well described by the van der Waals equation of state. The molar internal energy is given by $E = \dfrac{3}{2}RT - a/V$, where a is a positive constant. The gas is thermally isolated and is allowed to expand from an initial volume V_1 to a final volume V_2 by opening a valve into an evacuated space. Obtain an expression for the final temperature in terms of the given quantities and show that the final temperature is lower than the initial temperature. Give a physical explanation for the temperature decrease. (This is known as the Joule experiment in honor of the scientist who first attempted to measure the small temperature change in a process of this kind.) For a particular gas, $a = 0.1$ Jm^3. Find the temperature change for one mole of the thermally isolated gas when the volume is doubled, starting at a pressure of 1 atm and a temperature of 300 K.

1.2   A mass of 14 g of diatomic carbon monoxide gas is contained in a vessel at 290 K. Use the equipartition of energy theorem to obtain the sum of the translational and rotational energy contributions to the total energy of the gas assuming ideal gas behavior. Explain why the vibrational energy need not be considered. Compare the total energy with that of an equal mass of helium gas at the same temperature.

1.3   The virial equation of state for n moles of a real gas has the form $P = (n/V)RT\left[1 + B(T)(n/V) + \cdots\right]$, where the higher-order correction terms have been omitted. One mole of a real gas for which the second virial coefficient $B(T) = -1.2 \times 10^{-1}$ L mol^{-1} is initially at 300 K. If the gas is caused to expand at a constant pressure from an initial volume of 0.8 L to a final volume of 1.2 L, by raising the temperature T, find the increase in T in this process assuming that $B(T)$ remains constant. Comment on this assumption for the process considered. Compare the temperature change with that which would occur for an ideal gas.

1.4   A paramagnetic crystal is to be used as a magnetic thermometer at temperatures less than
      1 K. Magnetic susceptibility measurements show that the material has a Curie–Weiss
      temperature of 2 mK. What is the useful range of the thermometer if errors in temperature
      are not to exceed 2%?

1.5   Argon gas in a container is at a temperature of 300 K. What is the RMS speed of the
      gas molecules at this temperature? If the gas pressure is 1 atm, estimate the mean time
      between collisions of the molecules and the average distance traveled between collisions.

1.6   Use the Maxwell speed distribution to obtain an expression for the most probable speed
      v_{mp} of a molecule in a gas. Give your result in terms of the absolute temperature. Find a
      relationship between v_{mp} and the RMS speed v_{RMS}.

2 Energy
The First Law

2.1 THE FIRST LAW OF THERMODYNAMICS

One of the fundamental laws of physics is the law of energy conservation. In classical mechanics, this leads to the work–energy theorem that may be written in the following form when frictional forces are absent, and all forces are conservative:

$$W = \Delta E = \Delta K + \Delta U. \tag{2.1}$$

W is the work done on the system, and ΔE is the change in energy that is generally made up of a change in kinetic energy ΔK and a change in potential energy ΔU. When friction is present, an allowance must be made for the work done against friction. Equation 2.1 provides a powerful means for solving many problems in mechanics.

The law of energy conservation is the basis of the first law of thermodynamics. Consider an ideal gas system on which work ΔW is done and to which heat ΔQ is added, as shown schematically in Figure 2.1.

According to Equation 2.1, we may expect the kinetic energy of the particles in the gas to increase as a result of the two processes. No change in potential energy occurs because the gas is ideal, with negligible interactions between the particles. Gravitational potential energy changes are zero because the center of mass of the container remains fixed. These considerations lead to the following relationship for the system:

$$\Delta E = \Delta W + \Delta Q = \Delta K. \tag{2.2}$$

ΔE is the total change in energy of the gas, which is equal to the total change in kinetic energy of the particles making up the gas. If each particle has mass m, this gives

$$\Delta K = \sum_i \frac{\Delta p_i^2}{2m}. \tag{2.3}$$

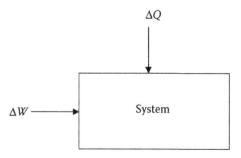

FIGURE 2.1 An ideal gas system to which heat ΔQ is added and on which work ΔW is done.

Δp_i^2 is the change in the momentum squared for the ith particle, and the sum is over all N particles in the system. If we put $\langle \Delta p^2 \rangle = 1/N \sum_i \Delta p_i^2$, where $\langle p^2 \rangle$ is the arithmetic mean change in momentum squared per particle, we get

$$\Delta K = N \frac{\langle \Delta p^2 \rangle}{2m} = N \Delta \varepsilon, \tag{2.4}$$

where $\langle \varepsilon \rangle$ is the mean change in kinetic energy per particle. Using the equipartition theorem from Chapter 1, we can relate $\langle \varepsilon \rangle$ to the change in absolute temperature of the gas. Returning to Equation 2.2, the relation $\Delta E = \Delta W + \Delta Q$ or infinitesimally,

$$dE = dW + dQ, \tag{2.5}$$

is the mathematical statement of the *first law of thermodynamics* and expresses the fact that the change in internal energy of the system is equal to the sum of the work done on the system and the heat added to the system. In SI units, all quantities in Equation 2.2 are measured in joules. Although the first law has been written down for an ideal gas system, it is a general law for *any* thermodynamic system. The energy change ΔE is usually made up of both potential energy and kinetic energy contributions. Note that the *sign convention* adopted in Equation 2.5 counts as *positive* for both the *work done on* the system and *heat added to* the system because these processes lead to an increase in internal energy. Our choice of sign convention is arbitrary but once it has been made must be used consistently.

2.2 APPLICATION OF THE FIRST LAW TO A FLUID SYSTEM

Consider a fluid, such as a gas, in a piston-cylinder container as shown in Figure 2.2.

If the pressure in the fluid is P, then a force $F = PA$ must be exerted on the piston to keep it fixed in position. If F is increased very slightly, the piston is caused to move inwards by an infinitesimal distance dx, and the infinitesimal work done on the system is

$$dW = -Fdx = -PA\ dx = -P\ dV. \tag{2.6}$$

A decrease in the volume of the system corresponds to work being done on the system or $dW > 0$ for $dV < 0$. This is consistent with the chosen sign convention for the work done. The first law for *fluids* becomes

$$dE = dQ - P\ dV. \tag{2.7}$$

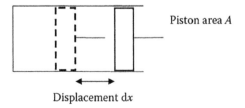

Displacement dx

FIGURE 2.2 A piston-cylinder container for a fluid system is shown. This arrangement permits work to be done on the system when the piston, of cross-sectional area A, is pushed into the cylinder.

In this form, the law can be applied to a variety of situations. It is often useful in considering thermal processes to introduce the idea of a large heat bath at a given temperature. The heat bath is sufficiently large that it can take in heat or give up heat to a system, with which it is in thermal contact, with negligible change in its temperature. Consider the following simple application as an exercise.

Exercise 2.1

An ideal gas contained in a piston-cylinder arrangement is at temperature T. If the piston is pushed in slowly so that the volume changes from V_i to V_f, how much heat must be rejected by the system to a heat bath in order for the temperature to remain constant?

We assume that the piston is pushed in sufficiently slowly that the system is always close to equilibrium. The ideal gas equation of state may be applied at all stages of the process. Figure 2.3 depicts the situation.

The first law (Equation 2.7) is given by $dE = dQ - P\,dV$. Because the gas is ideal, the internal energy does not change in an isothermal process, as can be seen with the aid of Equation 1.12. Because T remains fixed, the average energy of the molecules does not change. With $dE = 0$, we have $dQ = P\,dV$, and integration leads to $\Delta Q = \int_i^f P\,dV$. The ideal gas equation gives $P = nRT/V$, and we obtain

$$\Delta Q = nRT \int_{V_i}^{V_f} \frac{dV}{V} = nRT \, \ln\left(\frac{V_f}{V_i}\right) \tag{2.8}$$

If $V_f < V_i$, it follows that $\Delta Q < 0$, and this shows that heat is rejected by the system in the compression process. The above result will be useful in our discussion of heat engines in Chapter 3.

Exercise 2.2

Consider an isothermal compression process for a gas that obeys the van der Waals equation of state. Obtain an expression for the heat *rejected* by the system. Assume as an approximation that for a fairly small isothermal compression, the internal energy of the gas remains almost constant. This corresponds to the kinetic energy staying constant, according to the equipartition theorem, and a negligible change in the potential energy. (We discuss the intermolecular potential for a van der Waals gas in Chapter 7.)

With the assumption that $\Delta E \approx 0$, we have $\Delta Q = \int_{V_i}^{V_f} P\,dV$. For a van der Waals gas, with $n = 1$ mol, $P = RT/(V - b) - a/V^2$.

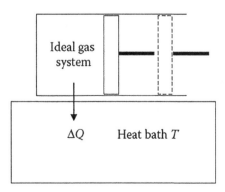

FIGURE 2.3 Isothermal compression of an ideal gas: heat ΔQ is rejected to the large thermal reservoir, or heat bath, which is maintained at a fixed temperature T.

This gives

$$\int_{V_i}^{V_f} P \, dV = RT \int_{V_i}^{V_f} \frac{dV}{V-b} - a \int_{V_i}^{V_f} \frac{dV}{V^2} = RT \ln\left(\frac{V_f - b}{V_i - b}\right) + a\left(\frac{1}{V_f} - \frac{1}{V_i}\right). \tag{2.9}$$

In this case, it is necessary to know the van der Waals constants a and b to obtain ΔQ. For $V_f < V_i$, the first term will be negative and the second positive. This reflects the fact that in the pressure term, which involves the constant a, a decrease in volume increases the importance of the interparticle attractive interaction, whereas in the volume term, the constant b simply reduces the effective volume available to the particles. If the internal energy change is not small, a more elaborate calculation is required. Expansion processes for a van der Waals gas are considered in Chapter 7.

2.3 TERMINOLOGY

The terminology used in the description of various processes that are carried out on systems is summarized in Table 2.1.

Other useful terminologies are given below.
State variable. Specifies the state of a system (e.g., P, V, and T).
Extensive variable. Value depends on the size of the system (e.g., V, N, or n).
Intensive variable. Value does not depend on the size of the system (e.g., T and P).
State function. A thermodynamic function of the state variables.

An example of a state function is the internal energy of a system. For an ideal gas, we see from Equation 1.13 that $E = \frac{1}{2} fNRT$, and in this case, E is simply a function of T, that is, $E = E(T)$. For a process in which a temperature change takes place, it does not matter how a particular final state is reached. The internal energy of the final state is specified completely by the final absolute temperature. In general, E will be a function of two state variables for systems that are not ideal so that we have $E = E(V, T)$. Note that because there is an equation of state that connects the state variables (P, V, and T for a gas), only two state variables are needed to specify the state of the system and hence the internal energy completely.

Although the first law relates dE, dQ, and dW, it is important to realize that although E is a state function, Q and W are *not* state functions. dE is an exact differential, which may be written in terms of partial derivatives of $E = E(V, T)$ as

$$dE = \left(\frac{\partial E}{\partial T}\right)_V dT + \left(\frac{\partial E}{\partial V}\right)_T dV. \tag{2.10}$$

TABLE 2.1
Terminology Used in Describing Thermodynamic Processes

Process	Condition
Isothermal	T constant – no change in temperature
Isobaric	P constant – no change in pressure
Isochoric	V constant – no change in volume
Adiabatic	$\Delta Q = 0$ – no heat energy transferred
Quasi-static reversible	Process carried out slowly; system is always very close to equilibrium

dQ and dW, on the other hand, are inexact differentials and cannot be written in a form similar to Equation 2.10. An infinitesimal change in internal energy may occur in various ways. This means that dQ and dW each depend on the path followed, but the sum gives the same value dE for specified initial and final states. Macroscopically, we find that different combinations of ΔQ and ΔW can lead to the same change &E. For future use, we note that for state variables such as $P = P(V, T)$, which are functions of other state variables, total differentials may be written in terms of partial derivatives as follows:

$$dP = \left(\frac{\partial P}{\partial V}\right)_T dV + \left(\frac{\partial P}{\partial T}\right)_V dT. \tag{2.11}$$

2.4   P–V DIAGRAMS

It is convenient to represent processes involving a fluid system on a pressure-volume or P–V diagram, and Figure 2.4 shows how this is done.

It is easy to see that the work done in a process, such as that shown in Figure 2.4, is given by the area under the curve that depicts the process on a P–V diagram, that is, $\Delta W = -\int_{V_i}^{V_f} P\,dV$.

In an expansion process, ΔW is negative, from our sign convention, because work is done *by* the gas. Isochoric and isobaric processes are represented on P–V diagrams by straight lines parallel to the vertical and horizontal axes, respectively. If P–V diagrams are drawn to scale, the area under a curve may be calculated directly using geometric results, and this gives the magnitude of the work done in the corresponding process. The sign is obtained by determining whether the process is an expansion (negative) or compression (positive).

2.5   QUASI-STATIC ADIABATIC PROCESSES FOR AN IDEAL GAS

An adiabatic process has been defined as one in which $\Delta Q = 0$, with the system thermally isolated so that no heat flow can take place. Experiment shows that in a quasi-static adiabatic process, the pressure and the volume of an ideal gas are related by the expression PV^γ = constant or

$$P_i V_i^\gamma = P_f V_f^\gamma, \tag{2.12}$$

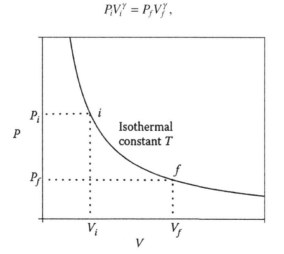

FIGURE 2.4   P–V diagram for a gas that shows an isothermal expansion process from an initial volume V_i to a final volume V_f.

where γ is a constant for a given system and has the value 5/3 = 1.66 for a monatomic gas. We derive Equation 2.12 later in this chapter. Because the ideal gas equation of state always holds in any such process, Equation 2.12 may be rewritten as

$$T_i V_i^{\gamma-1} = T_f V_f^{\gamma-1} \qquad (2.13)$$

The temperature of a gas will change in an adiabatic expansion or compression as follows: $T_f = T_i (V_i/V_f)^{\gamma-1}$. Adiabatic expansion processes are used to cool gases such as nitrogen in liquefier plants.

2.6 MAGNETIC SYSTEMS

In application of the first law to magnetic systems, it is necessary to consider the magnetic work done on a system when an applied magnetic field H is changed from some initial value to a final value. For a magnetic material, the magnetic induction in the system may be written as

$$\boldsymbol{B} = \mu_0 \left(\boldsymbol{H} + \boldsymbol{M} \right), \qquad (2.14)$$

with $\mu_0 = 4 \times 10^{-7}$ Hm^{-1} the permeability of free space. M is the magnetization or magnetic moment per unit volume of the material and has units ampere per meter. H is produced by an external electric current in a circuit and also has units ampere per meter. The magnetic susceptibility χ is defined by the relationship $\boldsymbol{M} = \chi \boldsymbol{H}$. Choosing the units of M as ampere per meter makes the magnetic induction B the force vector for a magnetic dipole rather than H. The systems dealt with in this book are paramagnetic and not ferromagnetic. As a simplification and as a good approximation for paramagnets, we assume that $M \ll H$ and that the relative permeability μ_r of any material of interest is close to unity with $\boldsymbol{B} \approx \mu_0 \boldsymbol{H}$.

Consider a long solenoid of N turns, cross section A, and volume V, carrying a current i and filled with magnetic material, as shown in Figure 2.5. The applied magnetic field produced by the current is given by $H = ni$.

The demagnetizing factor is assumed small for the chosen geometry. When the current through the solenoid changes with time, an emf ε is induced, according to Faraday's law of induction:

$$\varepsilon = -NA\left(\frac{dB}{dt} \right). \qquad (2.15)$$

The infinitesimal work done on the system of solenoid plus magnetic material by the external energy source in transferring a charge $i\,dt$ through the solenoid is

$$dW = \varepsilon i\,dt, \qquad (2.16)$$

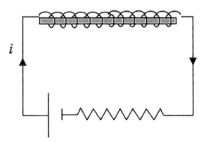

FIGURE 2.5 A rod of magnetic material is contained in a long solenoid through which a current i is passed.

with ε the *applied* potential difference across the solenoid. Bearing in mind our sign convention that the work done on a system is positive, combination of Equations 2.15 and 2.16 gives $dW = NAi\ dB$ and with $H = Ni/L$, where L is the solenoid length, we obtain

$$dW = VH\ dB = \mu_0 V (H\ dH + H\ dM). \tag{2.17}$$

The first term in Equation 2.17 corresponds to the work done in the establishment of the increased magnetic field in vacuum and the second term to that associated with the increase in magnetization of the magnetic material. If the system is taken to consist of the solenoid plus magnetic material (with $\mu_r \approx 1$), the first law may be written to a good approximation as

$$dE = dQ + VBdH + VB\ dM. \tag{2.18}$$

It is convenient to rewrite Equation 2.18 in the form

$$dE^* = d\left(E - \frac{1}{2}\mu_0 VH^2 \right) = dQ + VBdM, \tag{2.19}$$

where $E^* = \left(E - \frac{1}{2}V\mu_0 H^2 \right)$ contains the vacuum field energy as a shift in the arbitrarily chosen zero of energy. This corresponds to exclusion of the vacuum field as part of the system of interest and retention of just the energy term associated with the magnetic material in the applied field. An alternative but equivalent expression to Equation 2.19 is obtained if we make use of the differential $d(MB) = M\ dB + B\ dM$, or

$$B\ dM = d(MB) - M\ dB. \tag{2.20}$$

Inserting Equation 2.20 in Equation 2.19 gives $dE^* = dQ + V\ d(MB) - VM\ dB$, and this equation is written as

$$dE^+ = d\left(E^* - VMB \right) = dQ - VM\ dB \tag{2.21}$$

In Equation 2.21, $E^+ = (E^* - VMB)$ is the self-energy of the material alone, with the vacuum field energy and the mutual field energy of the sample field and the applied field subtracted from the total energy E. Both dE^* and dE^+ are exact differentials of state functions and correspond to different viewpoints of the system of interest.

The expressions for the infinitesimal work done in the two cases that are usually considered are

i. Magnetic material plus mutual field

$$dW = VB\ dM \tag{2.22}$$

ii. Magnetic material only

$$dW = -VM\ dB \tag{2.23}$$

Depending on the situation which is being considered, either of these two expressions may be used. It is important to note that the signs are different in the two expressions for dW. Examples of the use of these expressions are given later. In the above discussion, it has been implied that any changes in volume V of the sample are small and may be ignored, but if V changes sufficiently, an additional term $-P\,dV$ should be included in Equations 2.22 and 2.23 to allow for the mechanical work done in the expansion. In most cases of interest, this additional term may be omitted because it is much smaller than the magnetic work term.

Additional insight into the various contributions to the magnetic energy is obtained by considering the energy density of the magnetic field which is given by

$$\frac{E}{V} = \frac{1}{2}\frac{B^2}{\mu_0}.$$ (2.24)

With Equation 2.14, we obtain

$$E = \frac{1}{2}V\mu_0\left(H^2 + 2\boldsymbol{H}\cdot\boldsymbol{M} + M^2\right).$$ (2.25)

The three terms in Equation 2.25 are the vacuum field energy, the mutual field energy, and the self-energy, respectively. The quantities $E^\star$ and E^+, defined above, are readily identified by rearrangement of Equation 2.25.

For many calculations, it will be convenient to exclude the vacuum field energy and the mutual field energy and to use E^+, the self-energy or sample energy of the magnetic system. In this chapter, and for many sections of the book, we shall limit the discussion to ideal linear magnetic systems, specifically paramagnetic systems that obey Curie's law. Ferromagnets and antiferromagnets that undergo transitions to spin ordered phases below their transition temperatures are excluded here but are considered in Chapter 17, which deals with nonideal systems.

2.7 PARAMAGNETIC SYSTEMS

Having distinguished the self-energy from the other energy contributions, we now write down the form for this energy that is widely used in thermal physics applications. Consider a fixed moment μ in a field of magnetic induction $\boldsymbol{B} = \mu_0\boldsymbol{H}$. The potential energy of the moment is given by the scalar product of μ and $\boldsymbol{B}$, that is,

$$E = -\boldsymbol{\mu}\cdot\boldsymbol{B}.$$ (2.26)

This follows from a consideration of the work done in a process in which an orientation change of the moment in the applied field is made. (We allow the moment to be macroscopic and ignore quantization effects for the present.) For a macroscopic system, such as a paramagnet, the expression for the potential energy is given by

$$E = -V\boldsymbol{M}\cdot\boldsymbol{B} = -\mathcal{M}\cdot\boldsymbol{B},$$ (2.27)

with $\mathcal{M}$ as the total magnetic moment of the specimen and $\boldsymbol{B}$ as the magnetic induction. To simplify the notation, the superscript dagger has been omitted from E. This simplified notation will, in general, be used to denote the self-energy unless otherwise indicated. For a paramagnet, the magnetization is given by Curie's law (Equation 1.4). Further discussion of Equations 2.26 and 2.27 is given in Chapter 4.

Exercise 2.3

A paramagnetic material that obeys Curie's law is situated in a magnetic field whose strength is increased from H_i to H_f in an isothermal process. Obtain an expression for the work done on the material in this process. How much heat energy is rejected by the system?

We have from Equation 2.23 $dW = -\mu_0 VM dH$. Curie's law (Equation 1.4) is $M = CH/T$, and substitution for M in Equation 2.23 followed by integration gives

$$\Delta W = -\frac{1}{2}\left(\frac{\mu_0 CV}{T}\right)\left(H_f^2 - H_i^2\right). \tag{2.28}$$

From Equation 2.27, the change in internal energy of the magnetic material is simply

$$\Delta E = -\mu_0 V\left[M_f H_f - M_i H_i\right],$$

which for an isothermal process gives

$$\Delta E = -\left(\frac{\mu_0 CV}{T}\right)\left[H_f^2 - H_i^2\right], \tag{2.29}$$

where use has again been made of Curie's law. From the first law and using Equations 2.28 and 2.29, it follows that

$$\Delta Q = -\frac{\mu_0 CV}{2T}\left[H_f^2 - H_i^2\right]. \tag{2.30}$$

This is the heat energy rejected in the isothermal magnetization process. This result is qualitatively similar to the result obtained for the heat rejected in the isothermal compression of a gas where work is done on the system. Note that we have expressed our results in terms of the applied field H rather than in terms of the magnetic induction B. This is because the form chosen for Curie's law involves the external field H applied to a paramagnetic material.

M–H diagrams: By analogy with gaseous systems, where processes are represented on P–V diagrams, magnetic processes may be represented on M–H diagrams. In the magnetic case, M is the magnetization or magnetic moment per unit volume. If we multiply M by V, we obtain $\mathcal{M}$, the total magnetic moment, which is an extensive quantity. Figure 2.6 shows the isothermal and adiabatic demagnetization processes on an M–H diagram for a paramagnetic system.

Note that in the adiabatic demagnetization process, M remains constant as shown below. Demagnetization in this context means reduction of the applied field H produced by the external magnetic circuit. From Equation 2.23, the work done on the material in any process is proportional to the area under the M–H curve. We again exclude the vacuum field energy and mutual field energy. The paramagnetic material is assumed to obey Curie's law, and it follows that isothermals are represented by straight lines of slope C/T, which extrapolate to the origin. Adiabatic processes are represented by horizontal paths such as $1 \rightarrow 2$, in which the magnetic field strength changes from H_1 to H_2. The statement that M is constant in an adiabatic process is based on the first law (Equation 2.21), which for $dQ = 0$ gives $dE = dW = -\mu_0 VM dH$. Together with the expression for the energy in Equation 2.27, this leads to the following relationship $\Delta E = -\mu_0 V\left[M_2 H_2 - M_1 H_1\right] = \Delta W = -\mu_0 V \int_{H_1}^{H_2} M\, dH$. Replacement of M by CH/T shows that this relationship is satisfied only if M remains constant, proving that $M_1 = M_2$ in an adiabatic process. It follows from Curie's law that

$$\frac{H_1}{T_1} = \frac{H_2}{T_2}. \tag{2.31}$$

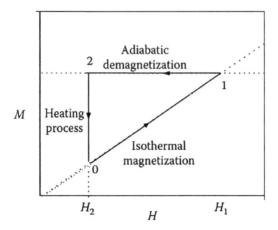

FIGURE 2.6 *M–H* diagram for a paramagnetic material that obeys Curie's law. The diagram shows a cycle that is made up of an isothermal magnetization process, an adiabatic demagnetization process, and a heating process at constant *H*.

Equation 2.31 is a useful relationship that expresses the initial and final temperatures of an ideal paramagnet in terms of the initial and final applied fields in an adiabatic process.

The work done in an adiabatic process is the area under the curve on an *M–H* diagram, with due regard to sign: $\Delta W_{12} = -\mu_0 VM (H_2 - H_1)$. Similarly, the work done in an isothermal process is readily obtained. Magnetic cooling, which is discussed in the next section, is an example of the practical importance of the processes shown in Figure 2.6.

2.8 MAGNETIC COOLING

Figure 2.7 depicts schematically an arrangement used in achieving cooling of a paramagnetic system by means of adiabatic demagnetization.

With the heat switch closed, the paramagnetic material is isothermally magnetized by an increase in the applied magnetic field, and this corresponds to path $0 \rightarrow 1$ in Figure 2.6. When equilibrium has been reached at T_1 in the high field H_1, the paramagnetic material is thermally isolated by opening

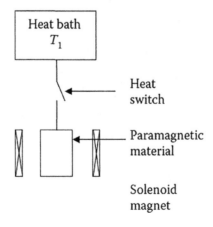

FIGURE 2.7 A schematic depiction of the arrangement used for magnetic cooling of a paramagnetic material by means of an adiabatic demagnetization process. The material is situated in a field supplied by a magnet. Thermal contact between the paramagnet and the heat bath at a fixed temperature T_1 is controlled by means of a heat switch. With the heat switch open, the applied field is gradually reduced and the temperature of the paramagnet decreases.

the heat switch. In practice, this may be done using a magnetic field operated switch made of super-conducting material, with high thermal conductivity in its normal state and low thermal conductivity in the superconducting state, or by using an exchange gas that can be pumped away to break thermal contact. Cooling of the paramagnetic material is achieved by adiabatic reduction of the applied magnetic field strength from H_1 to H_2 along path $1 \rightarrow 2$ in Figure 2.6. From Equation 2.31, the final temperature reached is given by

$$T_2 = \left(\frac{H_2}{H_1} \right) T_1. \tag{2.32}$$

This is easily understood from Curie's law because the magnetization reached in a large magnetic field H_1 at temperature T_1 is retained, at the completion of the adiabatic process, in the much lower field H_2. It might be thought that absolute zero temperature can be reached when H_2 is decreased to zero. This is not so because, for any real paramagnetic system, there are always some interactions between the spins that lead to a breakdown of Curie's law at very low temperatures. This point is discussed in greater detail in Chapter 6.

In addition to the breakdown of Curie's law, a complete treatment must consider interactions between the system of spins in the paramagnetic material and the host lattice in which they are located. The lattice is cooled by the spins through spin-lattice relaxation processes. At very low temperatures, spin-lattice relaxation times can become very long, and this prevents equilibrium being reached for the entire system of spins plus lattice. These details are not pursued further here.

2.9   GENERAL EXPRESSION FOR THE WORK DONE

For a fluid system, the infinitesimal work done is given by Equation 2.6 $dW = -P\,dV$. In the case of a magnetic system, we have two expressions that depend on what the system is taken to be. For the magnetic material plus field case, we have from Equation 2.22 $dW = \mu_0 VH\,dM = \mu_0 H\,d\mathcal{M}$, whereas for the case of magnetic material alone, Equation 2.23 is $dW = {}_-\mu_0 VM\,dH = -\mu_0 \mathcal{M}\,dH$. Two other processes that are often considered are stretch of a wire and increase in the surface area of a liquid. For a stretched wire,

$$dW = F\,d\ell, \tag{2.33}$$

where F is the applied force, and $d\ell$ is the increase in length of the wire. For the liquid surface,

$$dW = S\,dA, \tag{2.34}$$

where S is the surface tension and dA is the increase in area of the surface. Other examples include the work done by an electric cell, with emf ε, in the transfer of charge dQ through a circuit $dW = -\varepsilon\,dQ$.

Equations for the work done on a system are of the general form

$$dW = Y\,dX,$$

where Y is a generalized force, which is an intensive variable, and X is a generalized displacement, which is an extensive variable. For a particular system, care must be exercised in choosing the sign in expressions for dW to comply with the convention we have adopted. The various processes may be represented graphically on a Y versus X diagram, as shown in Figure 2.8. The generalized force, which is an intensive quantity, is usually plotted along the vertical axis, whereas the generalized displacement, which is an extensive quantity, is plotted along the horizontal axis.

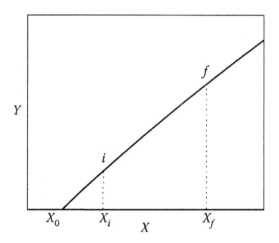

FIGURE 2.8 The Y–X diagram for a representative work process on some system. The generalized force (intensive variable) Y is plotted versus the generalized displacement (extensive variable) X for the process. X_0 is the value of the variable X for $Y = 0$.

The work done in a change of the extensive variable X from X_i to X_f is given by

$$W = \int_{X_i}^{X_f} Y \, dX. \tag{2.35}$$

This is the area under the curve between the limit values X_i and X_f. For reasons discussed in Section 2.6, care must be exercised in specifying the system in magnetic situations. Caution must similarly be exercised in expressions for the work done on polarizable media in electric fields.

2.10   HEAT CAPACITY

Consider a situation in which heat ΔQ is added to a system resulting in an increase in temperature ΔT. We define the heat capacity of the system as

$$C(T) \lim_{\Delta T \to 0} \left(\frac{\Delta Q}{\Delta T} \right) = \frac{dQ}{dT}. \tag{2.36}$$

The SI unit of C is joules per kelvin. Note that C may be a function of T. It is convenient to introduce the specific heat of a substance defined as the heat capacity either per unit mass

$$c(T) = \frac{1}{m} \left(\frac{dQ}{dT} \right), \tag{2.37}$$

or per mol

$$c(T) = \frac{1}{n} \left(\frac{dQ}{dT} \right). \tag{2.38}$$

Note the use of lower case c for the specific heat. The unit is J kg^{-1}K^{-1} or J mol^{-1}K^{-1}. For water at room temperature, $c = 4.18 \times 10^3$ J kg^{-1} K^{-1}. Specific heats for other substances are available in tables.

From the equipartition of energy theorem, it is a straightforward matter to obtain an expression for the heat capacity of an ideal gas. The internal energy of a monatomic ideal gas is from

Equation 1.13 $E = N\left(\frac{3}{2}k_BT\right) = \frac{3}{2}nRT$, and an infinitesimal change in E with temperature

is $dE = \frac{3}{2}nR\ dT$. This expression with the first law gives $\frac{3}{2}nR\ dT = dQ - P\ dV$, and with

rearrangement, we obtain

$$dQ = \frac{3}{2}nR\ dT + P\ dV. \tag{2.39}$$

To proceed, it is important to recognize that the heat capacity of a gas or, more generally, a fluid can be measured with two different constraints: either the volume or the pressure may be held constant. The corresponding heat capacities are denoted C_V and C_P, respectively. From Equation 2.39, it follows that

$$C_v = \left(\frac{dQ}{dT}\right)_V = \frac{3}{2}nR \tag{2.40}$$

and

$$C_P = \left(\frac{dQ}{dT}\right)_P = \frac{3}{2}nR + P\left(\frac{\partial V}{\partial T}\right)_P. \tag{2.41}$$

The ideal gas equation of states gives $(\partial V/\partial T)_p = nR/P$, which, with Equation 2.41, leads to

$$C_p = \frac{5}{2}nR. \tag{2.42}$$

The molar-specific heats are obtained immediately on division by n. From Equations 2.41 and 2.42, we see that

$$c_P - c_V = R. \tag{2.43}$$

The physical reason for c_P being greater than c_V is that work is done in the constant P case, which is not done at constant V. Some of the heat is converted into mechanical work when P is constant and V increases. Additional heat energy is therefore required to produce a given rise in temperature for constant P compared to constant V.

More generally, for an ideal gas made up of particles with f degrees of freedom to which the equipartition of energy theorem applies, the heat capacity expressions are $C_V = \frac{1}{2}fnR$ and $C_P = \left(\frac{1}{2}f + 1\right)nR$. The infinitesimal change in internal energy is from the first law and our definition of C_V given by $dE = dQ = C_V\ dT$. The first law may conveniently be written in the form

$$C_V\ dT = dQ - P\ dV. \tag{2.44}$$

Equation 2.44 is useful in calculations that involve ideal gas processes, as illustrated in the following example.

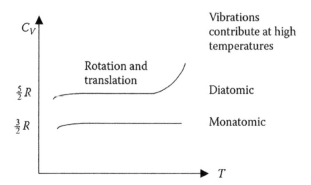

FIGURE 2.9 Schematic representation of the specific heats of monatomic and diatomic gases as a function of temperature. At sufficiently low T, gases liquefy and/or solidify. At very high temperatures, vibrational motions contribute to the specific heat for diatomic molecules.

Exercise 2.4

A monatomic ideal gas undergoes an adiabatic expansion from volume V_i to V_f. Obtain an expression for the ratio of the initial to the final temperature of the gas.

For $dQ = 0$ and with $C_V = \dfrac{3}{2}nR$, Equation 2.44 reduces to $\dfrac{3}{2}nR\ dT = -P\ dV$. With use of the ideal gas law, this equation may be integrated to give $\dfrac{3}{2}\ln\left(T_i/T_f\right) = \ln\left(V_i/V_f\right)$ from which we obtain $(T_i/T_f) = (V_f/V_i)^{2/3}$. More generally, for a gas that is not monatomic, it is easy to see that

$$\left(\frac{T_i}{T_f}\right) = \left(\frac{V_f}{V_i}\right)^{2/f},$$
(2.45)

which is an alternative form of Equation 2.13. Figure 2.9 shows, schematically, the form of the specific heat behavior with temperature for monatomic and diatomic gases. At low temperatures, gases liquefy and low temperatures are therefore excluded from the plot. The specific heats of liquids and solids are discussed in Chapter 8. For diatomic gases, the vibrational degrees of freedom contribute to the specific heat at sufficiently high temperatures, typically well above room temperature. More detailed consideration of the various contributions to the specific heats of polyatomic gases is given in Chapter 16.

The heat capacity of an ideal paramagnetic solid may be considered in a similar way to that just given for an ideal gas. Heat capacities may be defined as follows: $C_H = (dQ/dT)_H$ and $C_M = (dQ/dT)_M$. A discussion of these heat capacities is deferred until later in this book when an expression for the internal energy of a paramagnetic system has been derived using a microscopic approach.

2.11   QUASI-STATIC ADIABATIC PROCESS FOR AN IDEAL GAS REVISITED

In Section 2.5, we introduced the relationship $PV^\gamma = $ constant for adiabatic processes. It is now possible to relate the quantity γ to the specific heats introduced in Section 2.10. From Equation 2.44, the first law for $n = 1$ mol of an ideal gas is

$$dQ = c_V\ dT + P\ dV.$$
(2.46)

The ideal gas equation of state is $PV = RT$. Forming the differential of PV, we have

$$d(PV) = P\ dV + V\ dP = R\ dT. \tag{2.47}$$

On substitution for $P\ dV$ from Equation 2.46 into Equation 2.47, we obtain $dQ = (c_V + R)\ dT - VdP$, and with $(c_P - c_V) = R$, this gives

$$dQ = c_P\ dT - V\ dP. \tag{2.48}$$

For an infinitesimal adiabatic process, $dQ = 0$, and from Equation 2.48, we get

$$V\ dP = c_P\ dT. \tag{2.49}$$

With Equation 2.47, it follows that

$$P\ dV = -c_V\ dT. \tag{2.50}$$

We divide Equation 2.49 by Equation 2.50 and rearrange the result to obtain the first-order differential equation $dP/P = -(c_P/c_V)dV/V$. For a quasi-static process, integration leads to $\ln P = -(c_P/c_V) \ln V + \ln(\text{constant})$, or with antilogs

$$PV^{c_P/c_V} = \text{constant}. \tag{2.51}$$

Comparison of Equation 2.51 with Equation 2.12 shows that $\gamma = c_P/c_V$. Not only is the form of Equation 2.12 obtained, but an explicit expression for γ in terms of the specific heats is found. Note that the ratio $c_P/c_V = 5/3 = 1.66$ for a monatomic ideal gas.

2.12 THERMAL EXPANSION COEFFICIENT AND ISOTHERMAL COMPRESSIBILITY

For condensed matter, it is convenient to introduce two coefficients that characterize a particular system. These are the isobaric thermal expansion coefficient,

$$\beta = \frac{1}{V}\left(\frac{\partial V}{\partial T}\right)_P \tag{2.52}$$

and the isothermal compressibility,

$$\kappa = \frac{1}{V}\left(\frac{\partial V}{\partial P}\right)_T. \tag{2.53}$$

Measured values of these coefficients have been tabulated for many solids and liquids. The relationship between the specific heats for such systems takes the form

$$c_P - c_V = \frac{VT\beta^2}{\kappa}. \tag{2.54}$$

A derivation of this relationship is given later in the book.

Exercise 2.5

Show that the relationships that connect the specific heats at constant volume and constant pressure as given in Equations 2.43 and 2.54 are consistent for an ideal gas.

The ideal gas equation of state $PV = RT$ (for $n = 1$ mol), $\beta = 1/V(\partial V/\partial T)_P = R/PV = 1/T$, and $\kappa = -1/V(\partial V/\partial P)_T = RT/P^2V = 1/P$. Substituting for β and κ in Equation 2.54, we obtain $c_P - c_V = PV/T = R$, which is the result in Equation 2.43.

We now have the first law and other relationships required to analyze a wide variety of thermodynamic processes. In the next chapter, we deal with heat engines and refrigerators and from the results obtained are led to introduce the entropy concept, which is of fundamental importance in thermal physics.

PROBLEMS CHAPTER 2

2.1   One mole of an ideal monatomic gas initially at a pressure of 1 atm and temperature 0°C is isothermally and quasi-statically compressed until the pressure has increased to 2 atm. How much work is done on the system in the compression process? How much heat is transferred from the gas to a heat bath in the process?

2.2   A monatomic ideal gas at 280 K is at an initial pressure of 30 atm and occupies a volume of 1.5 L. If the gas is allowed to expand (a) isothermally and (b) adiabatically to a final volume of 3 L, what is the final temperature and pressure in each case? How much work is done by the gas in the two processes?

2.3   Obtain expressions for the slopes of curves on a P–V diagram representing isothermal and adiabatic expansion processes for an ideal gas. Show that for any value of V the slope of the adiabatic curve is greater than that of the isothermal curve by the factor γ that occurs in the adiabatic P–V relationship.

2.4   In an adiabatic quasi-static process, the initial and final pressures and volumes are related by the expression $P_1V_1^\gamma = P_2V_2^\gamma$, where γ is a constant for a particular gas. Show that the work done on an ideal gas in a process of this kind is given by $W = \left[1/(\gamma - 1)\right]\left[P_2V_2 - P_1V_1\right]$.

2.5   Argon gas in a piston-cylinder container has initial volume V_1 of 5 L and pressure P_1 of 1 atm at a temperature T_1 of 300 K. The gas is heated so that volume and pressure both increase in such a way that the path is represented by a *straight line* on a P–V diagram. The final pressure P_2 is 1.1 atm, and the final volume V_2 is 10 L. Sketch the P–V diagram for this process and obtain an expression for the work done by the gas in the expansion. How much heat is required for the process?

2.6   A monatomic ideal gas, initially at 3 atm in a volume of 5 L, is taken through a cycle involving the following processes: the volume is doubled under isothermal conditions at 300 K, the gas is then heated isochorically until the pressure is equal to the initial pressure, and finally the gas is cooled isobarically back to its initial condition. Sketch the P–V diagram for this cycle and obtain the work done and the heat transferred in each segment. By how much does the temperature of the gas change in the isochoric process?

2.7   Consider a cyclic quasi-static process for 0.1 mol of monatomic gas that has the form of a circle on a P–V diagram. Sketch the form of the cycle such that the diameters parallel to the P and V axes are from 1 to 3 atm and from 1 to 3 L, respectively. Obtain the net work done on the system per cycle. What is the internal energy change in the system in traversing a half cycle starting at a point where the pressure is 2.0 atm and the volume is 1.0 L on the horizontal diameter (i.e., between two points joined by a diameter drawn parallel to the V-axis or volume axis)? How much heat is absorbed in this process?

2.8   A paramagnetic sample that obeys Curie's law has mass 0.4 g and density 4 g cm^{-3}. The material is in good thermal contact with a heat bath at 2 K, permitting isothermal magnetization processes to be carried out. If an applied magnetic field is increased quasi-statically from zero to 10 T, find the work done on the sample. The Curie constant for the sample material is $C = 0.5$ K m^{-3}. If after magnetization the paramagnetic sample is thermally isolated from the heat bath and the applied field is quasi-statically reduced from 10 to 0.1 T, find the change in temperature of the sample material. How much work is done on the system in the demagnetization process? Compare your result with the work done in the isothermal magnetization procedure. Sketch the isothermal and adiabatic process on an M–H plot.

2.9 An object of mass m_o is heated to a temperature $t_h°C$ and is then dropped into a well-lagged vessel containing a mass m_w of water at $t_c°C$. If the final equilibrium temperature is $t_f°C$, obtain an expression for the specific heat of the object in terms of the initial and final temperatures and the specific heat of water.

2.10 The molar heat capacity of diamond has the following form at sufficiently low temperatures $c_V = 3R(4\pi/5)(T/\theta_D)^3$, where R is the gas constant and $\theta_D = 2200$ K is a parameter called the Debye temperature for diamond. How much heat is required to heat a 1-carat diamond from 4 to 300 K? (1 carat = 200 mg)?

2.11 A piece of copper of mass 200 g is isothermally compressed at 290 K by increasing the external pressure from 1 to 1000 atm. Find the volume change and the work done in the compression process. The isothermal compressibility of copper is $\kappa = 7.3 \times 10^{-5}$ atm^{-1} and the density is 8.9 g cm^{-3}. Express the work done as an integral over the pressure change using the average volume in the compression process. Justify the use of the average volume in your calculation.

3 Entropy
The Second Law

3.1 INTRODUCTION

The concept of entropy is of central importance in thermal physics. Historically, it was through the study of heat engine processes that the need for this new concept was first appreciated. We shall see that entropy, which is traditionally represented by the symbol S, is an extensive quantity and a state function like the internal energy E of a system. The second law of thermodynamics is related to the concept of reversibility in processes and expresses the reversible or irreversible nature of the evolution of the state of a system with time. The law is stated most compactly in terms of entropy, although alternative statements exist. In Chapters 4 and 5, we introduce a microscopic description for large assemblies of particles and find that entropy is related to the number of microstates available to a system. The microscopic approach provides deep insight into the nature of entropy, and through its connection with microstates, we establish a bridge between the microscopic and the macroscopic descriptions. In this chapter, we focus on entropy in thermodynamics.

3.2 HEAT ENGINES: THE CARNOT CYCLE

Heat engines operate by conversion of heat into useful work. It is found that only part of the heat taken in by an engine can be converted into work, and the remainder is rejected to a heat bath at a temperature lower than that of the bath from which heat is derived. As mentioned in Chapter 2, the term heat bath is used to describe a thermal reservoir, with a very large heat capacity, at a temperature T that remains essentially constant regardless of the heat transferred to or from the bath. The operation of a heat engine is illustrated schematically in Figure 3.1.

Heat engines operate in a cyclic way by taking in heat Q_1 from a high-temperature reservoir, converting part of this energy into work output W, and rejecting the remainder Q_2 to a low-temperature reservoir. Figure 3.1 shows the heat transferred and the work done in one complete cycle of a heat engine. The efficiency η of a heat engine is defined as

$$\eta = \frac{W_{\text{out}}}{Q_{\text{in}}} = \frac{W}{Q_1}. \tag{3.1}$$

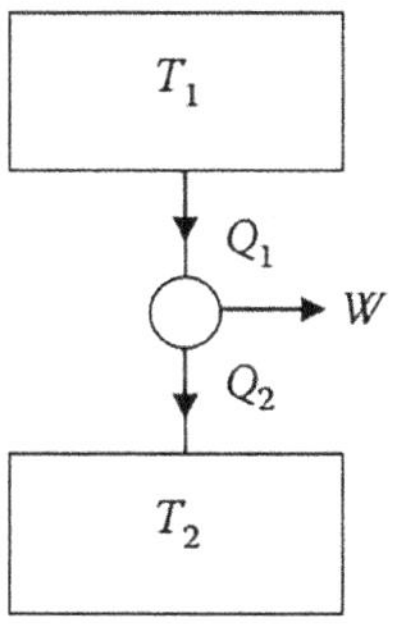

FIGURE 3.1 The figure gives a schematic representation of the operation of a heat engine. In each complete cycle, heat Q_1 is absorbed from the high-temperature bath at T_1, and heat Q_2 is rejected to the low-temperature bath at T_2. An amount of work W is produced by the engine in each cycle.

A natural question that arises concerns the maximum possible efficiency of a heat engine. This question was addressed in the nineteenth century by Sadi Carnot, who considered a reversible cyclic engine with an ideal gas as the work substance. In honor of Carnot, the ideal engine is called a Carnot engine. Figure 3.2 is a P–V diagram that depicts the cyclic operation of an ideal gas Carnot engine. The gas is contained in a cylinder fitted with a frictionless piston that can be moved very slowly and reversibly to change the volume. The Carnot cycle consists of two reversible quasi-static isothermal processes that involve the gas system in contact with heat baths at temperatures T_1 and T_2, respectively, and two reversible adiabatic processes in which the gas and the container are thermally isolated from the surroundings.

The efficiency of the Carnot engine may be obtained using the first law and the expressions derived in Chapter 2 for isothermal and adiabatic ideal gas processes. Application of the first law (Equation 2.2) to a complete cycle gives the internal energy change of the ideal gas as

$$\Delta E = Q_1 + Q_2 - W. \tag{3.2}$$

Q_1 is the heat input along isothermal *ab,* and Q_2 is the heat output along isothermal *cd.* Note that the signs of Q_1 and Q_2 must be explicitly evaluated, while W has been chosen as negative to represent work *output.* In a complete cycle, ΔE is zero because the system is returned to its initial state at temperature T_1. Equation 3.2 may therefore be rewritten for a cycle as

$$W = Q_1 + Q_2, \tag{3.3}$$

and the efficiency (Equation 3.1) is given by

$$\eta = \frac{Q_1 + Q_2}{Q_1}. \tag{3.4}$$

To proceed, it is necessary to obtain expressions for Q_1 and Q_2. From Equation 2.8, we have for the ideal gas $Q_1 = nRT_1 \ln (V_b/V_a)$ with $Q_2 = nRT_2 \ln (V_d/V_c)$. Substitution in Equation 3.4 gives

$$\eta = 1 + \frac{T_2}{T_1} \left[\ln\left(\frac{V_b}{V_a}\right) \right] \left[\ln\left(\frac{V_d}{V_c}\right) \right]^{-1}. \tag{3.5}$$

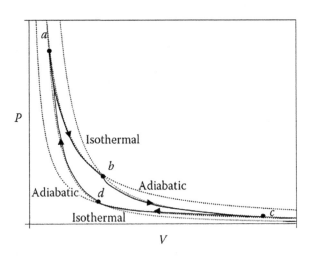

FIGURE 3.2 P–V diagram for a reversible Carnot ideal gas engine that operates between heat baths at temperatures T_1 and T_2. The cycle is made up of two isothermal processes and two adiabatic processes.

For the four paths that make up the complete cycle, four P–V relations hold:

$$\text{Isothermal} \quad a \to b \quad P_a V_a = P_b V_b; \qquad \text{Adiabatic} \quad b \to c \quad P_b V_b^\gamma = P_c V_c^\gamma;$$

$$\text{Isothermal} \quad c \to d \quad P_c V_c = P_d V_d; \qquad \text{Adiabatic} \quad d \to a \quad P_d V_d^\gamma = P_a V_a^\gamma.$$

Equation 2.12 has been used for the adiabatic processes. The four equations may be combined to show that $(V_b / V_a)^{\gamma-1} = (V_c / V_d)^{\gamma-1}$, and this result with Equation 3.5 leads to the required result for the efficiency

$$\eta = 1 - \frac{T_2}{T_1}. \tag{3.6}$$

W is obtained on substitution of the expressions for Q_1 and Q_2 in Equation 3.3 and use of $\ln (V_b/V_a) = - \ln (V_d/V_c)$. This procedure gives

$$W = nR[T_1 - T_2]\ln\left(\frac{V_b}{V_a}\right). \tag{3.7}$$

The efficiency in Equation 3.6 has a maximum value of unity, and this is achieved only at $T_2 = 0$ K, provided we disregard $T_1 = \infty$ as a practical temperature. The work output depends on T_1 and T_2 together with the volume ratio V_b/V_a. The larger the temperature difference between the heat baths, and/or the larger the V_b/V_a, which is the compression ratio for the engine, the greater the work output. Real engines have lower efficiencies than the ideal Carnot engine. An example of an approximation to a real engine cycle, the Otto cycle, is given in the problem set at the end of this chapter.

3.3 CARNOT REFRIGERATOR

Because all the processes that make up the cycle are reversible, a Carnot heat engine can be made to run backward as a refrigerator. This device takes in heat at a low temperature and rejects heat at a high temperature, provided work is supplied. For a refrigerator, the coefficient of performance is defined as follows: $\kappa = Q_2/W$, where W is now the work input per cycle and Q_2 the heat extracted from the low-temperature bath per cycle. Adapting the first law result (Equation 3.2) by making a change in sign to allow for work input gives $\kappa = -Q_2 / (Q_1 + Q_2)$. For an ideal gas refrigerator, we substitute for Q_1 and Q_2 from Equation 2.8 and, with the volume ratios as determined above, obtain for the coefficient of performance

$$\kappa = \frac{T_2}{T_1 - T_2}. \tag{3.8}$$

From Equation 3.8, it is clear that κ can be much larger than unity if the temperature difference $(T_1 - T_2)$ is small. This implies that more heat energy is extracted from the cold thermal reservoir than the work input in each cycle. The smaller the temperature difference between hot and cold reservoirs, the larger κ becomes. A refrigerator or an air conditioner unit typically operates between reservoirs that are not very different in temperature on the kelvin scale so that $T_2 \gg (T_1 - T_2)$ and $\kappa \gg 1$.

Exercise 3.1

A Carnot engine that uses 0.2 mol of an ideal gas as working substance operates between heat baths at 200°C and 60°C. If the expansion ratio along the high-temperature isothermal is 10, find the efficiency of the engine and the work output per cycle.

The efficiency is obtained with use of Equation 3.6 after converting the temperatures to kelvins. This gives $\eta = 1 - T_2/T_1 = 0.3$.

From Equation 3.7, the work output per cycle is

$$W = nR[T_1 - T_2]\ln\left(\frac{V_b}{V_a}\right) = (0.2 \times 8.314 \times 140)\ln 10 = 536 \text{ J}.$$

Exercise 3.2

Obtain the coefficient of performance for a Carnot refrigerator using an ideal paramagnet as the working substance.

We make use of the results obtained in Chapter 2, with the first law in the form $dE = dQ - \mu_0 V M\, dH$ (Equation 2.21) together with Curie's law, $M = CH/T$. On an $M - H$ diagram, a paramagnetic Carnot refrigerator has the cycle shown in Figure 3.3.

The coefficient of performance is $\kappa = -Q_2/(Q_1 + Q_2)$. From Equation 2.30, we obtain $Q_1 = \Delta Q_{ba} = -[(\mu_0 CV)/(2T_1)][H_a^2 - H_b^2]$ and $Q_2 = \Delta Q_{dc} = -(\mu_0 CV/2T_2)[H_c^2 - H_d^2]$. These expressions give

$$\kappa = -\left(\frac{T_2\left(H_a^2 - H_b^2\right)}{T_1\left(H_c^2 - H_d^2\right)} + 1\right)^{-1}. \tag{3.9}$$

Along the adiabatic paths cb and ad, $H_b/T_1 = H_c/T_2$ and $H_a/T_1 = H_d/T_2$, where use has been made of Equation 2.31. It follows that $H_d/H_a = T_2/T_1$ and $H_b/H_a = H_c/H_d$. Substitution in Equation 3.9 results in

$$\kappa = \frac{T_2}{T_1 - T_2}. \tag{3.10}$$

This is the same result that we obtained for an ideal gas Carnot refrigerator. Figure 3.4 gives a schematic representation of a paramagnet-based Carnot refrigerator.

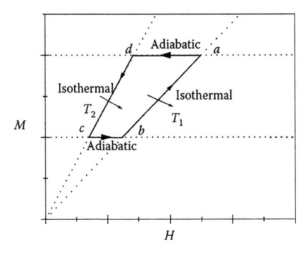

FIGURE 3.3 *M–H* diagram for a Carnot refrigerator with an ideal paramagnet as the work substance. Note that $T_1 > T_2$ with the slopes of the isothermals given by *C/T*. The adiabatic processes correspond to constant *M*. For the isothermal processes, the small arrows depict the heat flow direction.

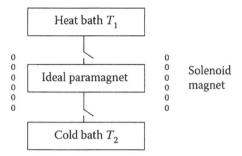

FIGURE 3.4   A diagram of a paramagnet-based Carnot refrigerator system that can provide cooling of specimens at low temperatures through magnetic work processes. Control of thermal contact between the paramagnet and the two heat baths is achieved with use of heat switches. Heat is pumped from T_2 to T_1.

3.4   ENTROPY

Equation 3.4 gives the efficiency of a Carnot engine in terms of heat transferred in a complete cycle, whereas Equation 3.6 gives η in terms of the temperatures of the heat baths used. When these two expressions are combined, we obtain $\eta = 1 + (Q_2 / Q_1) = 1 - T_2 / T_1$, which, on rearrangement, gives $Q_1 / T_1 + Q_2 / T_2 = 0$ or expressed as a summation

$$\sum_{i=1,2} \frac{Q_i}{T_i} = 0. \tag{3.11}$$

The result $\sum_i Q_i / T_i = 0$ can be generalized to any reversible cycle, which may be regarded as made up of a large number of elementary Carnot cycles, as shown schematically in Figure 3.5.

To replicate the arbitrary reversible cycle, a large number of heat baths at carefully chosen temperatures T_i ($i = 1,2,3,...$) are needed. Note that the adiabatic path contributions will cancel in pairs except for the first and last adiabatics. The elementary isothermal processes are chosen to match the path of the arbitrary cycle as closely as possible. In considering an arbitrary reversible cycle in which large numbers of elementary heat transfer processes occur at various temperatures, it is natural to replace the summation in Equation 3.11 by an integral around the cycle (denoted by $\oint$), and this gives

$$\oint \frac{dQ}{T} = 0. \tag{3.12}$$

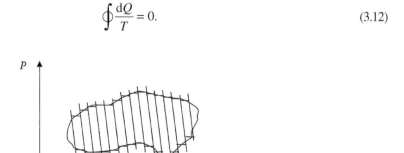

FIGURE 3.5   An arbitrary reversible cycle can be replaced by a large number of elementary Carnot cycles that together approximate the original cycle. The steep adiabatic paths cancel in pairs except for the two end cycles. A large number of heat baths are necessary for the reversible isothermals.

Equation 3.12 is an extremely important result for reversible cyclic processes. For the complete cycle, it is clear that the total heat transferred $\oint dQ \neq 0$ because some heat energy is converted to work. In Chapter 2, it was pointed out that the heat energy transferred in a process from some initial state to a final state depends on the path followed or, stated in an alternative way, dQ is not an exact differential. Equation 3.12 suggests that $1/T$ is the integration factor for dQ and leads to an exact differential, which is represented by dS, that is, $dS = dQ/T$. For a reversible cyclic process, we then have

$$\oint_R dS = \int_R \frac{dQ}{T} = 0. \tag{3.13}$$

This is *Clausius's theorem*, named in honor of the nineteenth-century physicist who first grasped its significance. S is called the *entropy*, and dS is the infinitesimal change in entropy associated with heat dQ transferred at temperature T. Equation 3.13 shows that the entropy is of fundamental importance and warrants a closer examination of its properties. The entropy change in a finite *reversible* process, where i and f represent the initial and final states, respectively, is

$$\Delta S = \int_i^f \frac{dQ}{T}. \tag{3.14}$$

This relationship is of central importance in thermal physics.

3.5 ENTROPY CHANGES FOR REVERSIBLE CYCLIC PROCESSES

In our discussion of the ideal gas Carnot cycle depicted in Figure 3.2, expressions for Q_1 and Q_2 were obtained with forms $Q_1 = nRT_1\ln(V_b/V_a)$ and $Q_2 = nRT_2\ln(V_d/V_c)$. It is instructive to consider the entropy changes in the two reversible isothermal processes. From Equation 3.14, the two entropy changes are $\Delta S_{ab} = 1/T_1 \int_a^b dQ = nR\ln(V_b/V_a)$ and $\Delta S_{cd} = 1/T_2 \int_a^b dQ = nR\ln(V_d/V_c)$. As shown in Section 3.2, $(V_d/V_c) = (V_a/V_b)$, and it follows that $\Delta S_{ab} = -\Delta S_{cd}$. The entropy changes for the ideal gas along the two reversible isothermal paths in the cycle are equal but opposite in sign, whereas the entropy changes along the reversible adiabatic paths are zero because no heat is transferred, that is, $\Delta S_{bc} = \Delta S_{da} = 0$. For the complete reversible cycle, it follows that

$$\oint dS = \Delta S_{ab} + \Delta S_{bc} + \Delta S_{cd} + \Delta S_{da} = 0. \tag{3.15}$$

The result in Equation 3.15 may be generalized to any reversible cycle using arguments similar to those in Section 3.4. The temperatures T_1 and T_2 and the volume changes involved in the cyclic process are arbitrary. The fact that the entropy change for the ideal gas in an arbitrary reversible cycle is zero leads us to conjecture that the entropy change in any process from some initial state i to a given final state f is independent of the path followed, provided it is reversible. This is demonstrated with the help of Figure 3.6 that shows states i and f on a P–V diagram, with arbitrary reversible paths 1 and 2 that connect i and f in a cyclic way.

The above results lead to the relation $\Delta S_1 + \Delta S_2 = 0$ or $\Delta S_1 = -\Delta S_2$ for the closed cycle. It follows that the entropy change in a process from i to f along *any* reversible path is the same and depends only on the initial and final states. This implies that entropy is a state function determined by the state variables. It follows that for an ideal gas we have $S = S(P, V, T)$. In fact, only two variables are needed to specify S completely because the equation of state connects P, V, and T.

In the above discussion, attention has been focused on a Carnot engine system that involves an ideal gas in a piston-cylinder arrangement. The entropy changes of the associated heat baths have not been considered, but these are readily obtained. For the high-temperature bath at fixed temperature T_1,

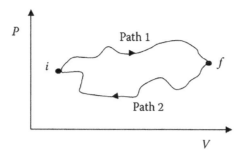

FIGURE 3.6 *P–V* diagram for an arbitrary closed reversible cycle made up of paths 1 and 2 between states *i* and *f* for a particular system. Because entropy is a state function, the entropy change along path 1 is equal and opposite to that along path 2.

$\Delta S_1^b = -Q_1/T_1$, and the use of the expression for Q_1 from Section 3.2 gives $\Delta S_1^b = -nR\ln(V_b/V_a)$. This is equal in magnitude and opposite in sign to the entropy change of the gas given above. Similarly, for the low-temperature reservoir, we have $\Delta S_2^b - nR\ln(V_d/V_c) = nR\ln(V_b/V_c)$. Comparison shows that $\Delta S_1^b = -\Delta S_2^b$ or $\Delta S_1^b + \Delta S_2^b = 0$. This is an important result. Not only is the entropy change of the working substance in the Carnot engine zero in a complete cycle, but the sum of the entropy changes of the two heat baths is also zero. This may be summarized as follows: *For reversible cyclic processes, the entropy change ΔS_u of the local universe is zero.* Local universe in this case means the system plus the heat baths. No other parts of the universe are involved in the operation of the heat engine. For any part of the cyclic process, it is easy to show that $\Delta S_u = 0$. Along the isothermal paths, as the entropy of the gas is increased, the entropy of the associated heat bath is decreased, whereas in the reversible adiabatic paths, where no heat is transferred, the entropy remains constant so that $\Delta S_u = 0$ again for these processes. These conclusions may be extended to any reversible paths.

3.6 ENTROPY CHANGES IN IRREVERSIBLE PROCESSES

The expansion of an ideal gas, as used in the operation of an engine, may be carried out suddenly and irreversibly instead of reversibly. With the cylinder and gas system thermally isolated, we imagine that the piston is quickly pulled outward, as shown in Figure 3.7.

In this situation, "suddenly" implies that the expansion process occurs in a time much shorter than the time taken for the gas to reach equilibrium. As an order of magnitude, the time should be shorter than the time it takes for a particle traveling at the average speed in the gas to traverse a distance comparable with the interparticle separation. In Chapter 1, this is seen to be of the order of nanoseconds for a gas such as He at ambient conditions. There is no heat bath involved, and the expansion is adiabatic. Because the expansion is sudden, no work is done on the system. Therefore, no change in temperature of the gas occurs, according to the equipartition of energy theorem, because the internal energy of the gas remains constant. Because entropy is a state function,

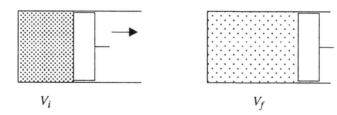

FIGURE 3.7 Irreversible sudden expansion of an ideal gas from initial volume V_i to final volume V_f. In the above representation, we imagine that the piston is withdrawn in a time much less than the mean collision time of particles in the gas.

the entropy change of the gas depends only on the initial and final states. In the calculation of the entropy change, we replace the irreversible process by a *reversible* isothermal process to carry out the integration. This change in entropy has already been obtained above in our discussion of the Carnot cycle given in Section 3.5. With initial state i and final state f, we obtain $\Delta Q = nRT\ln\left(V_f/V_i\right)$, and hence, $\Delta S = nR\ln\left(V_f/V_i\right)$. In this irreversible process, there is no change in entropy of the heat bath. This means that the entropy of the universe has increased as a result of the process, or

$$\Delta S_u > 0. \tag{3.16}$$

The increase in entropy cannot be removed once it has occurred. The system can, of course, be restored to its initial state i, with entropy S_i, either reversibly or irreversibly, but the entropy increase of the universe, which occurred in the original irreversible process, remains for all time. When the system is restored to its initial state, the entropy increase that accompanied the irreversible process is passed on to another system, such as a heat bath, and is never cancelled out. All irreversible processes are accompanied by an increase in entropy of the universe. As a further illustration of this result, consider an irreversible transfer of heat energy between two finite heat reservoirs, with heat capacities C_a and C_b, respectively, as shown in Figure 3.8.

At some instant, the reservoirs a and b are brought into thermal contact when the heat switch is closed. No work is done, and the first law applied to the combined system gives $\Delta E = 0 = \Delta Q_a + \Delta Q_b$, where $\Delta Q_a = C_a\left(T_1 - T_f\right)$ and $\Delta Q_b = C_b\left(T_2 - T_f\right)$. C_a and C_b are the heat capacities of the two baths. For simplicity, let $C_a = C_b = C$ so that $T_f = (T_1 + T_2)/2$. The change in entropy of the combined system is given by the sum of the entropy changes for each reservoir, $\Delta S = \Delta S_a + \Delta S_b = \int_{T_1}^{T_f} dQ_a/T_a + \int_{T_2}^{T_f} dQ_b/T_b$. Because $dQ = C\,dT$, the integrations are readily carried out, and we obtain

$$\Delta S = C\left[\ln\left(\frac{T_f}{T_1}\right) + \ln\left(\frac{T_f}{T_2}\right)\right] = C\,\ln\left[\left(\frac{T_1+T_2}{2T_1}\right)\left(\frac{T_1+T_2}{2T_2}\right)\right] = C\,\ln\left[\frac{(T_1+T_2)^2}{4T_1T_2}\right].$$

It is straightforward to show that the quantity in square brackets is larger than unity. (The condition is simply $(T_1 - T_2)^2 > 0$.) It follows that $\Delta S > 0$.

The increase in entropy of the universe that occurs in irreversible processes corresponds to a loss of opportunity to obtain useful work from a system. For example, in the case of the two heat reservoirs just considered, before the closure of the heat switch, a heat engine could be operated between the reservoirs to provide work for a time until the reservoirs reach a common temperature. However, when the reservoirs exchange heat energy irreversibly after the heat switch is closed,

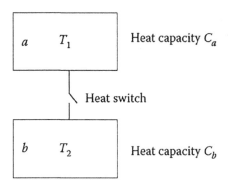

FIGURE 3.8   The two finite heat reservoirs a and b, initially at temperatures T_1 and T_2, are brought into thermal contact by closing the heat switch. The final equilibrium temperature is T_f.

the opportunity for operation of a heat engine between them is lost. Similar analyses may be carried out for all irreversible processes. Consider, for example, a large mass at the top of a high building. The mass can be lowered slowly using a system of low-friction pulleys that are arranged to provide useful work output. Alternatively, the mass can be allowed to fall freely with the initial potential energy converted into kinetic energy and then finally largely into heat, when the mass strikes the ground, without useful work being done. The entropy of the universe is increased much more in the second irreversible process than in the first approximately reversible process.

3.7   THE SECOND LAW OF THERMODYNAMICS

The discussion of reversible and irreversible processes and the associated entropy changes given above can be summarized in a compact statement,

$$\Delta S_u \geq 0. \tag{3.17}$$

This is known as the *second law of thermodynamics*, which is written as follows: *In any process, the entropy change of the local universe, considered to consist of a system and adjacent bodies, is either zero or greater than zero.* The equality holds, provided all processes are carried out reversibly. The inequality holds when irreversible processes are involved.

Although it appears to be very simple, the second law has profound implications, which will be dealt with in later chapters. As with any physical law, it is based on a large body of experimental observations. Alternative statements of the second law may be given. These are of historical importance but can be shown to be equivalent to the statement given in Equation 3.17. A brief discussion of these alternative statements is given later in this chapter. Most processes that occur in nature are irreversible. It follows that the entropy of the universe must steadily increase. The direction of time is linked to entropy increase, and the second law provides what has been called *time's arrow*. The forward flow of time is accompanied by an increase in entropy of the universe.

3.8   THE FUNDAMENTAL RELATION

If we combine the first and second laws, we obtain what is called the *fundamental relation* of thermodynamics. The first law in terms of infinitesimal quantities is $dE = dQ + dW$. For fluids, this is written as $dE = dQ - P\,dV$ (Equation 2.7). We now have $dQ = T\,dS$, for an infinitesimal heat transfer process, as introduced above, and this leads to $dE = T\,dS - P\,dV$, or equivalently,

$$T dS = dE + P\,dV. \tag{3.18}$$

This is the fundamental relation, and as its name implies, it is of great importance in thermodynamics. Inspection of Equation 3.18 shows that it involves the state variables P, V, and T and the state functions E and S. The inexact differentials dQ and dW have been replaced by exact differentials. Many applications of Equation 3.18 are given later. Because it involves state functions that depend only on the initial and final state, Equation 3.18 is valid for *any* process and not only for reversible processes.

3.9   ENTROPY CHANGES AND *T–S* DIAGRAMS

For many purposes, it is sufficient to calculate changes in entropy, and it is not necessary to determine absolute values of the entropy. In entropy calculations, use is made of Equation 3.14, $\Delta S = \int_i^f dQ/T$.

The integral *must* be carried out along a reversible path, but we can choose any reversible path we like because entropy is a state function. The SI entropy unit is joules per kelvin. It is convenient to represent processes in a *T–S* diagram, as shown in Figure 3.9. Following convention, we plot T (intensive) as

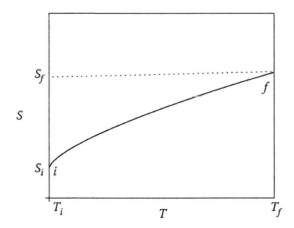

FIGURE 3.9   A temperature entropy or T–S diagram that represents a process from an initial state i to a final state f.

the ordinate and S (extensive) as the abscissa. This is different to the convention used in PV and MH diagrams introduced in Chapter 2 in which the extensive variable is plotted as the ordinate. In those cases, the work done is given by the area under the curve.

The area $\Delta Q = \int_i^f T dS$, which can be obtained from a T–S diagram, gives the total amount of heat transferred in a reversible process. Adiabatic processes, which correspond to constant S, are represented by horizontal isentropic lines on the T–S diagram, whereas isothermals are represented by vertical lines. The Carnot cycle has the simple rectangular form shown in the T–S diagram in Figure 3.10.

The heat energy absorbed at T_1 along isothermal ab is $Q_1 = T_1(S_2 - S_1)$. Similarly, along path cd, $Q_2 = T_2 (S_1 - S_2) = -T_2 (S_2 - S_1)$. From Equation 3.4, the efficiency is $\eta = (Q_1 + Q_2)/Q_1$, and substitution for Q_1 and Q_2 gives, as before, $\eta = (T_1 - T_2)/T_1$. The coefficient of performance of a refrigerator may be obtained just as easily using this graphical T–S representation with paths reversed.

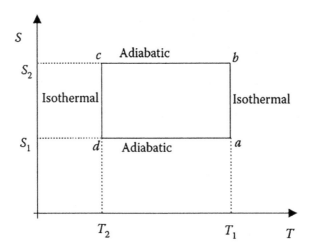

FIGURE 3.10 T–S diagram representation of a Carnot cycle that shows two isothermal processes at temperatures T_1 and T_2, respectively, and two adiabatic processes. The cycle represents either an engine or a refrigerator depending on the direction of operation.

3.10   THE KELVIN TEMPERATURE SCALE

For an ideal gas Carnot cycle, Equation 3.11 relates T_1 and T_2 as follows: $Q_1/T_1 + Q_2/T_2 = 0$ or $T_1/T_2 = -Q_1/Q_2$. This shows that it is possible to compare the temperatures of two heat baths from measurements of the amounts of heat transferred to or from the heat baths when a Carnot engine is operated between them. Although Equation 3.11 has been derived for an ideal gas heat engine, other Carnot cycles, for example, an ideal paramagnet engine, can be used. It follows that temperatures on the basis of Carnot cycle heat measurements are not dependent on the properties of any particular substance. The scale obtained is called the absolute or kelvin temperature scale. Absolute temperatures are specified in kelvin. It is, of course, necessary to fix a particular temperature as a reference temperature. The triple point of water is chosen to have a temperature of 273.16 K, as described in Chapter 1. Because Equation 3.11 was derived using the ideal gas equation of state, it follows that the ideal gas scale must be the same as the kelvin scale.

3.11   ALTERNATIVE STATEMENTS OF THE SECOND LAW

As noted previously in Section 3.8, the second law may be stated in the compact form $\Delta S_u \geq 0$ (Equation 3.17). This statement means that, in any process, the entropy of the local universe either remains constant or increases. Local universe in this context again means all systems and reservoirs involved in the process are considered. Although the entropy of a particular component system may decrease, the entropy of the entire assembly will either remain unchanged or increase.

Earlier statements that are equivalent to Equation 3.17 are known as the Clausius statement and the Kelvin–Planck statement. Both are concerned with heat transfer and, like other physical laws, are laws of experience. The classical statements involve processes that experience shows to be impossible.

The Clausius Statement
It is impossible to construct a device that allows the transfer of heat from a reservoir at a low
  temperature to a reservoir at a higher temperature with no other effect.

The impossible Clausius process is depicted in Figure 3.11.

Exercise 3.3

From a consideration of the net entropy change in a complete cycle of the refrigerator shown in Figure 3.11, prove that the process violates the second law as stated in Equation 3.17.

In a complete cycle, the entropy change of the local universe (the two heat baths) is given by $\Delta S_u = Q/T_1 - Q/T_2 = Q(T_2 - T_1)/T_1 T_2 < 0$, which violates the second law.

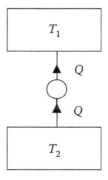

FIGURE 3.11   The Clausius statement of the second law of thermodynamics asserts that the process shown in the diagram, in which heat Q is transferred from a cold reservoir at T_2 to a hot reservoir at T_1, with no work input, is impossible.

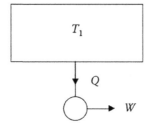

FIGURE 3.12 The Kelvin–Planck statement of the second law asserts that the process shown in the diagram, in which heat Q is converted entirely into work W, is impossible.

The Kelvin–Planck Statement
It is impossible to construct a device that allows the absorption of heat energy from a reservoir and the production of an equivalent amount of work and no other effect.

The impossible Kelvin–Planck process is shown in Figure 3.12.

Exercise 3.4

Show that the Clausius and Kelvin–Planck statements are equivalent, as is to be expected.
    The equivalence can be shown by considering the arrangement in Figure 3.13, which involves two coupled cyclic devices.
    The refrigerator A in Figure 3.13 violates the Clausius statement of the second law. The refrigerator A plus Carnot engine B together violate the Kelvin–Planck statement because zero heat energy is transferred to the low-temperature reservoir while work is produced. It is therefore apparent that the two second law statements can be linked, and violation of one statement implies violation of the other.

An axiomatic basis for thermodynamics, and in particular the second law, was developed by Carathéodory in the early part of the twentieth century. In a general approach, he considered systems with three or more independent thermodynamic variables. An example of a system with five state variables is a paramagnetic gas for which the variables are P, V, T, H, and M. In the case of five variables, there are two equations of state linking the variables, and any three specify the state of the system completely.
    By considering reversible adiabatic processes starting from a particular initial state, Carathéodory generated surfaces in the three variable space. He then argued that such surfaces

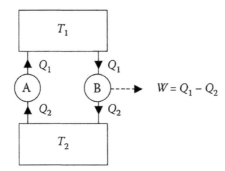

FIGURE 3.13 A refrigerator A and a Carnot engine B operate in a coupled cycle between heat baths at temperatures T_1 and T_2, with $T_1 > T_2$. It is impossible for the combined refrigerator–engine system to produce work output with zero net heat input from the high-temperature bath. The coupled system illustrates the equivalence of the Clausius and the Kelvin–Planck statements of the second law.

could not intersect and that it was impossible to move from one surface to another by means of a reversible adiabatic process. This became the axiomatic foundation of the second law. In fact, the Carathéodory axiomatic approach, although mathematically elegant, is equivalent to the Kelvin–Planck and Clausius statements of the second law. We do not discuss these historical formulations of the second law in greater detail here. Instead, a detailed microscopic discussion of entropy is given in Chapter 4. The microscopic approach provides a general procedure for the calculation of the entropy of a system and leads to great physical insight into reversible and irreversible processes. The formulation of the second law given in Equation 3.17 is readily understood in terms of the microscopic approach.

3.12 GENERAL FORMULATION

Thermodynamics is traditionally developed using the laws of thermodynamics as the starting point. The fundamental relation given in Equation 3.18, $dE = T\,dS - P\,dV$ or, equivalently, $T\,dS = dE + P\,dV$, effectively combines the first and second laws. The third law, which is discussed in Chapter 6, concerns the way in which the entropy behaves at low temperatures and is expressed in terms of the unattainability of absolute zero temperature. The three laws of thermodynamics are laws of experience on the basis of experimental observation.

An alternative fundamental approach uses three basic and reasonable postulates as a point of departure. The success of the theory constructed in this way is determined by comparison of the predictions with experiment. The approach is briefly presented below. Again, only systems at, or very close to, equilibrium are considered. The first postulate is concerned with specification of equilibrium states, whereas the second and third postulates are concerned with the entropy function. To be as general as possible, multicomponent systems are considered. This has the benefit of introducing an important new quantity, called the chemical potential, into our discussion.

Postulate 1

A complete specification of the equilibrium state of a system may be given in terms of extensive variables, specifically the internal energy E, the volume V, and the numbers N_i of the different constituent particles or molecules.

The second postulate introduces the entropy, which plays a key role in the subject.

Postulate 2

A function called the entropy S exists and is specified completely, for all equilibrium states, in terms of the extensive variables E, V, and N_i.

$$S = S(E, V, N_1, \ldots, N_i, \ldots, N_N).\tag{3.19}$$

The entropy function is continuous and differentiable.

To allow for systems that are made up of various subsystems, a third postulate is stated.

Postulate 3

The entropy S of a composite system is the sum of the entropies S_α of the constituent subsystems.

$$S = \sum_\alpha S_\alpha.\tag{3.20}$$

A corollary to this postulate is that S is a homogeneous first-order function of the extensive variables, which satisfies the mathematical relationship.

$$S(\lambda E, \lambda V, \lambda N_i) = \lambda S(E, V, N_i). \tag{3.21}$$

The proof of this corollary is obvious. In many cases of interest, the entropy S is a monotonically increasing function of the energy E, but this is not universally true. Systems with an upper energy bound, such as spin systems, show a maximum in S for some value of E. S is found to decrease at higher E, and this feature is discussed in detail in Chapters 5 and 6. An important consequence of the extremum in the entropy is the need to allow for negative temperatures at high energies in systems with an upper energy bound. Again, this topic is briefly discussed in Chapter 6.

The relationship, $S = S(E, V, N_1,\ldots, N_n)$, is an alternative statement of the fundamental relation, which we can understand through consideration of the total differential of S, given by

$$dS = \left(\frac{\partial S}{\partial E}\right)_{V,N_i} dE + \left(\frac{\partial S}{\partial V}\right)_{E,N_i} dV + \sum_i \left(\frac{\partial S}{\partial N_i}\right)_{E,V,N_j \neq N_i} dN_i. \tag{3.22}$$

The coefficients may be formally defined as variables. To make contact with our earlier discussion, we compare Equation 3.22 with the fundamental relation for fluids Equation 3.18, and identify

$$\left(\frac{\partial S}{\partial E}\right)_{V,N_i} = \frac{1}{T}, \tag{3.23}$$

$$\left(\frac{\partial S}{\partial V}\right)_{E,N_i} = \frac{P}{T}. \tag{3.24}$$

For a single component system with only one type of particle, the third term on the right-hand side of Equation 3.22 is zero. This gives $dS = (1/T)dE + (P/T)\,dV$ or $T dS = dE + PdV$, which is the fundamental relation given in Equation 3.18.

The partial derivative of S with respect to the N_i, given by $\left(\partial S/\partial N_i\right)_{E,V,N_j \neq N_i}$, introduces a new quantity into the subject, related to the particle concentration, and we define

$$\left(\frac{\partial S}{\partial N_i}\right)_{E,V,N_j \neq N_i} = -\frac{\mu_i}{T}, \tag{3.25}$$

where, for historical reasons, μ_i is called the chemical potential of species i. Note that in the partial derivative, the variables E, V, and all the N_j are kept fixed while only N_i is allowed to change. The partial derivatives of S with respect to the extensive variables give $1/T$, P/T, and $-\mu_i/T$, as shown above. These intensive quantities are functions of the extensive variables E, V, and N_i. For example, we have $T = T(E, V, N_i)$, which may be regarded as an equation of state for the system. Other equations of state exist for each of the intensive variables. Knowledge of all the equations of state is equivalent to knowledge of the fundamental relation for a system.

The general fundamental relation, which is based on Equation 3.22, may now be written, with use of the identifications made in Equations 3.23–3.25, as

$$T\,dS = dE + P\,dV - \sum_i \mu_i\,dN_i. \tag{3.26}$$

This general form is used for processes in multicomponent systems. For single component systems, the final term that involves the chemical potential can be omitted when N is fixed.

In considering multicomponent systems, we have introduced the chemical potential μ into our thermodynamic formalism. This important new quantity determines, for example, how particles in a single component or multicomponent gas in which there is a nonequilibrium concentration gradient reaches equilibrium through particle diffusion. Further discussion of the chemical potential is given in Chapters 5 and 7. In Chapter 5, an explicit expression for the chemical potential of a single component ideal gas is obtained on the basis of the microscopic definition of the entropy that is given there. The chemical potential is very important in the discussion of phase transitions and chemical equilibrium and in the development of quantum statistics later in the book. As a specific illustration of the role of the chemical potential in establishing the final equilibrium state of a system, consider a cylindrical container of a gas that is separated into parts 1 and 2 by movable piston that is thermally conducting and has small channels through which the particles can diffuse. If at some instant it is arranged, through various devices, that the temperatures, pressures, and particle concentrations on the two sides of the piston are all unequal and the system is then allowed to reach equilibrium, so that the entropy of the two part system is a maximum, we shall later show (Section 11.1) that in equilibrium $P_1 = P_2$, $T_1 = T_2$, and $\mu_1 = \mu_2$. When particle diffusion plays a role in a process, the chemical potential gradient drives the diffusion.

The general formulation outlined above is given in what is called the entropy representation. An alternative formulation may be given in the energy representation. The continuity, differentiability, and normally monotonic properties of the entropy function imply that it can be inverted to give $E = E(S, V, N_i)$. By forming the total differential of E, various partial derivatives are obtained, which may be defined as thermodynamic variables, using an approach equivalent to that used in the entropy formulation. We leave this development as an exercise. The entropy formulation is similar to the microscopic statistical formulation developed in Chapters 4 and 5.

3.13   THE THERMODYNAMIC POTENTIALS

In the development of thermodynamics, it is found useful to introduce new quantities that involve both state functions and state variables in combinations that are termed thermodynamic potentials. In Chapters 7 and 8, the potentials are used, for example, in considering processes in gases and condensed matter. In Chapter 9, the potentials are of central importance in the thermodynamic description of phase transitions and critical phenomena. In addition, in Chapters 10 and 11, we shall use certain of the potentials to establish extremely important bridge relationships that connect the microscopic statistical and macroscopic thermodynamic descriptions of large systems. In view of their importance, we give a short introduction to the thermodynamic potentials here.

The three potentials that we shall be most concerned with are the enthalpy represented by H, the Helmholtz potential F, and the Gibbs potential G. The names of the last two of these quantities are chosen to honor scientists who made major contributions to the development of the subject.

The three potentials are defined as follows:

$$\text{Enthalpy } H \ = \ E \ + \ PV,$$

$$\text{Helmholtz potential } F \ = \ E \ - \ TS,$$

$$\text{Gibbs potential } G \ = \ E \ - \ TS \ + \ PV.$$

All three involve the internal energy E together with products of intensive and extensive variables that result in quantities with energy dimensions. Insight into the seemingly arbitrary definitions can be gained with use of the Legendre transform, details of which are given in Appendix D. The underlying idea of the Legendre transform can be grasped by considering alternative ways of representing a function $F(x)$ of a single variable x. In addition to giving values of the function for various x, the function may be specified by a complete set of tangents to the function, in a graphical representation, together with corresponding intercepts. The tangents form the envelope of the curve. Following the notation used in Appendix D, the Legendre transform function $L(s)$ of the function $F(x)$ is given by $L(s) = F(x(s)) - s(x) x(s)$ or more simply $L = F - sx$, where $S(x) = \partial F(x)/\partial x$ is the slope of the function at some point x and $F(x(s))$ is the value of the function at that point. The transform $L(s)$ is related to the intercept of the tangent line in the geometrical representation. The Legendre transform can be defined for functions of more than one variable as discussed in Appendix D.

In the energy representation introduced in Section 3.13, we have for a homogeneous system with a fixed number of particles $E = E(S, V)$. The partial Legendre transform with S held constant is $L = E - (\partial E/\partial V)_S V = E + PV = H$, where use has been made of the fundamental relation given in Equation 3.18 and the definition of the enthalpy H given above. The partial Legendre transform of $E(S, V)$ with V held constant gives the Helmholtz potential F and the full transform, allowing both S and V to vary, leads to the Gibbs potential G. Details are given in Appendix D. As pointed out there in connection with the enthalpy, the Legendre transform of $E(S, V)$ replaces S, which is an extensive quantity that is not readily controlled, with the intensive variable P that can be controlled as an independent variable. This is the important change that the Legendre transform facilitates.

The fundamental relation given in Equation 3.18 is $T\,dS = dE + P\,dV$, and combining this relation with the differential of F, it is readily shown that $S = -(\partial F/\partial T)_V$ and $P = -(\partial F/\partial V)_T$. similarly from the differential of G, it follows that $S = -(\partial G/\partial T)_P$ and $V = (\partial G/\partial P)_T$. These useful results are presented in greater detail in Chapter 7 in the discussion of the thermodynamics of gaseous systems.

PROBLEMS CHAPTER 3

3.1   A Carnot engine operates between a hot reservoir at 250°C and a cold reservoir at 30°C. If the engine absorbs 500 J of heat from the hot reservoir, how much work does it deliver per cycle? If the same engine were run in reverse as a heat pump, how much heat would be removed from the cooler reservoir per cycle if the work input is 200 J? If the engine were operated as a heat pump between an outside temperature of 0°C and an inside temperature of 22°C, how much heat would be delivered per cycle assuming the same work input as before?

3.2   A system consisting of 1 mol of an ideal monatomic gas is initially at pressure P and volume V The gas is allowed to expand isothermally until its volume has doubled and is then compressed isochorically to a pressure $2P$. Sketch the process on a P–V diagram. Obtain expressions for the change in internal energy and entropy of the gas for the complete process.

3.3   In a cyclic process, n moles of a monatomic ideal gas are isothermally expanded at high-temperature T_1 from volume V_1 to V_2 followed by an isochoric lowering of pressure at V_2 then an isothermal compression process at low-temperature T_2 from volume V_2 to V_1 and finally an isochoric compression back to the starting point. Construct a P–V diagram for the cycle. Indicate in which processes work is done and obtain expressions for W in each of these processes. Determine for which processes heat is transferred and obtain expressions for Q in each of these processes. If the isochoric processes are carried out reversibly, briefly describe how this is done. Obtain an expression for the efficiency of the cycle. For the gas, $C_V = 3/2\,nR$ and $C_P = 5/2\,nR$, and the ideal gas equation holds.

3.4   The Otto cycle provides an approximate description of the operation of a gasoline-powered engine. An ideal gas is used as working substance, and all processes are carried out quasi-statically. The combustion of gasoline is replaced by an isochoric process in which heat Q_1 is added using a series of heat baths at steadily increasing temperatures. Heat rejection Q_2

is again accomplished using a series of heat baths. Sketch a P–V diagram for the Otto cycle involving the following four processes. (The initial gas intake and final exhaust processes that occur at atmospheric pressure may be ignored.)

a. Adiabatic compression from P_1, V_1 to P_2, V_2;

b. Isochoric (V_2) increase in pressure from P_2 to P_3 corresponding to intake of heat Q_1;

c. Adiabatic expansion from P_3, V_2 to P_4, V_1; and

d. Isochoric (V_1) drop in pressure from P_4 to P_1 corresponding to heat rejection Q_2.

   Give an expression for the efficiency of the Otto cycle in terms of Q_1 and Q_2. Express Q_1 and Q_2 in terms of the temperature changes in the two isochoric processes and relate these temperature changes to the volumes V_1 and V_2. Show that the efficiency of the Otto cycle depends on the compression ratio $C_R = V_1/V_2$. Find the efficiency of an Otto cycle with a compression ratio of nine and using a monatomic ideal gas as working substance.

3.5 A clamped frictionless conducting piston separates two unequal quantities of ideal gas into equal volumes in a thermally insulated cylindrical container so that the pressures on the two sides of the piston are in the ratio of 2:1. If the piston is unclamped, obtain expressions for the final equilibrium temperature of the combined system and the net entropy change in reaching equilibrium.

3.6 A Carnot engine is operated between two heat reservoirs, of equal heat capacity C, initially at temperatures T_1 and T_2, respectively. What is the net entropy change for the process in which the engine produces work before the two heat reservoirs reach a common final equilibrium temperature T_f? Obtain an expression for T_f in terms of T_1 and T_2. Show that the engine will deliver an amount of work $W = C(T_1 + T_2 - 2T_f)$ before the two heat reservoirs reach final equilibrium at T_f.

3.7 If in Question 3.6, the two heat baths were simply placed in contact without the heat engine, what would the final equilibrium temperature be? Obtain an expression for the entropy change of the combined system in this case and show that this change is positive. Contrast your result for the entropy change with that of Question 3.6.

3.8 Two thermally insulated reservoirs each contain 2 kg of water at temperatures of 80°C and 20°C, respectively. The two quantities of water are mixed in one of the containers without heat transfer to the surroundings and are allowed to reach equilibrium. Determine the final equilibrium temperature and find the change in entropy of the local universe in the mixing process. The specific heat of water is 4.18 J g^{-1} K^{-1}.

3.9 A paramagnetic salt that obeys Curie's law is situated in a magnetic field H at temperature T. Show that the heat capacity of a volume V of the material at constant H is given by $C_H = \mu_0 CVH^2/T^2$, where C is the Curie constant. Obtain an expression for the entropy change in a constant H cooling process.

3.10 A paramagnetic solid of volume V that obeys Curie's law, $M = CH/T$, is subjected to an isothermal magnetization process in which an applied field is increased from H_1 to H_2. Sketch an M–H diagram for the process. How much magnetic work is done? What is the change in internal energy of the system? Obtain expressions for the heat transfer and the entropy change in the process.

3.11 The susceptibility of a magnetic solid follows the Curie–Weiss law. Obtain an expression for the entropy change of the solid in an isothermal process that involves a quasi-static increase in an applied magnetic field from H_1 to H_2.

3.12 Consider the 1-carat diamond described in Question 2.10 for which the molar specific heat is given as $c_V = 3R(4\pi/5)(T/\theta_D)^3$ with $\theta_D = 2200$ K. What is the entropy increase of the diamond in heating from 4 to 300 K? Give a schematic plot of the entropy of the diamond for the temperature range considered in a T–S diagram. Describe the behavior you expect for the entropy as T tends toward very high temperatures and the specific heat tends to a constant value of $3R$ as given by the law of Dulong and Petit.

3.13 The boiling point of a monatomic liquid is at T_b. What is the entropy change of 1 mol of the substance in transforming from the liquid to the vapor phase at T_b? The latent heat of vaporization is L J mol⁻¹. Give a sketch plot of the vaporization process on a T–S diagram showing regions on both sides of the transition. Assume that the specific heat of the liquid is approximately constant just below the transition and that the vapor phase may be treated as an ideal gas at constant pressure. Include expressions for the entropy changes with T below and above T_b.

3.14 A system consisting of 2 kg of ice is heated from an initial temperature of −10°C through the melting point at 0°C to a final temperature of 20°C. The specific heat of ice c_I is 2.1×10^3 J kg⁻¹ K⁻¹, that of water c_W is $2\,c_I$, and the latent heat of melting L is 3.3×10^5 J kg⁻¹. Calculate the heat absorbed by the system in this melting process. What is the entropy change of the system? Sketch the T–S diagram for the process.

Section IB

Microstates and the Statistical
Interpretation of Entropy

4 Microstates for Large Systems

4.1 INTRODUCTION

In the nineteenth century, Boltzmann explored the connection between entropy and statistical probability. This linkage leads to methods for calculating the entropy of a system and provides a great deal of insight into the entropy concept. In the development of the statistical approach, we use a microscopic description on the basis of quantum mechanics and enumerate the states of systems of interest, specifically for the ideal gas and ideal spin systems. A classical phase space description of fluids, which is complementary to the quantum mechanical description, is also discussed.

To grasp the essential connection between entropy and the probability of finding a system in a particular state, consider the irreversible adiabatic expansion of a gas from an initial volume V_1 to a final volume $V_1 + V_2$ when a valve between two containers is opened, as shown in Figure 4.1. Initially, container 1 is filled with gas at some pressure P, and container 2 is evacuated. The process is similar to the sudden expansion of a gas shown in Figure 3.7, but in this case, no piston is involved.

As shown in Section 3.5, the entropy increase in an irreversible expansion of this kind is

$$\Delta S = nR \ln\left[\frac{V_1 + V_2}{V_1}\right].\qquad(4.1)$$

We obtained this result by replacing the irreversible path with a reversible isothermal path joining the initial and the final states of the system to carry out the necessary integration. Because $(V_1 + V_2) > V_1$, it follows from Equation 4.1 that $\Delta S > 0$, as expected from the second law for an irreversible process. When the tap is opened, it is more probable that the gas will occupy all of the space available to it rather than part of the space. Experience shows that the probability of finding the system back in its initial state is essentially zero, and the system will never return to that state once the expansion has occurred. The entropy increase is linked to a probability increase, and it is therefore natural to seek a formal connection between entropy and statistical probability.

It is necessary to place the above ideas on a firm quantitative basis in order that a microscopic statistical theory can be developed. The way in which this is done is based on the knowledge of the number of microstates in which the system may be found. We shall see that very large numbers of microstates are associated with a given macrostate specified by E, V, and N.

4.2 MICROSTATES—CLASSICAL PHASE SPACE APPROACH

In statistical calculations, it is necessary to specify the various states that are available for the particular system considered. For example, we specify the outcome when a single die is thrown by the number

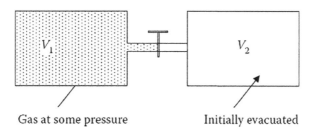

Gas at some pressure   Initially evacuated

FIGURE 4.1 An irreversible expansion process for an ideal gas occurs when the tap that joins the containers is opened. The gas expands freely from an initial volume V_1 to a final volume $(V_1 + V_2)$.

on the face that is uppermost after the throw. The die has six states in which it can be found. Similarly, for a coin, there are two possible states, heads or tails. For collections of dice, a composite state may be specified in terms of the numbers of dice in particular states. Statistical methods and probabilities are discussed in some detail in Chapter 10. To specify the various microstates of a system, it is convenient to use quantum mechanics. Historically, a classical approach was used before the development of quantum mechanics. For completeness, the classical approach is briefly discussed first in this section.

Consider N particles of mass m in a container of volume V. The classical Hamiltonian is

$$\mathcal{H} = K + U = \sum_i \left(\frac{p_i^2}{2m} \right) + \sum_{i<j} u(\boldsymbol{r}_{ij}), \tag{4.2}$$

with p_i the momentum of the ith particle and $u(\boldsymbol{r}_{ij})$ the potential energy due to the interaction of particles i and j separated by $\boldsymbol{r}_{ij}$. For an ideal gas, we have $u(\boldsymbol{r}_{ij}) = 0$. At a particular instant, the position $\boldsymbol{q}_i$ and momentum $\boldsymbol{p}_i$ coordinates for all N particles may be specified as the set $(\boldsymbol{q}_1, ..., \boldsymbol{q}_N; \boldsymbol{p}_1, ..., \boldsymbol{p}_N)$. (The limitations on how precisely we can specify the position and momentum coordinates as a result of the quantum mechanical uncertainty principle are introduced below.) The vectors $\boldsymbol{q}_i$ and $\boldsymbol{p}_i$ will each have three components, which in Cartesian coordinates are written q_{ix}, q_{iy}, q_{iz}, and p_{ix}, p_{iy}, p_{iz}. Altogether, $6N$ coordinates are needed to specify the positions and momenta of N particles in three-dimensional space. Over time, the position and momentum coordinates of particles will change, subject to the constraints on energy, $\sum_{iv} (p_{iv}^2/2m) = E$ and volume, $0 \le q_{iv} \le L_v$ ($v = x$, y, z), where E is the total energy, which is fixed within limits $\pm \delta E$, and L_v is the edge length of the container in direction v. Each different set of position and momentum coordinates for all N particles corresponds to a different microstate.

The position and momentum coordinates for each particle cannot be given as precisely as we choose because of the Heisenberg uncertainty principle in quantum mechanics. For the ith particle's position and momentum in direction x, for example, we have $\delta q_{ix} \delta p_{ix} \ge \hbar$ with $\hbar$ Planck's constant divided by 2π. The uncertainty principle should be borne in mind in the discussion that follows. For N particles, a geometrical representation of all the position and momentum coordinates involves a $6N$-dimensional hyperspace, which is called phase space. Although it is not possible to draw a $6N$-dimensional space (with $N \sim 10^{23}$), it is nevertheless a useful construct in formulating a microscopic description of a large system. In the $6N$-dimensional space, a single point represents all the position and momentum coordinates for the particles in a system, and this special point, which is called the *representative point*, specifies the microstate of the system, at a given instant, in terms of the set $(q_{1x}, q_{1y}, q_{1z}, ..., q_{Nx}, q_{Ny}, q_{Nz}; p_{1x}, p_{1y}, p_{1z}, ..., p_{Nx}, p_{Ny}, p_{Nz})$.

As time passes, the representative point will traverse the $6N$-dimen-sional hyperspace subject to the constraints given above. To illustrate these ideas, consider a single particle in a one-dimensional container for which the phase space, or $q - p$, representation is shown in Figure 4.2.

The representative point traverses the accessible regions of phase space that consist of the two shaded bands shown in Figure 4.2. The momentum is given by $p = \pm\sqrt{2m\varepsilon}$, where ε is the kinetic energy of the particle. To specify or label microstates for this one-dimensional system, it is useful to subdivide the two-dimensional phase space into cells of size $\delta p \delta q = h_0$, with $h_0 \ge \hbar$, so that a given microstate may be specified by giving the particular cell in phase space in which the representative point for the system is located. These ideas may be extended in an obvious way to our $6N$-dimensional hyperspace. It is clear that the number of cells involved will become extremely large, but over time, the representative point will traverse all the accessible regions of phase space. This is known as the *ergodic* hypothesis or, more strictly, the *quasi-ergodic* hypothesis. The hypothesis is of fundamental importance in establishing the foundations of statistical mechanics. We introduce a related hypothesis in considering accessible quantum states in Chapter 5.

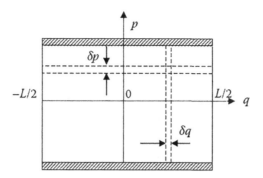

FIGURE 4.2 Phase space representation of the position and momentum coordinates for a single particle in a one-dimensional container. The particle has momentum in the range from p to $p + dp$ and can be anywhere in the one-dimensional available space from $-L/2$ to $L/2$.

The phase space representation may be extended to a large set or *ensemble* of similar systems, all of which are subject to the same constraints. The set of representative points for the whole ensemble will constitute a type of fluid moving through phase space with time. By considering this new kind of fluid, it is possible to obtain what is called Liouville's equation for the rate of change of the density of representative points in a given region of phase space. These ideas are pursued further in Chapter 18. We now consider an alternative specification of microstates using simple quantum mechanical ideas.

4.3 QUANTUM MECHANICAL DESCRIPTION OF AN IDEAL GAS

Schrödinger's equation for a nonrelativistic system of particles is

$$-\frac{\hbar^2}{2m}\nabla^2\psi\left(r_1, \ldots, r_N\right) + V\left(r_1, \ldots, r_N\right)\psi = E\psi\left(r_1, \ldots, r_N\right), \tag{4.3}$$

where $\psi\left(r_1, \ldots, r_N\right)$ is the wave function for the entire system of particles and $V\left(r_1, \ldots, r_N\right)$ is the potential energy operator. For an ideal gas, the potential energy is zero, and the potential energy term may be dropped. The energy eigenvalues are, in this case, those for N noninteracting particles in a box. For a single particle in a box with sides L_x, L_y, and L_z, the energy eigenvalues obtained from the solution of the Schrödinger equation given in Appendix C are

$$\varepsilon_n = \frac{\pi^2\hbar^2}{2m}\left(\frac{n_x^2}{L_x^2} + \frac{n_y^2}{L_y^2} + \frac{n_z^2}{L_z^2}\right), \tag{4.4}$$

where the quantum numbers n_x, n_y, and n_z take integer values, that is, n_x, n_y, $n_z = 1, 2, 3, \ldots$. For a cubical container of edge length L and volume $V = L^3$, the eigenvalues are simply

$$\varepsilon_n = \left(\frac{\pi^2\hbar^2}{2mV^{2/3}}\right)\left(n_x^2 + n_y^2 + n_z^2\right). \tag{4.5}$$

A microstate of the single particle system is designated by the set of quantum numbers (n_x, n_y, n_z) and it follows that for N particles, the state of the entire system is specified by the set of $3N$ quantum numbers $(n_{1x}, n_{1y}, n_{1z}; n_{2x}, n_{2y}, n_{2z}; \ldots; n_{Nx}, n_{Ny}, n_{Nz})$. Figure 4.3 shows the single particle energy eigenvalues for increasing values of the quantum number n for a one-dimensional box.

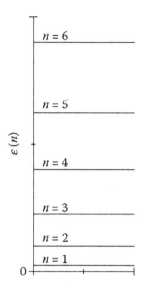

FIGURE 4.3 Energy states specified by the quantum number n for a single particle in a one-dimensional box.

When there are N particles in the box of volume V, they occupy single particle states that are identical for all particles. If the particles are fermions, they obey the Pauli exclusion principle, which states that no two fermions may have the same set of quantum numbers, and if spin quantum numbers are ignored for the moment, this prevents two fermions from being in the same particle-in-a-box state. For bosons, no such restriction applies. These cases are dealt with in detail in the quantum statistics section of the book, particularly in Chapters 12–14.

In the high-temperature and low-particle-density limit, the particle statistics become unimportant. This is because the energy levels are sparsely populated, and the probability that two particles will be in the same state becomes extremely small. Although the number of particles N may be very large ($\sim 10^{23}$), the number of accessible quantum states is very much larger than this. It follows that at any instant, most states are likely to be unoccupied, with some states occupied by just one particle and, if the statistics allows this, a few states occupied by more than one particle. This situation corresponds to the classical limit. In Chapter 12, it is shown that these assumptions about occupancy of states are justified in the high-temperature, low-density limit of the quantum distribution functions.

To determine whether quantum effects are important or if the classical limit applies, it is helpful to use wave packet ideas. The average de Broglie wavelength of particles with mean momentum $\langle p \rangle$ is $\langle \lambda \rangle = h/\langle p \rangle$. If $\langle \lambda \rangle$ is much less than the average interparticle spacing d, it is permissible to neglect quantum effects because there is little overlap of the particle wave functions. The mean spacing may be estimated using the relation $d \sim (V/N)^{1/3}$. In the classical limit, we assert $\langle \lambda \rangle \ll d$. The equipartition of energy theorem for a monatomic gas, such as helium, gives for the mean particle momentum $\langle p \rangle = m\langle v \rangle \approx (3mk_BT)^{1/2}$ and, with the de Broglie relation, given above, we obtain $\langle \lambda_T \rangle = h/(3mk_BT)^{1/2}$, where T is the absolute temperature. (The subscript indicates that $\langle \lambda_T \rangle$ is the thermal de Broglie wavelength.) As an estimate of the particle separation, in an ideal gas, we use $d \approx (V/N)^{1/3} = (k_BT/P)^{1/3}$.

Exercise 4.1

Show that the condition $\langle \lambda \rangle \ll d$ is well satisfied for helium gas at $T = 100$ K and $P = 1$ atm $= 1.01 \times 10^5$ Pa.

The mass of a helium atom is $4.003 \times 1.66 \times 10^{-27}$ kg. Making use of the expressions given above, we obtain $\langle \lambda_T \rangle = 0.10$ nm and $d \approx 2.4$ nm. The condition $\langle \lambda \rangle \ll d$ is well satisfied for helium gas for the given conditions of T and P.

We introduce the quantum volume as $V_Q = (h/\langle p \rangle)^3$. For future reference, it is convenient to write the quantum volume as $V_Q = (h^2/2m\langle\varepsilon\rangle)^{3/2} = (h^2/3mk_BT)^{3/2}$. The mean volume per particle is $V_A = V/N$, and the condition for negligible overlap of particle wave functions is $V_Q \ll V_A$. We show later that, in the classical limit, many important results can be expressed in terms of the ratio (V_Q/V_a). When (V_Q/V_a) approaches unity for a gas, the classical approximation breaks down.

For an ideal gas system, there are no interactions between the particles, and no collisions that involve an exchange of energy between particles can occur. Furthermore, we assume that collisions with the container walls are elastic. In this limit, the quantum numbers do not change with time, and the system is in a *stationary* state. If weak interactions between particles are switched on, collisions will occur, and the set of quantum numbers will gradually be shuffled. Over a sufficiently long time, much longer than the mean time τ between collisions discussed in Chapter 1, the system will explore all of the accessible states subject to the constraints of fixed E, V, and N. We have now established a procedure for designating microstates of an ideal gas in terms of the corresponding set of $3N$ quantum numbers. The procedure permits us to enumerate microstates for the two model systems, the ideal gas and the ideal spin system. We first consider the ideal spin system.

4.4   QUANTUM STATES FOR AN IDEAL LOCALIZED SPIN SYSTEM

Consider a solid paramagnetic system of N particles, each with spin angular momentum $\hbar S$ and associated magnetic dipole moment $\boldsymbol{\mu} = -g\mu_B \boldsymbol{S}$, given in terms of the g factor and the Bohr magneton $\mu_B = e\hbar/2m$. This expression applies to electron spins, and a similar relation may be written for nuclear spins. For electrons in our ideal spin system, we take $g = 2$. The minus sign in the expression that relates μ to $\boldsymbol{S}$ shows that the magnetic dipole vector is antiparallel to the spin angular momentum vector for negatively charged electrons.

We assume that the spins are located on fixed lattice sites in a solid, and there are negligible interactions between the localized moments. *Localization* of the spins is important, as we shall see. For a single spin in applied magnetic field of induction $\boldsymbol{B}$ along the z-direction, the Hamiltonian operator is

$$\mathcal{H} = -\boldsymbol{\mu} \cdot \boldsymbol{B}, \tag{4.6}$$

where B is the local field seen by the spin. This is the operator form of Equation 2.27. For a paramagnetic system, the local field is very nearly the same as the applied field $B = \mu_0 H$ along z, and we have

$$\mathcal{H} = -\mu_z B = g\mu_B B S_z. \tag{4.7}$$

The energy eigenvalues of this Hamiltonian are simply multiples of the eigenvalues of S_z, with

$$E_{ms} = g\mu_B B m_s, \tag{4.8}$$

where $m_S = S, S-1, \ldots, -S$. For $S = \frac{1}{2}$, there are two eigenstates, $|+\frac{1}{2}\rangle$ and $|-\frac{1}{2}\rangle$, with energies

$$E_\pm = \pm\frac{1}{2} g\mu_B B = \mp\mu_z B. \tag{4.9}$$

These states are depicted in Figure 4.4 for $g = 2$. The energy gap ΔE increases linearly with B.

For nuclei, it is usual to represent the spin operator by $\boldsymbol{I}$, and the nuclear magnetic dipole moment is given by $\boldsymbol{\mu} = \gamma\hbar\boldsymbol{I}$, where γ, the magnetogyric ratio for the nucleus of interest, can be positive or negative. For the present discussion, the form Equation 4.7 for the Hamiltonian and Equation 4.9 for the eigenvalues will be adopted.

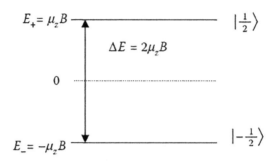

FIGURE 4.4 Energy levels for a single electron spin $\frac{1}{2}$ in a magnetic field B. The energy gap between the two levels is $2\mu_z B$.

For N noninteracting dipoles, the total energy is the sum of the individual energies. The state of the system may be specified by giving the set of quantum numbers, $(m_1, m_2, ..., m_n)$ or in Dirac notation $|m_1, m_2, ..., m_n\rangle$. The subscript S has been dropped to simplify the notation. If the energy of the system E is fixed corresponding to fixed B and N, only certain combinations of quantum numbers are allowed subject to the constraint

$$E = -n_+\mu_z B + n_-\mu_z B, \tag{4.10}$$

where n_+ denotes the number of spins in the lower energy $\left|-\frac{1}{2}\right\rangle$ state and n_- the number in the higher energy $\left|+\frac{1}{2}\right\rangle$ state. (It is convenient from now on in our general discussion to denote the number of spins in the lower energy state, with their dipole moment μ oriented along B, as n_+.) For weak interactions between the spins, the individual quantum numbers can change with time through mutual spin energy exchange. However, the *total numbers* in each state n_+ and n_- will stay constant because of the fixed total energy constraint.

4.5   THE NUMBER OF ACCESSIBLE QUANTUM STATES

The labeled quantum states for our two model systems permit us to count states in a particular energy range. The number of accessible states is denoted by $\Omega(E)$, and this is interpreted as the number of states in the energy range E to $E + \delta E$, where δE is a small interval that allows for some uncertainty in the fixed total energy E of the system. In terms of the energy density of state $\rho(E)$, the number of accessible states is

$$\Omega(E) = \rho(E)\delta E. \tag{4.11}$$

In the calculation of $\rho(E)$, allowance must be made for any degeneracy of states with the same total energy.

a. *The Density of States for a Single Particle in a Box.* In Section 4.3, the quantum states for N particles in a box of volume V are considered. The single particle states are from Equation 4.5 given by $\varepsilon_n = \left(\pi^2\hbar^2/2m\ V^{2/3}\right)\left(n_x^2 + n_y^2 + n_z^2\right)$, with $n_x, n_y, n_z = 1, 2, 3, ...$. It is convenient to introduce a geometrical representation of the states in a space spanned by the values of the quantum numbers n_x, n_y, and n_z, and this representation is shown in Figure 4.5.

The number of states $N(\varepsilon)$ in the range from 0 to ε is easily obtained. Each state corresponds to a cell of volume unity in the quantum number representation. The total number

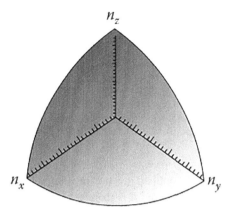

FIGURE 4.5 Representation of particle in a box states in quantum number space specified by n_x, n_y, $n_z = 1$, 2, 3, ... The octant of the sphere shown encloses states with particular values of $n_x^2 + n_y^2 + n_z^2$.

of states is therefore given by the volume of the octant of a sphere as shown in Figure 4.5, with radius $R = \left(n_x^2 + n_y^2 + n_z^2\right)^{1/2}$, where the selected set of quantum numbers (n_x, n_y, and n_z) generates the spherical surface corresponding to energy ε. This enumeration procedure gives the total number of states as

$$N(\varepsilon) = \frac{1}{8}\left[\frac{4\pi}{3}\left(n_x^2 + n_y^2 + n_z^2\right)^{3/2}\right]. \tag{4.12}$$

From Equation 4.5, we obtain $\left(n_x^2 + n_y^2 + n_z^2\right)$ in terms of ε and other quantities, and substitution in Equation 4.12 gives

$$N(\varepsilon) = \frac{1}{6}\left(\frac{V}{\pi^2\hbar^3}\right)(2m\varepsilon)^{3/2}. \tag{4.13}$$

The density of states follows directly by differentiation of $N(\varepsilon)$ with respect to ε

$$\rho(\varepsilon) = \frac{dN(\varepsilon)}{d\varepsilon} = \left(\frac{V}{4\pi^2\hbar^3}\right)(2m)^{3/2}\,\varepsilon^{1/2}. \tag{4.14}$$

This is an important result, which shows that $\rho(\varepsilon) \propto \varepsilon^{1/2}$ for three-dimensional single particle states. The number of states in a shell of thickness $\delta\varepsilon$ at ε is

$$\Omega(\varepsilon) = \rho(\varepsilon)\delta\varepsilon = \left(\frac{V}{4\pi^2\hbar^3}\right)(2m)^{3/2}\,\varepsilon^{1/2}\delta\varepsilon \tag{4.15}$$

In Chapter 5, we make use of this expression for the number of accessible states.

Exercise 4.2

Estimate the number of *single particle* states $N(\varepsilon)$, with energies in the range from 0 to $\varepsilon \approx \frac{3}{2}\,k_B T$, for helium atoms in a box of volume 1 L at a temperature of 300 K. Compare the number of states $N(\varepsilon)$ with Avogadro's number N_A.

We have shown in Equation 4.13 that the number of single particle states in the range from 0 to ε is given by the expression $N(\varepsilon) = \frac{1}{6}(V/\pi^2\hbar^3)(2m\varepsilon)^{3/2} = (4\pi/3)(V/V_Q)$. (Note that the number $N(\varepsilon)$ increases as V increases, for constant quantum volume V_Q.) The upper energy limit is, for conservative estimate purposes, based on the equipartition of energy theorem. For helium, the atomic mass is $m = 4.002 \times 1.660 \times 10^{-27}$ kg. For these values for ε and m, we get $N(\varepsilon) \simeq 6 \times 10^{30}$. This is much larger than Avogadro's number $N_A = 6.02 \times 10^{23}$ mol^{-1}. The large ratio $N(\varepsilon)/N_A$ supports the claim made in Section 4.2 that the single particle states are sparsely populated in the classical high-temperature, low-density limit.

b. *The Density of States for N Noninteracting Particles in a Box.* In the determination of $\rho(E)$ for N particles, with total energy E, confined in a box of volume V, great care must be exercised. The particles are *indistinguishable*, and it is therefore physically impossible to count states as distinct that simply involve particle interchange or, in simple terms, it is not possible to label particles. A further complication is that the particles may be either fermions or bosons. Fermions obey the Pauli exclusion principle whereas bosons do not. However, in the classical limit of low densities and high temperatures, where the states are sparsely populated, it is legitimate to ignore the quantum statistics features, as we show in detail in Chapter 12.

For N particles, the total number of states with energy in the range from 0 to E, if we ignore indistinguishability for the moment, is estimated as $N(E) \simeq [N(\varepsilon)]^N$ where $\varepsilon \simeq E/N$. This estimate combines all states for particle 1 with all the states for particle 2 and so on. Because of the indistinguishability of particles, the estimate is not correct and overcounts states by a large factor. To a good approximation, the indistinguishability problem can be overcome by multiplying by $1/N!$ to allow for those permutations of particles that give rise to indistinguishable states of the system. Our revised estimate of the number of states is $N(E) \simeq (1/N!)[N(\varepsilon)]^N$. This correction factor is introduced in a natural way in the classical limit of the quantum distribution functions, as discussed in Chapter 12.

With $N(\varepsilon)$ from Equation 4.13, we get

$$N(E) \simeq \left(\frac{V^N}{N!}\right)\left[\left(\frac{1}{6\pi^2}\right)\left(\frac{2m}{\hbar^2}\right)^{3/2}\right]^N \varepsilon^{3N/2}, \qquad (4.16)$$

and using $d\varepsilon/dE \simeq 1/N$, we obtain the density of states from Equation 4.16 as

$$\rho(E) = \frac{dN(E)}{dE} = \left(\frac{dN(E)}{d\varepsilon}\right)\frac{d\varepsilon}{dE}$$

or

$$\rho(E) = \left(\frac{V^N}{N!}\right)\left[\left(\frac{1}{6\pi^2}\right)\left(\frac{2m}{\hbar^2}\right)^{3/2}\right]^N \left(\frac{3}{2}N\varepsilon^{(3N/2)-1}\right)\frac{1}{N}.$$

This gives the density of states to a good approximation as

$$\rho(E) \approx \left(\frac{V^N}{N!}\right)\left[\left(\frac{1}{6\pi^2}\right)\left(\frac{2m}{\hbar^2}\right)^{3/2}\right]^N \left(\frac{E}{N}\right)^{3N/2} \qquad (4.17)$$

The number of states in the range from E to $E + \delta E$ follows immediately as

$$\Omega(E) = \frac{3}{2}\left(\frac{V^N}{N!}\right)\left[\left(\frac{1}{6\pi^2}\right)\left(\frac{2m}{\hbar^2}\right)^{3/2}\right]^N\left(\frac{E}{N}\right)^{3N/2}\delta E. \tag{4.18}$$

For future reference, it is convenient to obtain $\ln\Omega$, (E) with use of an extremely useful approximation known as Stirling's formula: $\ln N! \simeq N\ln N - N$ as given in Appendix A. With use of this result, we obtain

$$\ln\Omega(E) = N - N\ln N + N\ln V + N\ln\left(\frac{E}{N}\right)^{3/2} + N\ln\left[\frac{1}{6\pi^2}\left(\frac{2m}{\hbar^2}\right)^{3/2}\right] + \ln\frac{3}{2}\delta E,$$

$$\approx N\ln\left(\frac{V}{N}\right) + N\ln\left[\frac{4\pi}{3}\left(\frac{2mE}{Nh^2}\right)^{3/2}\right] + N.$$

As a good approximation, the term $\ln\frac{3}{2}\delta E$ is ignored in comparison with the other retained terms that involve $N \sim 10^{23}$ and are therefore very large.

A number of observations may be made concerning $\Omega(E)$. First, $\Omega(E)$ corresponds to a very large number of states. $\ln\Omega(E)$ is of the order of N, and if for our estimate purposes we choose N of order Avogadro's number, this implies that $\Omega(E) \sim e^{10^{23}}$, an extremely large number. Second, $\Omega(E)$ is a very rapidly increasing function of E because $\Omega(E) \propto E^{3N/2}$. These two features play an important role in the development of the statistical theory. The expression for $\ln\Omega(E)$ obtained above involves a number of approximations. However, a reliable expression on the basis of results obtained in Chapter 16 for a classical ideal gas leads to the very similar result

$$\ln\Omega(E) = N\ln\left(\frac{V}{N}\right) + N\ln\left[\left(\frac{2\pi}{3}\right)^{3/2}\left(\frac{2mE}{Nh^2}\right)^{3/2}\right] + \frac{5}{2}N$$

or

$$\ln\Omega(E) = N\ln\left[\frac{V}{N}\left(\frac{4\pi mE}{3Nh^2}\right)^{3/2}\right] + \frac{5}{2}N = N\left[\ln\left(\frac{V_A}{V_Q}\right) + \frac{5}{2} + \frac{3}{2}\ln\left(\frac{2\pi}{3}\right)\right]. \tag{4.19}$$

In Equation 4.19, $V_A = V/N$ is the volume per particle, and V_Q is the *quantum* volume introduced in Section 4.3. We take $\langle\varepsilon\rangle = E/N$ in introducing the quantum volume. In the classical limit, we have $V_A \gg V_Q$, and $\ln\Omega(E)$ is clearly very large, of the order of the number of particles N, as expected. The expression for $\ln\Omega(E)$ will prove useful in a later discussion of the entropy of an ideal gas in Chapter 5.

c. *Accessible States for a System of N Noninteracting Spins.* In Section 4.4, it is shown that a spin $\frac{1}{2}$ particle (electron or nucleus) with magnetic moment μ in a magnetic field B has two energy levels with separation $2\mu_z B$. (To simplify the notation, we drop the subscript on μ in the following discussion.) For N spins, the possible energies are shown in Figure 4.6, where the convention for specifying the numbers of spins in a state as given in Section 4.4 has been adopted.

The lowest energy state corresponds to all *moments* being aligned parallel to the field (or up) with $n_+ = N$ and the highest energy state, in turn, to all moments being aligned antiparallel to the field (or down) so that $n_- = N$. The density of states is constant for the N spin system and is given by

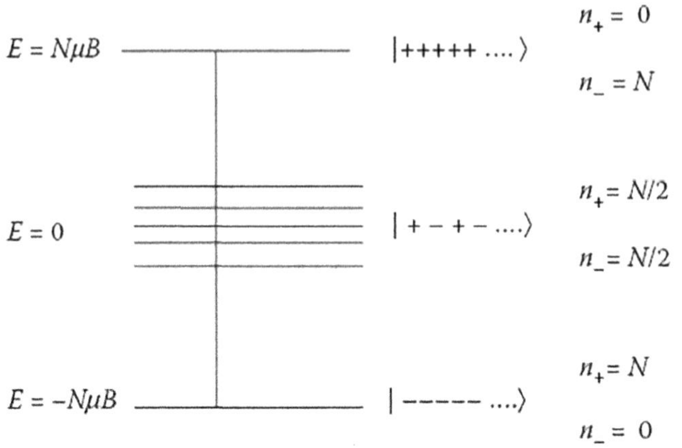

$$E = N\mu B \quad \underline{\hspace{4cm}} \quad |+++++\ ....\rangle \qquad \begin{array}{l} n_+ = 0 \\ \\ n_- = N \end{array}$$

$$E = 0 \quad \begin{array}{l} \underline{\hspace{2.5cm}} \\ \equiv\!\!\equiv\!\!\equiv \\ \underline{\hspace{2.5cm}} \end{array} \quad |+-+-\ ....\rangle \qquad \begin{array}{l} n_+ = N/2 \\ \\ n_- = N/2 \end{array}$$

$$E = -N\mu B \quad \underline{\hspace{4cm}} \quad |------\ ....\rangle \qquad \begin{array}{l} n_+ = N \\ \\ n_- = 0 \end{array}$$

FIGURE 4.6   Energy levels for N electron spins $\dfrac{1}{2}$ in a magnetic field B. The energy states are specified by the *spin* quantum number sets as shown. The number of up moments aligned parallel to B is denoted by n_+, with $n_- = N - n_+$. Successive energy levels correspond to turning over a single spin with a change in n_+, and consequently n_-, by plus or minus 1, respectively.

$$\rho(E) = \frac{1}{2\mu B}, \tag{4.20}$$

which corresponds to just one state in the energy range $2\mu B$. For localized moments in a rigid lattice, the spins are, in principle, *distinguishable*. Using clever imaging techniques involving particle beams, for example, it is feasible that we can determine the orientation of any given spin. Although this might be extremely difficult in practice, the fact that it is possible in a *gedanken* experiment means that the localized spins must be considered in a different way to the delocalized particles in a gas. The factor $1/N!$ introduced to overcome overcounting of states in the case of the gas is not needed for localized, distinguishable spins.

The number of states in the range from E to $E + \delta E$ is $\Omega(E) = g(N)\,\rho(E)\,\delta E$, with $g(N)$ a degeneracy factor that allows for the permutation of up and down spins without altering n_+ or n_-. Note that

$$E = -n\mu B + (N - n)\mu B = (N - 2n)\mu B. \tag{4.21}$$

The factor $g(N)$ is easily obtained and is simply the number of ways of arranging N objects, where n are of one kind with moment up (or spin down) and $(N-n)$ are of another kind with moment down. The result is

$$g(N) = \frac{N!}{n!(N-n)!} = \binom{N}{n}, \tag{4.22}$$

where $\binom{N}{n}$ denotes the binomial coefficient. The binomial distribution, which involves the binomial coefficient, is discussed in Chapter 10 and Appendix B. Using Equation 4.22 for $g(N)$ and Equation 4.20 for $\rho(E)$ gives

$$\Omega(E) = \binom{N}{n}\left(\frac{\delta E}{2\mu B}\right). \tag{4.23}$$

For future reference, it is useful to obtain an expression for ln Ω(E), that is, ln $\Omega(E) = \ln N! - \ln$
$(N-n)! - \ln n! - \ln(\delta E/2\mu B)$, and using Stirling's formula, we get to a good approximation

$$\ln \Omega(E) \simeq N\ln N - (N-n)\ln(N-n) - n\ln n. \tag{4.24}$$

The term $\ln(\delta E/2\mu B)$ has been omitted because $(\delta E/2\mu B)$ is of order unity, and the logarithm of
this quantity will be negligible compared with the other terms, which are of order N. The right-
hand side of Equation 4.24 does not contain E explicitly. From Equation 4.21, it is readily seen that
$n = \frac{1}{2}\left[N - (E/\mu B)\right]$ and $(N-n) = \frac{1}{2}\left[N + (E/\mu B)\right]$.

Substituting for n and $(N-n)$ in Equation 4.24 leads to

$$\ln \Omega = N \ln N - \left(\frac{N}{2} - \frac{E}{2\mu B}\right)\ln\left(\frac{N}{2} - \frac{E}{2\mu B}\right) - \left(\frac{N}{2} + \frac{E}{2\mu B}\right)\ln\left(\frac{N}{2} + \frac{E}{2\mu B}\right). \tag{4.25}$$

Exercise 4.3

Differentiating the expression for ln $\Omega(E)$, given in Equation 4.25, with respect to E, and equating
the result to zero show that an extremum (maximum) occurs at $E = 0$.
   From Equation 4.25, $d \ln \Omega(E)/dE = \ln[(N/2 - E/2\mu B)/(N/2 + E/2\mu B)] = 0$, and solving for E gives
$E/2\gamma B = 0$. The extremum occurs at $E = 0$ as predicted. By differentiating again, it is readily found
that the extremum at $E = 0$ is a maximum.
   A plot of ln $\Omega(E)$ as a function of E is given in Figure 4.7.

For $E = 0$, the function ln $\Omega(E)$ has a maximum value of $N \ln 2$, as may be seen from Equation 4.25
or Equation 4.24 with $n = N/2$. $\Omega(E)$ increases, from a value of 1 (ln $\Omega(E) = 0$) when $n = N$ to a
value of 2^N for $n = N/2$ and then decreases to 1 at the upper energy bound, $N\mu B$, which corresponds
to $n = 0$ with all moments down (spins up). For $N \sim N_A$, the maximum value for $\Omega(E)$ is very large. It
is clear that $\Omega(E)$ increases extremely rapidly with E for $E < 0$ and decreases again extremely rapidly
for $E > 0$. This is an important characteristic that is used in establishing a connection between the
entropy S and ln $\Omega(E)$. Figure 4.8 gives a schematic representation of the variation of $\Omega(E)$ with E.
   The curve, shown in Figure 4.8 for $N = 100$, tends to Gaussian form and peaks more and more
sharply at $E = 0$ as N is increased. The sharply peaked feature of the distribution is of central importance in the discussion of accessible states given in Chapter 5.

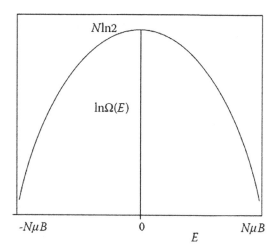

FIGURE 4.7   Plot of ln Ω(E) versus E for a system of N spins $\frac{1}{2}$.

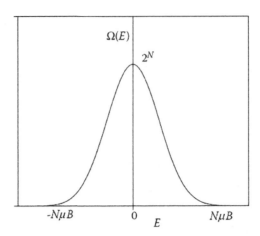

FIGURE 4.8 Plot of $\Omega(E)$ versus E for a spin system. The rapid increase near $E = 0$ is due to the explosive increase in the degeneracy factor $g(n)$ given in Equation 4.22.

PROBLEMS CHAPTER 4

4.1 Two containers of equal volume are connected by a valve that is initially closed. One container contains 100 molecules, whereas the other is empty. Describe what happens when the valve is opened, and give the situation that you expect when equilibrium has been reached. Find the entropy increase for the system in this irreversible process. What is the probability of finding a particular molecule in one of the two containers after equilibrium has been reached? Use the binomial distribution to calculate the probability of finding the molecules back in the initial state with all molecules in one container.

4.2 Consider a system consisting of 0.1 mol of argon gas in a container of volume 1 L at 300 K. Compare the thermal average value of the de Broglie wavelength $\langle \lambda_t \rangle$ of the atoms with the average interparticle spacing. Find the ratio of the atomic volume $V_A = V/N$ to the quantum volume V_q given by the cube of the de Broglie wavelength. Is argon a classical gas under these conditions? Argon molar mass = 39.95 g/mol.

4.3 Argon gas is in a container of volume 2 L at a temperature of 290 K and at a pressure of 1 atm. Estimate the total number of available single particle states $N(\varepsilon)$ with energies in the range from 0 to $\varepsilon = \dfrac{3}{2} k_B T$ for the system and compare with the number of argon atoms. How would the ratio change if the volume of the gas were reduced by two orders of magnitude? Comment on the physical significance of your results.

4.4 A paramagnetic salt contains unpaired electron spins with $g = 2$. Give the Hamiltonian for this spin system when situated in a magnetic field B. Assuming that interactions between the spin are negligible, calculate the energy eigenvalues for the electron spins in a magnetic field of 1 T. Obtain the frequency of electromagnetic radiation that would be needed to induce transitions between the states. The Bohr magneton is 9.27×10^{-24} J T^{-1}.

4.5 A system of noninteracting nuclear spins with $I = 1$ is situated in a field B. Give the Hamiltonian and the energy eigenvalues for each spin. Sketch the energy levels for a single spin. Calculate the energy difference between levels for $B = 1$ T, taking the nuclear moment to be two nuclear magnetons, that is, $g_N = 2$. Compare the energy splitting with that for the electrons in Question 4.4. The nuclear magneton is 5.05×10^{-27} J T^{-1}.

4.6 Obtain an expression for the density of states for the nuclear spin $I = 1$ system in Question 4.5. Compare your result with the density of states for a spin $I = 1/2$ system in the same magnetic field. Give a numerical value for the density of states for the spin 1 case using the numbers given in Question 4.5.

4.7   Noninteracting nuclear spins with $I = 3/2$ are situated in a field B. Write down the spin Hamiltonian and give a sketch of the energy eigenvalues for each spin. Obtain the energy difference between levels for $B = 10$ T, taking the nuclear moment to be two nuclear magnetons, that is, $g_N = 2$. If the spin system is at a temperature of 50 mK, how does the energy gap between the levels compare with the thermal energy $k_B T$?

4.8   Two containers with a volume ratio 4:1 are connected by a valve that is closed. The larger volume contains N_A molecules of type A, whereas the smaller volume contains N_B molecules of type B. The pressures in the two containers are the same. If the valve is then opened, what are the average numbers of each type of molecule in the two containers when the system has reached equilibrium? Assume that the molecules interact very weakly. Obtain an expression for the probability of finding the molecules in the original configuration before the valve was opened. Evaluate the logarithm of this probability for the special case $N_A = 100$ and $N_B = 25$.

4.9   Assuming monatomic molecules in Question 4.8, obtain an expression for the logarithm of the number of accessible states $\ln {}'\Omega$ after mixing in terms of the total volume V, the temperature T, and the molecular masses.

4.10 Calculate $\ln {}'\Omega$ for 0.1 mol of helium gas in a container of volume $10^{-3}\,\mathrm{m}^3$ at a temperature of 300 K. First calculate the volume per helium atom and the thermal de Broglie wavelength and use the ratio in your calculation. Show that the classical approximation applies in this case.

4.11 A monatomic gas adsorbed on a surface may be considered to be a two-dimensional ideal gas if the molecules are free to move on the surface. Write down the energy eigenvalues for each atom and show that the single particle density of states is independent of energy ε. Obtain the density of states as a function of total energy E for the N particle system.

4.12 A system consists of N particles in a cubical container of edge length L. Assuming that the force exerted by a particle in a collision with a wall is given by $F_i = -\partial \varepsilon_i / \partial L$, where ε_i is the energy of the particle in a given quantum state specified by i, show that the pressure in the container may be written as $P = (2N/3V)\,\langle \varepsilon \rangle$, where $\langle \varepsilon \rangle$ is the average energy of a particle. Give an explicit expression for $\langle \varepsilon \rangle$ in terms of the particle in a box quantum numbers. By comparing your expression for the pressure with the ideal gas equation of state, obtain the equipartition of energy result for the average energy of a particle in terms of the temperature T.

4.13 Show that for two interacting spin systems, each consisting of $N/2$ spins of magnetic moment μ in a field B, the number of accessible states for the composite system may be written as the product of Gaussian functions of the form $\Omega_1(E_1) = \Omega_1(0) \exp[(-2/N)(E_1/2\mu B)^2]$, where E_1 is in the range from $-(N/2)\,\mu B$ to 0, with a similar form for $\Omega_2(E_2)$. Use your result to sketch (or computer plot) the product function versus energy of one of the systems.

5 Entropy and Temperature
Microscopic Statistical Interpretation

5.1 INTRODUCTION: THE FUNDAMENTAL POSTULATE

In Chapter 4, we introduced the number of accessible microstates $\Omega(E)$ for a system obtaining expressions for this quantity for an ideal spin system and an ideal gas. It is now possible to use statistical methods to establish various results such as the equilibrium conditions for two systems that interact. To do this, it is necessary to consider the probability for a system to be in any one of its accessible states. A helpful analogy is to consider the probability of various outcomes when a die is thrown. If the die is good, we expect the probability for any one of the six faces to land uppermost to be equal to 1/6. This postulate may be tested by experiment. Similarly, for a large system with $\Omega(E)$ accessible microstates, we expect that no one microstate is to be preferred over any other microstate, and the probability of the system to be found in any one of its accessible microstates should be the same and equal to $1/\Omega(E)$.

The fundamental postulate of statistical physics is based on these considerations and may be stated as follows: A system in equilibrium in a given macrostate is equally likely to be found in any one of its accessible microstates.

As with any postulate on which theory is based, the predictions made by the theory are tested through experiment. The fundamental postulate has been extensively tested in this way, as we shall see, and is found to be a good postulate. The postulate is the quantum description analog of the classical ergodic hypothesis introduced in Section 4.2 of Chapter 4 in considering the path of the representative point through accessible regions in classical phase space. Over a sufficient time, all accessible regions are traversed by the representative point.

Because the probabilities are all equal, the probability of finding a system in a particular microstate i is simply $P_i = 1/\Omega(E)$. We require that the sum of the probabilities for all accessible microstates is equal to unity or $\sum_{i \in \Omega(E)} P_i = 1$. As a consequence of the fundamental postulate, the most probable macrostate, specified by appropriate state variables (e.g., V and T), will correspond to that state which has the largest number of accessible microstates. This permits us to understand how a system will approach equilibrium. The macrostate will change with time until the number of accessible microstates reaches a maximum, with the system equally likely to be found in any one of these microstates. For example, in the situation shown in Figure 4.1 in Chapter 4, opening the tap increases the volume available to the particles of the gas. Equation 4.19 shows that $\ln \Omega(E) \propto V^N$, and it is clear that the number of accessible microstates increases enormously as a result of the increase in V. The gas particles fill the whole volume once equilibrium has been reached. Fluctuations in the number of particles per unit volume in any region of the container will be very small. In the next section, these ideas are explored in a quantitative way for two interacting spin systems. Although spins constitute a particular type of system, the approach used and some of the results obtained are quite general.

5.2 EQUILIBRIUM CONDITIONS FOR TWO INTERACTING SPIN SYSTEMS

Consider two spin systems, labeled 1 and 2, with fixed total energy E, which interact at the microscopic level by exchanging heat. System 1 contains N_1 spins, whereas system 2 has N_2 spins. The systems are situated in a magnetic field B, and all spins have magnetic moment μ. Figure 5.1 depicts the situation.

Let the number of microstates accessible to system 1 be $\Omega_1(E_1)$ and the number accessible to system 2 be $\Omega_2(E_2)$. Because of the thermal interaction, E_1 and E_2 can change, subject to the constraint that the total energy remains fixed, that is,

$$E = E_1 + E_2 = \text{constant.} \tag{5.1}$$

The total number of microstates available to the combined system is clearly

$$\Omega(E) = \Omega_1(E_1)\Omega_2(E_2). \tag{5.2}$$

This follows because each accessible microstate of system 1 can be combined with any of the accessible microstates of system 2. On the basis of the fundamental postulate, we require that in equilibrium, $\Omega(E)$ should be a maximum and consequently the energy will be shared between the two systems in such a way as to achieve this. In equilibrium, we therefore require that $d\Omega(E)/dE_1 = 0$ and $d\Omega(E)/dE_2 = 0$. These considerations lead to the condition $d\Omega(E)/dE_1 = \Omega_2(E_2)[d\Omega_1(E_1)/dE_1] + \Omega_1(E_1)[d\Omega_2(E_2)/dE_1] = 0$.

Dividing by $\Omega_1(E_1)\,\Omega_2(E_2)$ and using the chain rule gives

$$\frac{d\ln\Omega_1(E_1)}{d(E_1)} + \frac{d\ln\Omega_2(E_2)}{d(E_2)}\left(\frac{dE_2}{dE_1}\right) = 0.$$

From Equation 5.1, $(dE_2/dE_1) = -1$, and the equilibrium condition becomes

$$\frac{d\ln\Omega_1(E_1)}{dE_1} = \frac{d\ln\Omega_2(E_2)}{dE_2}. \tag{5.3}$$

The argument up to this point is perfectly general and could apply to any two interacting systems.

Exercise 5.1

Show that two thermally interacting spin systems 1 and 2, with numbers of spins N_1 and N_2, respectively, share the total energy in such a way that in equilibrium, the fractional number of up moments is the same in the two systems.

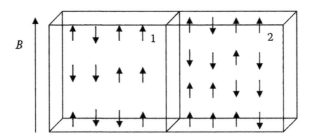

FIGURE 5.1 Two interacting spin systems 1 and 2 containing N_1 and N_2 spins, respectively, situated in a field B. The total energy of the two spin systems is fixed, but the subsystems 1 and 2 can exchange energy.

For spin systems, Equation 4.25 gives the following expression for $\ln \Omega_1(E_1)$:

$$\ln \Omega_1(E_1) = N_1\ln N_1 - \left(\frac{N_1}{2} - \frac{E_1}{2\mu B}\right)\ln\left(\frac{N_1}{2} - \frac{E_1}{2\mu B}\right) - \left(\frac{N_1}{2} + \frac{E_1}{2\mu B}\right)\ln\left(\frac{N_1}{2} + \frac{E_1}{2\mu B}\right),$$

with a similar expression for $\ln \Omega_2(E_2)$. After differentiation and rearrangement, this leads to

$$\frac{d\ln \Omega_1(E_1)}{dE_1} = \frac{1}{2\mu B}\ln\left(\frac{N_1}{2} - \frac{E_1}{2\mu B}\right) - \frac{1}{2\mu B}\ln\left(\frac{N_1}{2} + \frac{E_1}{2\mu B}\right), \tag{5.4}$$

again with a similar expression for $d\ln \Omega_2(E_2)/dE_2$. From Equation 5.3, these two expressions must be equal in equilibrium, and after simplification we obtain the result,

$$\frac{E_1}{E_2} = \frac{N_1}{N_2}. \tag{5.5}$$

Because $E_1 = -n_1\mu B + (N_1 - n_1)\mu B$, we have $E_1 = (N_1 - 2n_1)\mu B$. Similarly, $E_2 = (N_2 - 2n_2)\mu B$, with n_1 and n_2 the number of up moments in systems 1 and 2, respectively. (Substituting for E_1 and E_2 in Equation 5.5 gives in equilibrium

$$\frac{\tilde{n}_1}{\tilde{n}_2} = \frac{N_1}{N_2} \text{ or } \frac{\tilde{n}_1}{N_1} = \frac{\tilde{n}_2}{N_2}, \tag{5.6}$$

where $\tilde{n}_1$ and $\tilde{n}_2$ are the equilibrium values for n_1 and n_2, respectively. The equilibrium energy values $\tilde{E}_1$ and $\tilde{E}_2$ follow immediately. These results show that in equilibrium, the energy is shared in such a way that the fractional number of up moments is the same in the two systems.

Fluctuations in n_1 and n_2 away from the equilibrium values will occur, but because $\Omega(E) = \Omega_1(E_1)$ $\Omega_2(E_2)$ is a very sharply peaked function, it follows that at the maximum, where $\tilde{E}_1 = -\tilde{E}_2$, the fluctuations are exceedingly small. Using Equation 4.25 and taking antilogs, we obtain

$$\Omega_1(E_1) = \frac{N_1^{N_1}}{\left[N_1/2 - E_1/(2\mu B)\right]^{\left[(N_1/2)-(E_1/2\mu B)\right]}\left[N_1/2 + E_2/(2\mu B)\right]^{\left[(N_1/2)+(E_1/2\mu B)\right]}} \tag{5.7}$$

with a similar expression for $\Omega_2(E_2)$.

It is possible to show that a Gaussian function provides a very good approximation to $\Omega(E)$ in the vicinity of the peak. This is discussed in Chapter 10 and Appendix B. In terms of the total energy E, the Gaussian form for a system of N spins with a maximum at $E = 0$ is

$$\Omega(E) = \Omega(0)e^{-(2/N)(E/2\mu B)^2}, \tag{5.8}$$

with $\Omega(0) = 2^N$ because for $E = 0$, each moment has equal probability to be in its up or down state.

Exercise 5.2

Using a Taylor expansion about the maximum in $\Omega(E)$, show that in the limit of large N, the form of Equation 5.8 is obtained for the number of accessible states.

For large N, it is permissible to treat $\Omega(E)$ as a continuous differentiable function in spite of the fact that E takes discrete values. This is because the discrete values are very closely spaced, in comparison with the range of E values involved, and form a quasi-continuum.

Expanding $\ln \Omega(E)$ in a Taylor series about $E = 0$, we obtain

$$\ln \Omega(E)_{E=0} = \ln \Omega(0) + \left(\frac{d\ln \Omega(E)}{dE}\right)_0 E + \frac{1}{2}\left(\frac{d^2 \ln \Omega(E)}{dE^2}\right)_0 E^2 + \cdots.$$

The first derivative is zero at the maximum ($E = 0$), whereas the second derivative gives $(d^2 \ln \Omega(E)/dE^2)0 = -(1/2\mu B)^2 (4/N)$. Insertion of this expression in the Taylor expansion and taking antilogarithms give the required form.

To examine the behavior of the product function $(\Omega_1)(E_1) \Omega_2(E_2)$ in detail near the maximum, consider two systems with equal numbers of spins $N_1 = N_2 = N/_2$, which interact thermally and are situated in a magnetic field B. Let the total energy of the two combined systems have the fixed value E. For this special case of equal numbers of spins in the two systems, Equations 5.2 and 5.8 give

$$\Omega(E) = \Omega_1(E_1)\Omega_2(E - E_1) = 2^N e^{-(2/N)(E_1/2\mu B)2} \, e^{-(2/N)(E - E_1/2\mu B)^2}. \tag{5.9}$$

Figure 5.2 depicts the functions $\Omega(E1)$ and $\Omega2 (E - E1)$ as a function of E1.

It is easy to show by differentiation of Equation 5.9 with respect to E1 that, at the extremum in $\Omega1 \ \Omega2$, the energy is shared equally between the two systems $E_1 = E_2 = \frac{1}{2}E$. By differentiating again, it can be seen that the extremum is a maximum. For this special case, Equation 5.9 becomes

$$\Omega(E) = 2^N \, e^{-2/N(E/2\mu B)2}. \tag{5.10}$$

Equation 5.10 has the form of Equation 5.8, as expected. In Figure 5.2, the energies E and E_1 have been chosen to be less than zero, corresponding to a greater number of *moments* in the parallel to applied field or up state than in the antiparallel or down state.

Exercise 5.3

Show that for two interacting spin systems 1 and 2, the number of accessible states $\Omega(E)$ is a sharply peaked function and that fluctuations in energy are extremely small.

Assume that $N_1 = N_2 = N/2$ so that Equation 5.9 applies. In equilibrium, it follows that $\tilde{E}_1 = \tilde{E}_2 = \frac{1}{2}E$. In examining the form of Equation 5.9, it is helpful to put $E_1 = E/2 - \alpha N\mu B/2$, with α a factor in the range from 0 to 1. In equilibrium, $\tilde{E}_1 = E/2 - \alpha(0)N\mu B/2$, where α (0) = 0 is the equilibrium value of α. This gives after some simplification

$$\Omega(E) = \left[2^N e^{-(2/N)(E/2\mu B)^2} \right] e^{-(N/4)\alpha^2}. \tag{5.11}$$

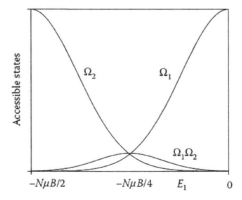

FIGURE 5.2 The number of accessible states Ω_1 and Ω_2 for two interacting spin systems 1 and 2, each containing $N/2$ spins in a field B as a function of the energy E_1 of system 1.

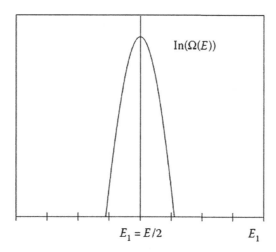

FIGURE 5.3 Plot of $\ln \Omega_1 (E)$ versus E_1 on the basis of Equation 5.11 for two interacting spin systems 1 and 2, with numbers of spins $N_1 = N_2 = 100$. $\tilde{E}_1$ is the most probable energy for system 1.

Because N is very large, $\Omega(E)$ decreases very rapidly as a takes values other than zero. From the fundamental postulate of equal probabilities for all accessible states, it is clear that it is highly improbable that the systems will be found in states that are not very close to the equilibrium state. Fluctuations in E_1 (and a) are extremely small. Figure 5.3 shows a plot of $\ln \Omega(E)$ as a function of E_1 for systems with $N_1 = N_2 = 100$ spins.

Comparison of Equation 5.11 with a standard Gaussian function shows that the half-height *width* of the peak is proportional to $1/\sqrt{N}$ so that, for $N \sim 10^{22}$ spins, fluctuations in E_1 will be given to order of magnitude by $\Delta E_1/\tilde{E}_1 \sim 10^{-11}$. Fluctuations are therefore very small and undetectable in conventional measurements.

The discussion of interacting spin systems given above has served to illustrate the usefulness of the accessible state formulation when considered together with the fundamental postulate. We now proceed to generalize the approach to include any thermally interacting system.

5.3 GENERAL EQUILIBRIUM CONDITIONS FOR INTERACTING SYSTEMS: ENTROPY AND TEMPERATURE

In Section 5.2, the specific case of two localized spin systems that exchange energy through thermal interaction is considered. Certain of the results obtained are perfectly general for any interacting systems subject to the constraint that the total energy E is fixed. Figure 5.4 depicts the general case.

If the additional constraints of fixed volume V and fixed particle number N are added to the energy constraint, Equations 5.2 and 5.3 may be taken over immediately and applied to systems other than spin systems. The number of accessible states for the combined system is $\Omega(E) = \Omega_1(E) \Omega_2(E_2)$, and in equilibrium, Equation 5.3 gives $d \ln \Omega_1 (E_1)/dE_1 = d \ln \Omega_2 (E_2)/dE_2$ as the condition for $\Omega(E)$ to be a maximum. These results on the basis of *microscopic* considerations may be compared with *macroscopic* equilibrium conditions given by thermodynamics. In equilibrium, the entropy S is a maximum, as given by the second law, and the temperatures are equal, that is, $T_1 = T_2$. This suggests a fundamental connection between the entropy and $\Omega(E)$ of the form $S \propto \ln \Omega(E)$. The proportionality constant is introduced below. We relate the entropy to $\ln \Omega(E)$ rather than $\Omega(E)$ because we require entropy to be an extensive quantity. For our interacting systems, we have $\Omega(E) = \Omega_1(E_1) \Omega_2(E_2)$, and it follows that $S = S_1 + S_2$. We now examine the proposed relationship between S and

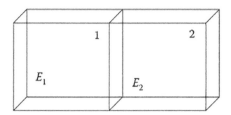

FIGURE 5.4   Two interacting large systems with fixed total energy $E = E_1 + E_2$.

ln $\Omega(E)$ in some detail to show that it leads to results consistent with the thermodynamic approach discussed in Chapter 3.

From the fundamental relation (Equation 3.18) in the form $T\,dS = dE + P\,dV$, it follows that $(\partial S/\partial E)_V = 1/T$, and comparison with the relationship between first derivatives in Equation 5.3 d ln $\Omega_1(E_1)\,dE_1 = $ d ln $\Omega_2(E_2)/dE_2$ provides strong support for the relationship $S \propto$ ln Ω as we shall see. To proceed, we need to introduce a proportionality constant, and for reasons given below, this is simply taken as Boltzmann's constant k_B,

$$S = k_B \ln \Omega. \tag{5.12}$$

This is an extremely important relationship. We show in Section 5.5 that identifying Boltzmann's constant as the constant of proportionality leads to consistency with our thermodynamic definition of entropy. Equation 5.12 expresses the entropy in terms of the logarithm of the number of accessible states and provides a direct means for obtaining the entropy. The definition Equation 5.12 shows immediately that the entropy for the composite system in Figure 5.4 is given by

$$S = k_B \ln \Omega_1 + k_B \ln \Omega_2 = S_1 + S_2. \tag{5.13}$$

This is consistent with entropy being an extensive quantity as pointed out above.

The use of Equation 5.12 in Equation 5.3 immediately leads to $1/k_B(\partial S_1/\partial E_1 = 1/k_B(\partial S_2/\partial E_2)$, or $T_1 = T_2$, which is the thermodynamic equilibrium condition. We can write

$$\frac{\partial \ln \Omega}{\partial E} = \frac{1}{k_B T} = \beta, \tag{5.14}$$

and this relationship defines β as the slope of the ln $\Omega(E)$ versus E plot. β is related to $1/T$ by Equation 5.14. The behavior of the reciprocal temperature β with E for ideal gases and ideal spin systems is discussed in Chapter 6. β is a convenient quantity to use in a number of expressions, as we shall see later, and is called the temperature parameter. Equation 5.12 is an important bridge relationship between the microscopic description and the macroscopic or thermodynamic description of a system. As noted above, knowledge of the number of accessible states immediately gives the entropy for a particular system of interest. Examples based on results obtained in Chapter 4 for our two ideal model systems are given in the following section.

The approach that we have used to arrive at Equation 5.12 involves statistical methods for a system with fixed total energy. We have considered time average properties for a single system. In Chapter 10, the concept of statistical *ensembles* or collections of identical systems is introduced. We shall see there that the statistical approach used in the present chapter involves the *microcanonical ensemble*, in which the energy E and the particle number N are kept fixed for each member of the ensemble. Two other ensembles, termed the *canonical ensemble* and the *grand canonical ensemble*, with different constraints on E and N, are considered in Chapters 10 and 11.

5.4 THE ENTROPY OF IDEAL SYSTEMS

In Chapter 4, expressions for $\ln \Omega(E)$ are obtained for an ideal gas, consisting of N particles in a box of volume V, and also for an ideal spin system, made up of N spins of magnetic moment μ, situated in a magnetic field B. From Equation 5.12, it is now a simple matter to write down entropy expressions for these two systems.

   a. *Ideal Gas.* Using Equation 4.19 for $\ln \Omega(E)$ in Equation 5.12 gives the following expression for the entropy of an ideal monatomic gas with no internal degrees of freedom:

$$S = Nk_B \ln\left[\frac{V}{N}\left(\frac{4\pi mE}{3Nh^2}\right)^{3/2}\right] + \frac{5}{2}Nk_B. \tag{5.15}$$

This important expression for the entropy, known as the Sackur–Tetrode equation, gives a properly extensive quantity that depends on the number of particles per unit volume N/V and on the energy of the system E. In terms of the atomic volume V_A and the quantum volume V_Q, the classical limit expression is

$$S = Nk_B\left[\ln\left(\frac{V_A}{V_Q}\right) + \frac{5}{2} + \frac{3}{2}\ln\left(\frac{2\pi}{3}\right)\right]. \tag{5.16}$$

Equation 5.16 is compact and useful.

   Differentiation of S in Equation 5.15 with respect to E gives

$$\left(\frac{\partial S}{\partial e}\right)_{V,N} = \frac{1}{T} = \frac{\partial}{\partial E}\left(\frac{3}{2}Nk_B \ln E\right)_{N,V} = \frac{3}{2}Nk_B\left(\frac{1}{E}\right)$$

or

$$E = \frac{3}{2}Nk_BT.$$

This is the classical equipartition theorem result given in Chapter 1.

   The heat capacity at constant volume follows immediately, $C_V = (dQ/dT)_V = (\partial E/\partial T)_V = (\partial E/\partial T)_V = \frac{3}{2}Nk_B$, or in terms of the gas constant, $C_V = \frac{3}{2}nR$, in agreement with Equation 2.41.

   The ideal gas equation of state may be obtained from Equation 3.24, which gives with Equation 5.15 $(\partial S/\partial V)_{E,N} = (P/T) = (\partial/\partial V)(Nk_B \ln V)_{E,N} = (Nk_B/V)$ or $PV = Nk_BT = nRT.$

   We see that knowledge of $\ln \Omega$ immediately leads to an expression for the entropy S and that other thermodynamic quantities may then be obtained with use of the fundamental relation. As stated previously, the relation $S = k_B \ln \Omega$ provides an important bridge between the *microscopic* and *macroscopic* descriptions for large systems, such as an ideal gas in a container.

   In the general formulation of thermodynamics in Section 3.13 of Chapter 3, we have introduced the chemical potential μ, with the fundamental relation (Equation 3.26) in the general form $T\,dS = dE + P\,dV - \mu\,dN$ for a single component system for which the energy, volume, and particle number are allowed to change through, for example, interaction with a large reservoir. As mentioned in Chapter 3, we shall see the importance of the chemical potential in discussing chemical reactions and phase transitions and in the derivation of the quantum distribution functions. For later use, we give an expression for μ obtained from our expression for the entropy. From $\mu = -T(\partial S/\partial N)_{E,V}$ and with $\langle \varepsilon \rangle = E/N = \frac{3}{2}k_BT$ as the mean energy per particle, it follows that

$$\mu = -k_BT \ln\left[\left(\frac{V}{N}\right)\left(\frac{4\pi m\langle \varepsilon \rangle}{3h^2}\right)^{3/2}\right]. \tag{5.17}$$

We note that μ depends on the density of particles and on the temperature. The equipartition theorem Equation 1.10 and the expression for the thermal de Broglie wavelength, introduced in Section 4.3 of Chapter 4, enable us to rewrite Equation 5.17 as $\mu = k_BT[\ln(V_Q/V_A) - C]$, with, as before, $V_A = V/N$ the atomic volume and $V_Q = (h\,/\langle p \rangle)^3$ the quantum volume. The numerical constant $C = \frac{3}{2}\left[\ln(2\pi/3) +\right]$ is of order unity. In the classical ideal gas case, we have $V_Q \ll V_A$, which gives $\mu < 0$ in this limit. We may expect the above expressions for μ to break down when this condition is not met, that is, for $V_Q \sim V_A$. Different expressions apply to Fermi and Bose gases in the quantum limit, as we shall find in Chapter 12.

b. *Ideal Spin System.* The entropy of an ideal system of N spins $\frac{1}{2}$ in a magnetic field B follows immediately from Equation 4.24 when inserted into Equation 5.12 $S = k_B \ln\Omega(E)$. This gives the useful form

$$S = Nk_B \ln N - \left(\frac{N}{2} - \frac{E}{2\mu B}\right)k_B \ln\left(\frac{N}{2} - \frac{E}{2\mu B}\right) - \left(\frac{N}{2} + \frac{E}{2\mu B}\right)k_B \ln\left(\frac{N}{2} + \frac{E}{2\mu B}\right). \tag{5.18}$$

Note that for $E = 0$, with $N/2$ moments up and $N/2$ down, we obtain $S = Nk_B \ln 2$, which corresponds to 2^N spin arrangements. This is the maximum value of S as a function of E. Equation 3.23 gives the reciprocal temperature for the N spin system as

$$\frac{1}{T} = \left(\frac{\partial S}{\partial E}\right)_{N,B} = \left(\frac{k_B}{2\mu B}\right)\ln\frac{(N\mu B - E)}{(N\mu B + E)}.$$

This equation may be solved for E by taking antilogs and multiplying top and bottom by $e^{-\mu B/k_BT}$. This procedure leads to the following expression:

$$E = -N\mu B\left[\frac{e^{\mu B/k_BT} - e^{-\mu B/k_BT}}{e^{\mu B/k_BT} + e^{-\mu B/k_BT}}\right] = -N\mu B\,\tanh\left(\frac{\mu B}{k_BT}\right). \tag{5.19}$$

At high temperature in magnetic fields that are not too large, we have $\mu B \ll k_BT$, and the approximation $\tanh x \approx x$ can be made. This gives the following simple expression for the energy, $E = -(N\mu^2B^2)/(k_BT)$. Now, as discussed in Section 2.7 of Chapter 2, $E = -MB$, where M is the magnetic moment of the system. It follows from Equation 5.19 that for $\mu B \ll k_BT$, the magnitude of the magnetic moment is $M = (N\mu^2B)/(k_BT)$ with magnetization

$$M = \frac{M}{V} = \frac{N\mu^2\mu_0H}{Vk_BT}, \tag{5.20}$$

where it is assumed that internal fields are negligibly small, so that $B = \mu_0H$. Equation 5.20 is Curie's law, and comparison with Equation 1.4 shows that the Curie constant is given by $C = N\mu^2\mu_0/Vk_B$.

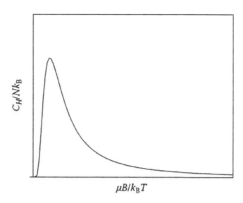

FIGURE 5.5 Reduced specific heat $C_H/N\,k_B$ of a paramagnetic material in a magnetic field B as a function of $\mu B/k_B T$. The curve shows a Schottky peak followed by a decrease in C_H at high T, for constant B, as the population difference between levels decreases.

The heat capacity C_H is readily obtained using $C_H = (dQ/dT)_H = (\partial E/\partial T)_H$, where E is the energy of the magnetic material in the applied field, as discussed in Sections 2.6 and 2.7 of Chapter 2. This gives

$$C_H = Nk_B \left(\frac{\mu B}{k_B T}\right)^2 \operatorname{sech}^2\left(\frac{\mu B}{k_B T}\right). \tag{5.21}$$

The behavior of C_H/Nk_B as a function of $\mu B/k_B T$ is shown in Figure 5.5.

A further discussion of spin systems is given in Chapter 10 as an illustration of the canonical ensemble approach. The peak in the specific heat is known as a Schottky peak and occurs for $\mu B/k_B T \sim 1$. Physically, the peak corresponds to a maximum change in the populations of the spin states and hence the energy of the system for a given temperature rise in a given field. For low B and high T, in the low B/T region of the curve, the heat capacity decreases toward zero as $(B/T)^2$. For high T, the spin populations tend to equality and the heat capacity therefore becomes very small.

The form of the specific heat for a paramagnetic material shown in Figure 5.5 is found to be in good agreement with experimental results for representative paramagnetic materials that obey Curie's law over the B/T range of interest.

5.5   THERMODYNAMIC ENTROPY AND ACCESSIBLE STATES REVISITED

For an ideal gas of N particles in a box, with total energy E, Equation 4.19 gives for the logarithm of the number of accessible states

$$\ln \Omega(E, V, N) = N\left[\ln\left(\frac{V}{N}\right)\left(\frac{4\pi mE}{3Nh^2}\right)^{3/2}\right] + \frac{5}{2}N.$$

Consider the ideal gas as the working substance of a reversible heat engine such as a Carnot engine. In a complete cycle, the change in the number of accessible states is zero because E and V return to their original values rand N is unchanged in the process. This may be expressed as $\oint d \ln \Omega = 0$, where, as before, the special integral sign represents the sum of all the infinitesimal changes in $\ln \Omega$ around the cycle. The total differential of $\ln \Omega$ may be written as $d \ln\Omega = (\partial \ln\Omega / \partial E)_{V,N}\,dE + (\partial\ln\Omega / \partial V)_{E,N}\,dV$.

If we put $\beta = (\partial \ln \Omega/E)_{V,N}$ and $\gamma = (\partial \ln \Omega/V)_{E,N}$, we have

$$\oint d \ln \Omega = \oint \beta\left(dE + \left(\frac{\gamma}{\beta}\right)dV\right) = 0. \tag{5.22}$$

For a reversible cyclic process, Equation 3.13 gives $\oint_R \mathrm{d}S = \oint_R \mathrm{d}Q/T = 0$. From the fundamental relation (Equation 3.18) for a gas, $T\,\mathrm{d}S = \mathrm{d}E + P\,\mathrm{d}V$, we obtain

$$\oint_R \frac{1}{T}(\mathrm{d}E + P\,\mathrm{d}V) = 0. \tag{5.23}$$

Comparison of Equations 5.22 and 5.23 leads to the relationships

$$\left(\frac{\partial \ln \Omega}{\partial E}\right)_{V,N} = \beta \propto \frac{1}{T} \tag{5.24}$$

and

$$\left(\frac{\partial \ln \Omega}{\partial V}\right)_{E,N} = \gamma \propto \frac{P}{T}. \tag{5.25}$$

If we choose $\beta = 1/k_B T$, with k_B Boltzmann's constant, we obtain the expression given previously for the temperature parameter in Equation 5.14. For consistency, the same proportionality constant k_B is chosen for $(\partial \ln \Omega/\partial V)_{E,N}$ in Equation 5.25:

$$\left(\frac{\partial \ln \Omega}{\partial V}\right)_{E,N} = \frac{P}{k_B T} = \beta P. \tag{5.26}$$

This discussion supports the arguments given previously in Section 5.4, in which the important bridge relationship (Equation 5.12) $S = k_B \ln \Omega$ was established. The microscopic and macroscopic descriptions are therefore shown to be completely consistent using the entropy identification we have made. The microscopic approach permits us to obtain explicit expressions for the entropy, provided we are able to calculate $\ln \Omega$ for the system of interest. This has been done in Section 5.4(a) for the ideal gas in what is termed the *microcanonical ensemble* approach with E, V, and N held constant. The approach has also been applied to an ideal spin system in Section 5.4(b) for which E, B, and N are kept fixed. We now apply the approach to a number of other systems in a series of exercises.

Exercise 5.4

Obtain the equation of state for an ideal lattice gas consisting of a lattice of sites, which may be either occupied or unoccupied by particles, as shown in Figure 5.6.

Assume that for the lattice gas there are n particles distributed among N lattice sites. From Equation 5.12, the entropy is given by $S = k_B \ln \Omega$. The number of accessible states is

$$\Omega = \frac{N!}{n!(N-n)!} = \binom{N}{n}, \tag{5.27}$$

which is the number of ways of arranging the n particles on the N possible sites. If we take the logarithm of Ω and apply Stirling's formula, we get

$$S = k_B \left[N \ln N - n \ln n - (N-n)\ln(N-n). \right] \tag{5.28}$$

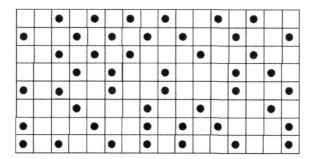

FIGURE 5.6 The figure gives a two-dimensional representation of a lattice gas. There are N lattice sites, with n particles distributed among the sites.

From the fundamental relation (Equation 3.18), the pressure is obtained as $P = T\,(\partial S/\partial V)_E$.

For the lattice gas, the particles do not interact with each other and interact very weakly with the lattice structure. As a result, there is negligible potential energy contribution to the total energy. The kinetic energy is also assumed to be negligible. In a quasi-static, adiabatic compression of the system, the energy E therefore remains constant.

Let the number of lattice sites per unit volume be ρ, so that $N = \rho V$, with V the volume of the lattice gas. From the chain rule, we can write $P = T\,(\partial S/\partial N)(\partial N/\partial V)$, and with Equation 5.28 this leads to $P = \rho k_B T[\ln N - \ln(N - n)] = -\rho k_B T \ln[1 - (n/N)]$.

For $n \ll N$, $\ln[1 - (n/N)] \simeq -n/N$ and $P \simeq \rho k_B Tn/N$ or $P = n k_B T/V$. This is the required equation of state. The entropy of the lattice gas increases as the volume is increased because of the increase in the number of accessible states. In the limit of small V, we have $n = N$, and Equation 5.28 gives $\Omega = 1$ and $S = 0$. The pressure exerted by the lattice gas is linked to the entropy change that accompanies any change in volume.

Exercise 5.5

Obtain an expression for the number of Schottky defects in a cubic crystal at temperature T. A Schottky defect corresponds to a vacancy in a crystal lattice. The ion from the site is assumed to have migrated to the surface of the crystal. Figure 5.7 schematically illustrates the nature of the defect.

If the crystal contains N lattice sites and there are n Schottky defects with $n \ll N$, the entropy of disorder of the crystal is to a good approximation given by

$$S = k_B \ln \Omega = k_B \ln \binom{N}{n} = k_B \left[N \ln N - n \ln n - (N - n)\ln(N - n) \right]. \qquad (5.29)$$

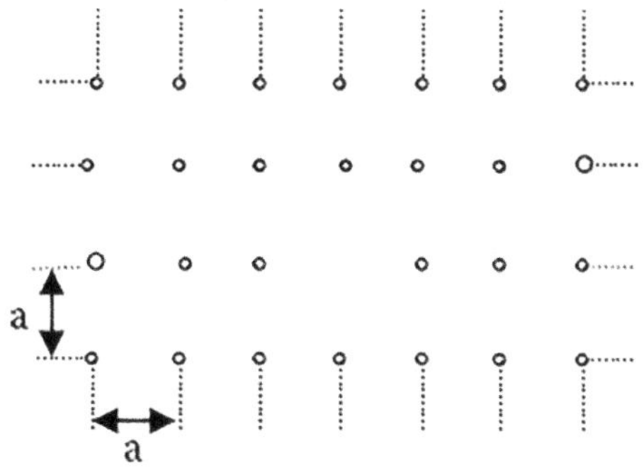

FIGURE 5.7 The two-dimensional sketch shows a Schottky defect in a crystalline solid. An ion has migrated from a lattice site to the crystal surface leaving a vacancy as shown.

The contribution to the entropy from ions that have migrated to sites at the surface is negligible because of the small surface-to-volume ratio for a crystal of macroscopic size. Any volume changes that accompany the formation of Schottky defects are extremely small.

From the fundamental relation (Equation 3.18), $1/T = (\partial S/\partial E)_V = (\partial S/\partial n)_V (\partial n/\partial E)_V$. The energy E associated with the defects is given by $E = n\varepsilon$, where ε is the energy of formation of a single defect. The use of Equation 5.29, with evaluation of the partial derivatives for $n \ll N$, leads to

$$\frac{n}{N} \simeq e^{-\varepsilon/k_B T}. \tag{5.30}$$

Equation 5.30 shows that the fraction of defects n/N depends very sensitively on ε via the exponential function. The formation energy ε of a Schottky defect is of the order of 1 eV. Because Boltzmann's constant has the value $k_B = 8.62 \times 10^{-5}\,\text{eVK}^{-1}$, it is necessary to reach fairly high temperatures $T \geq 1000\,\text{K}$ (provided the material does not melt) for a significant number of defects to be formed.

Exercise 5.6

Obtain an expression for the force required to extend a polymer chain by a small amount. Use a simple model for a polymer in which the polymer molecules consist of long chains of segments in which adjacent segments can take up different orientations with respect to one another.

To simplify the problem, we choose a particularly simple model for a polymer, which allows just two relative orientations, either parallel (0°) or antiparallel (180°), for two representative segments of the long chain as depicted in Figure 5.8.

Let the polymer chain consist of N segments, each of length l. If n segments are aligned parallel and the other $(N - n)$ are antiparallel, the length of the chain is $L = ml$, where $m = n - (N - n) = (2n - N)$. This expression gives $n = \frac{1}{2}(N + m)$ and $(N - n) = \frac{1}{2}(N - m)$. The number of accessible microstates is clearly

$$\Omega = \frac{N!}{\left[\frac{1}{2}(N+m)\right]!\left[\frac{1}{2}(N-m)\right]!}. \tag{5.31}$$

The entropy of the chain is therefore, in compact notation, given by

$$S = k_B \left[N \ln N - N - \frac{1}{2}(N \pm m)\ln\left[\frac{1}{2}(N \pm m)\right] + \frac{1}{2}(N \pm m) \right],$$

where use has been made of Stirling's formula (Appendix A). (The ± notation implies that both sets of signs must be used to give the complete expression.) The above expression simplifies to give

$$S = k_B \left[N \ln N - \frac{1}{2}(N \pm m)\ln\left\{\frac{1}{2}(N \pm m)\right\} \right]. \tag{5.32}$$

(a) (b)

FIGURE 5.8 Two segments of a polymer chain with relative orientations (a) parallel (0°) and (b) antiparallel (180°) to each other.

For a polymer subjected to a force F with associated extension dL, the fundamental relation is $T dS = dE - F\, dL$, which gives $F = -T(\partial S/\partial L)_E$ or, for the present case, $F = -T(\partial S/\partial m)_E(\partial m/\partial L)$. This expression with Equation 5.32 leads to

$$F = -\left(\frac{k_\mathrm{B} T}{2\ell}\right)\ln\left[\frac{(N-m)}{(N+m)}\right]. \qquad (5.33)$$

For $m \ll N$, we approximate the ln function by $-2m/N$ (with use of the binomial theorem) and obtain $F \simeq (2 m k_\mathrm{B} T)/(2 l N) = k_\mathrm{B} T[L/(N l^2)]$.

The force required to extend the polymer is proportional to the temperature and to $L/N l2$. An increase in temperature increases the entropy, which corresponds to increased disorder in the chain, and this tends to decrease the length L.

In Chapter 6, we use the results obtained in the present chapter to gain insight into the concept of absolute temperature. Our two model systems, the ideal gas and the ideal spin system, are used for illustrative purposes. We are led to a formulation of the third and final law of thermodynamics.

5.6 STATISTICAL ENSEMBLES

In Chapters 4 and 5, important connections are established between the microscopic and macroscopic descriptions of thermodynamic systems. Chapter 4 introduces the microstates for two representative ideal cases, the ideal gas and the ideal spin system. Expressions are obtained for the total number of accessible microstates $\Omega(E)$, as a function of energy for both cases. Note that the systems are taken to be in equilibrium with the number of particles and the total energy kept fixed. Use of the fundamental postulate as stated in Section 5.1 leads to the relationship $S = k_\mathrm{B} \ln \Omega$ for the entropy, as given in Equation 5.12. This important bridging relationship between the macroscopic and macroscopic formalisms is named the Boltzmann equation in honor of Ludwig Boltzmann who did pioneering work in the field in the nineteenth century. The energy and particle number constraints correspond to the conditions used in what is called the microcanonical ensemble statistical approach.

In applying statistical methods to situations involving the probabilities of obtaining various event outcomes, it is convenient to introduce statistical ensemble averaging to replace time averaging over a sequence of events. An ensemble consists of a very large number of identical systems prepared in identical ways. As an illustration, consider the tossing of a coin with outcomes heads or tails. To find the probability of a particular outcome, we could toss the same coin many times and keep a record of each outcome. Alternatively, we could use an ensemble of identical coins each of which is tossed just once. The probability of a particular outcome is obtained by counting up the individual outcomes and determining the fraction of each possible result for the ensemble. Further details of probability theory are given in Chapter 10.

Statistical physics makes use of three different ensembles called the microcanonical, the canonical, and the grand canonical ensembles. The microcanonical ensemble consists of a very large number of thermally isolated systems with fixed energy E and fixed particle number N as mentioned above. In the canonical ensemble, the fixed energy constraint is changed to fixed temperature T while N remains fixed. Finally, in the grand canonical ensemble, the two constraints become fixed temperature T and fixed chemical potential μ. Fixed temperature is achieved by having ensemble members in thermal contact with identical constant temperature heat baths, while fixed chemical potential involves each ensemble member being in diffusive contact with a large particle reservoir. The probability of a system being found in a particular microstate is defined as the fraction of ensemble members found in that state.

For systems which contain a macroscopically large number of particles, and which are in equilibrium, the three statistical ensembles are equivalent, and a particular choice can therefore be made to suit a particular situation. This ensemble optionality comes about because fluctuations in energy and/ particle number are vanishingly small for macroscopic systems in contact with large reservoirs. The freedom to choose an ensemble is helpful, for example, in deriving the quantum distribution functions as shown later in this book.

In view of the equivalence of the three statistical ensembles for macroscopic systems, it is possible to use Boltzmann's equation for the entropy in situations in which the microcanonical ensemble constraints do not apply. In Chapter 10, another general equation for the entropy, called the Gibbs equation, is introduced. The Gibbs equation, named after J.U. Willard Gibbs, who played a major role in developing statistical mechanics in the classical era, expresses the entropy in terms of the probabilities of finding a system in its various microstates. The Boltzmann and Gibbs equations are shown to be equivalent.

PROBLEMS CHAPTER 5

5.1 Calculate the entropy of 0.1 mol of helium gas at 300 K in a container of volume $2 \times 10^{-3}\,\mathrm{m}^3$. Express your answer in terms of the gas constant R.

5.2 A mixture of two monatomic gases consists of N_A molecules of gas A and N_B molecules of gas B in a container of volume V at temperature T. Obtain an expression for the entropy of the mixture in terms of the total energies E_A and E_B for the two molecular species, the numbers of molecules of each kind, and the volume of the container. Make use of the equipartition of energy theorem to write your expression in terms of the temperature, and hence, derive an expression for the specific heat of the gas.

5.3 By considering the number of accessible states for an ideal two-dimensional gas made up of N adsorbed molecules on a surface of area A, obtain an expression for the entropy of a system of this kind. Use the entropy expression to obtain the equation of state in terms of N, A, and the force per unit length F. What is the specific heat of the two-dimensional gas at constant area?

5.4 A solid contains N lattice sites and an equal number of interstitial sites. Energy ε is required to transfer an atom from a lattice site to an interstitial site. If at temperature T there are n atoms at interstitial sites, obtain an expression for the entropy of the system in terms of N and n. Use your expression to obtain the temperature parameter β in terms of ε and hence the energy and specific heat as a function of T. Assuming that the solid does not melt as T is raised, sketch the specific heat versus T curve and give a qualitative explanation for its form.

5.5 Derive an expression for the entropy of a system that consists of N noninteracting localized particles each of which has two energy states 0 and ε. Express your result in terms of the total energy E. Use your expression to obtain the temperature parameter β in terms of E and hence the specific heat as a function of T. Sketch the specific heat versus T curve and give a qualitative explanation for its form.

5.6 In this chapter, it is shown that the energy of a spin system may be written as $E = N\mu B$ $\tanh[\mu\beta/k_BT]$. Consider a system of electron spins $S = 1/2$ and $g = 2$ in a magnetic field of 1 T. Find the temperature at which the magnetization of the system reaches 90% of the saturation value. Repeat the calculation for solid helium-3 nuclear spins for which $I = 1/2$ and $\mu = 2.2$ nuclear magnetons.

5.7 A paramagnetic salt containing 10^{20} electron spins with $S = 1/2$ and $g = 2$ is situated in a magnetic field of 4 T. If the sample temperature is 2 K, determine the spin contribution to the energy and the entropy of the system using microstate considerations.

5.8 A d-dimensional ideal gas of N particles is contained in an L^d-sized container where L is the length of a side. Obtain an expression for the density of states for the system and hence the entropy. Express your results in terms of d. Determine the equation of state for the system, the total energy, and the heat capacity as a function of temperature.

6 Zero Kelvin and the Third Law

6.1 INTRODUCTION

The kelvin or absolute temperature scale is introduced in Chapter 1 in terms of the constant volume ideal gas thermometer and is further discussed in Section 3.11 of Chapter 3, in a general way, on the basis of reversible heat engines. Through the choice of a fixed reference point, the triple point of water, to which a particular temperature of 273.16 K is assigned, the absolute temperature or kelvin scale is completely specified. The choice of the particular temperature value for the triple point is based on historical considerations and ensures that 1 K is equal in magnitude to 1°C.

Because the analysis of the Carnot cycle makes use of the ideal gas equation of state, it follows that the kelvin scale and the ideal gas thermometer scale must be the same. An ideal gas thermometer can be used to measure absolute temperatures provided they are not too low. Helium-4 (^{4}He) gas liquefies at 4.2 K, at a pressure of one atmosphere, and other gases liquefy at higher temperatures at atmospheric pressure. Transition to another phase therefore limits the range of use of gas thermometers. Secondary thermometers such as resistance thermometers, which use metals or semiconductors as sensors, are generally used at temperatures down to 0.05 K (or 50 mK). A variety of other thermometers, such as ideal paramagnetic thermometers, have been developed for use at still lower temperatures.

This chapter is concerned with the low-temperature behavior of systems of many particles and with the third law of thermodynamics. Figure 6.1 shows the temperature range from 10^{-8} to 10^7 K, plotted on a log scale, with an indication of various important phase transition and other temperatures that include superfluidity and superconductivity transition temperatures. The temperature of the cosmic microwave background radiation and the interior temperature of a representative black hole are also shown. With the temperature plotted on a log scale rather than a linear scale, it is apparent that it becomes increasingly difficult to reach lower and lower temperatures.

Temperatures below 1 nK have been achieved in nuclear spin systems that are not in equilibrium with the lattice in which they are located. The nuclear spin-lattice relaxation times are generally very long, and this allows experiments to be performed on the "thermally isolated" nuclear spins at temperatures in the nanokelvin or even picokelvin range. Adiabatic demagnetization techniques are used to cool the nuclear spins. The coldest places in the universe are inside the horizons of black holes where the temperature depends on the inverse mass of the black hole. A representative temperature is shown for a black hole of five solar masses.

6.2 ENTROPY AND TEMPERATURE

In Section 3.4 of Chapter 3, it was pointed out that the inexact differential dQ may be converted into an exact differential on division by the absolute temperature T. This gives the infinitesimal change in entropy, dS = dQ/T, that accompanies a process in which heat energy dQ is transferred at temperature T. Entropy is a state function and, as pointed out previously, is of central importance in thermodynamics.

The fundamental relation of thermodynamics for fluids, introduced in Section 3.9 of Chapter 3, with a fixed number of particles, may be written as T dS = dE + PdV (Equation 3.18). It follows that $1/T = (\partial S/\partial E)_V$. From the definition of entropy introduced in Chapter 5, $S = k_B \ln \Omega(E)$, where $\Omega(E)$ is the number of accessible states for a system, the fundamental relation leads to Equation 5.14 $\beta = 1/k_B T = (\partial \ln \Omega/\partial E)_{X,N}$. The symbol X represents variables that are held fixed, such as the volume V for a fluid or the applied magnetic field H for a paramagnet. To gain further insight into the temperature concept, it is instructive to use the expressions for $\ln \Omega(E)$ obtained in Chapter 4 to examine the behavior of β as a function of energy for the ideal spin system and the ideal gas system.

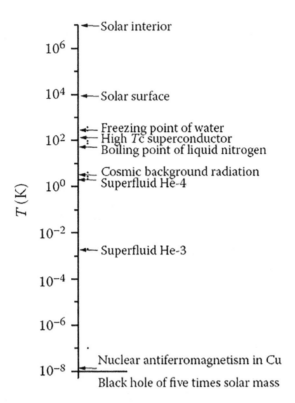

FIGURE 6.1 Representative temperatures that occur in nature or that have been achieved in the laboratory, plotted on a logarithmic scale. Some important phase transition and other significant temperatures are shown.

6.3 TEMPERATURE PARAMETER FOR AN IDEAL SPIN SYSTEM

For N noninteracting spins $\dfrac{1}{2}$ situated in a magnetic field B, Equation 4.25 gives

$$\ln \Omega(E) = N \ln N - \left(\frac{N}{2} - \frac{E}{2\mu B}\right)\ln\left(\frac{N}{2} - \frac{E}{2\mu B}\right) - \left(\frac{N}{2} + \frac{E}{2\mu B}\right)\ln\left(\frac{N}{2} + \frac{E}{2\mu B}\right).$$

A plot of $\ln \Omega(E)$ versus E is shown in Figure 4.7, and this is reproduced in Figure 6.2a. The energy of the spin system has both a lower energy bound $(-N\mu B)$ and an upper energy bound $(N\mu B)$. The maximum in $\ln \Omega(E)$ occurs at $E = 0$ and has the value $N \ln 2$, which corresponds to $\Omega(E) = 2^N$. Each spin can point up or down with equal probability so that there are two orientations for each of the N spins and therefore 2^N arrangements for all of the spins. The temperature parameter β that corresponds to a particular energy is given by the slope of the tangent to the curve, in a plot of $\ln \Omega(E)$ versus E, for that particular energy value. Figure 6.3b shows the behavior of β as a function of E, whereas Figure 6.3c shows the temperature $T \propto 1/\beta$ as a function of E.

From Figure 6.2c, we see that, for $-N\mu B \leq E \leq 0$, the temperature increases from 0 to $+\infty$ as the energy increases. Infinite temperature corresponds to each spin having equal probabilities for up or down orientations. Zero temperature corresponds to all spins pointing parallel to the applied field, and in this situation, there is only one arrangement for the N spins so that $\Omega(E) = 1$ and $\ln \Omega(E) = 0$.

A highly interesting region occurs for $E > 0$, where the temperatures are negative, as shown by the negative slope of the plot of $\ln \Omega(E)$ versus E. A population bar diagram is helpful in understanding the significance of negative temperatures. Figure 6.3 shows the populations of the two states for

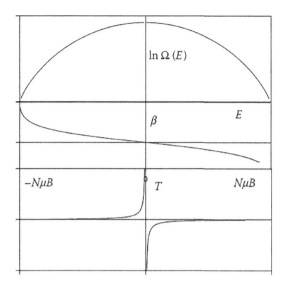

FIGURE 6.2 Plots of ln $\Omega(E)$, β, and T versus E for an ideal system of spins with moment μ in a field B. The temperature passes from $+\infty$ to $-\infty$ as E passes through zero.

N spins $\frac{1}{2}$ in a field B. Note that the spins are not necessarily electrons, and in this case, we have shown the $|+\frac{1}{2}\rangle$ state as the lower energy state, which implies that the magnetic dipole moment and the spin are parallel.

In Figure 6.3, the length of the bar for each of the two spin states is proportional to the population of that state. Positive temperatures correspond to the lower energy state having a larger population than the higher energy state, whereas the reverse applies for negative temperatures. Infinite temperature occurs when the populations are equal. The negative temperature region involves what is termed *population inversion* and requires manipulations of the spin system by some means to achieve a high-energy configuration. For example, we could in principle suddenly reverse the applied field, although this is difficult to accomplish sufficiently quickly in practice. Special methods have been developed for inverting spin populations in particular systems. Spin systems at negative temperatures are very hot. They cool down, with the loss of energy to other degrees of freedom of the system, and pass from negative to positive temperatures via infinite T, with $T = +\infty$ and $T = -\infty$ taken as being equivalent. It is not possible to move from positive to negative temperatures through 0 K, as we shall see.

Many experiments have been carried out to verify the physical importance of negative temperatures. Spin calorimetry experiments can be carried out between two spin systems: one at a positive temperature and the other at a negative temperature. These results verify the negative temperature concept. It must be emphasized that negative temperatures can only occur in systems where there is an upper energy bound and the plot of ln $\Omega(E)$ versus E has a region of negative slope.

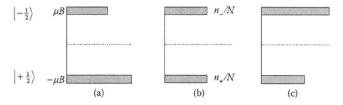

FIGURE 6.3 Population bar diagrams for various spin configurations n_+ and n_- for three temperatures: (a) positive, $T > 0$ K; (b) infinite, $T = \infty$; and (c) negative, $T < 0$ K. The particles are not electrons and have their magnetic dipole moments parallel to the spin.

6.4 TEMPERATURE PARAMETER FOR AN IDEAL GAS

For an ideal gas of N particles in a box of volume V, Equation 4.19 gives $\ln \Omega (E)$ as follows:

$$\ln \Omega(E) = N \ln \frac{V}{N} + \frac{3}{2} N \ln \frac{E}{N} + \frac{3}{2} N \ln \left(\frac{4\pi m}{3h^2} \right)^{3/2} + \frac{5}{2} N.$$

This shows that $\ln \Omega(E)$ increases steadily with E, as depicted schematically in Figure 6.4a. There is no upper energy bound in this case, and the energy can increase indefinitely in principle. Using the equipartition theorem, we obtain $\beta = \frac{3}{2}(N/E)$, which shows that β decreases as E increases. It is important to examine the behavior of $\ln \Omega(E)$ as $E \to 0$. In this limit, it is no longer appropriate to ignore the fact that the particles may be either fermions or bosons. Such considerations will be dealt with carefully later in the discussion of quantum statistics. For the present, we simply assume that as $E \to 0$, $\Omega(E) \to 1$ and $\ln \Omega(E) \to 0$. In the lowest energy state, there is only one state accessible. (We shall see that bosons can condense into a single ground state, but because of the exclusion principle, fermions cannot.)

The temperature T shown in Figure 6.4c increases steadily with the energy E. This type of behavior is found for many systems, including solids and liquids, for which the energy has no upper bound, although the details for the condensed phases are different. Solid and liquid systems may, of course, change phase, with the possibility of a sudden change in $\ln \Omega(E)$ and S. Changes of this sort are discussed in Chapter 9.

6.5 THE APPROACH TO $T = 0$ K

In our discussion in the previous two sections, we have seen that $\ln \Omega(E)$ and correspondingly the entropy S go smoothly to zero as the energy approaches some lower energy bound value, which, for the present discussion, will be designated E_0. T–S diagrams are helpful in consideration of the approach to absolute zero. Figure 6.5 shows a T–S diagram for an ideal paramagnet, where the two curves correspond to different applied magnetic fields.

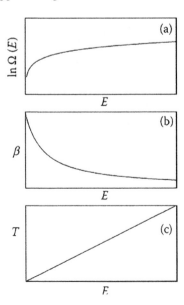

FIGURE 6.4 Plots of (a) $\ln \Omega(E)$, (b) β, and (c) T versus E for an ideal gas system. The temperature increases linearly with E, as required by the equipartition theorem.

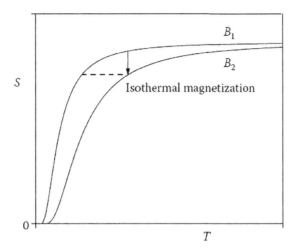

FIGURE 6.5 *T–S* diagram for a paramagnet. The two curves correspond to different magnetic fields B_1 and B_2, with $B_2 > B_1$. The vertical arrow represents an isothermal magnetization process, whereas the dashed horizontal line represents an isentropic (adiabatic) demagnetization process.

The curves in Figure 6.5 may be generated using the entropy expression (Equation 5.18) together with the energy of a spin system given in Equation 5.19.

Exercise 6.1

Discuss the form of the entropy for a system of *N* spins in a field *B* for the low-temperature case where $\mu B \gg k_B T$.

From Equations 5.18 and 5.19, the entropy has the form

$$S = Nk_B \ln N - \left[\frac{1}{2}N + \frac{1}{2}N\tanh\left(\frac{\mu B}{k_B T}\right)\right]k_B \ln\left[\frac{1}{2}N + \frac{N}{2}\tanh\left(\frac{\mu B}{k_B T}\right)\right]$$

$$-\left[\frac{1}{2}N - \frac{1}{2}N\tanh\left(\frac{\mu B}{k_B T}\right)\right]k_B \ln\left[\frac{1}{2}N - \frac{N}{2}\tanh\left(\frac{\mu B}{k_B T}\right)\right]. \qquad (6.1)$$

From Equation 6.1, it can be seen that as $T \to 0$, the entropy $S \to 0$ because $\tanh(\mu B/k_B T) \to 1$ in this limit. Furthermore, it is clear that entropy curves, which correspond to different magnetic fields *B*, tend to zero together, as shown in Figure 6.5.

An important question that arises is whether it is possible to reach the absolute zero of temperature $T = 0$ K? To answer this question, it is necessary to examine ways in which the entropy of a system may be reduced or, in other words, how the order of a system may be increased. Processes for reducing the entropy are often referred to as *entropy-squeezing processes*. Examination of Figure 6.1 shows that approaching absolute zero represents a formidable challenge.

6.6 ENTROPY-SQUEEZING PROCESSES

a. *Adiabatic Demagnetization.* Adiabatic demagnetization processes were discussed in Section 2.7 of Chapter 2 for paramagnetic systems that obey Curie's law. The temperature reached in such a process is given by Equation 2.31: $T_2 = T_1(H_2/H_1)$, where T_2 is the final temperature in field $H_2 = B_2/\mu_0$, whereas T_1 is the initial temperature in field $H_1 = B_1/\mu_0$. Equation 2.31 suggests that if the final field $H_2 = 0$, then we expect $T_2 = 0$. It was pointed

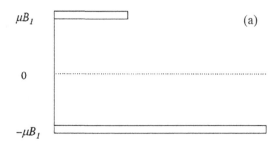

μB_1

0

$-\mu B_1$

(a)

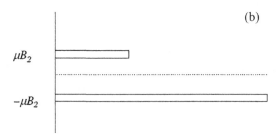

μB_2

$-\mu B_2$

(b)

FIGURE 6.6 Population bar diagram for an ideal spin $\dfrac{1}{2}$ paramagnetic system (a) in high-field B_1 and (b) in low-field B_2 after adiabatic demagnetization. The populations of the levels shown are unchanged in this process.

out in Section 2.8 that the presence of local fields in paramagnets causes departures from Curie's law and prevents the attainment of $T = 0$ K. For real systems, there are always small interactions between the spins, resulting in departures from the ideal spin system behavior at sufficiently low temperatures. These effects may produce spontaneous ordering into a ferromagnetic or antiferromagnetic state at some finite temperature. This complication prevents the attainment of lower temperatures in these systems by means of demagnetization processes. Equation 2.31 may be modified as follows to allow for local fields H_L:

$$T_2 = T_1 \left(\frac{\left(H_2^2 + H_L^2 \right)^{1/2}}{\left(H_1^2 + H_L^2 \right)^{1/2}} \right) \tag{6.2}$$

Because $H_L \neq 0$, it follows that it is not possible to reduce T_2 to 0 K.

It is helpful to use a population bar diagram to understand adiabatic demagnetization processes. Figure 6.6 shows the populations in the energy states before and after the demagnetization process.

The entropy is unchanged in the adiabatic demagnetization process, and the populations of the two states remain unchanged in the lower field. From Equation 6.1, we see that constant S implies constant B/T. Lowering B at constant S results in reduced T so that the final state is at a lower temperature than the initial state. In Figure 6.5, an adiabatic demagnetization process, which is isentropic, is represented by a horizontal path.

b. *Adiabatic Expansion of a Gas.* Reduction of temperature may also be achieved by the adiabatic expansion of a gas. Equation 2.13 may be written in the form

$$T_2 = T_2 \left(\frac{V_1}{V_2} \right)^{\gamma - 1}, \tag{6.3}$$

with $\gamma = c_P/c_V$ as the ratio of the specific heats as shown in Equation 2.51. For the condition $V_2 > V_1$, the gas cools. Processes of this sort are used in Carnot engines and refrigerators. It is clear from Equation 6.3 that, for a finite volume change, a finite drop in temperature will occur. However, weak interactions between particles will lead to condensation into a liquid phase when the temperature is sufficiently low, and this prevents further use of the method.

Exercise 6.2:

By what factor should the volume of nitrogen be increased to cool the gas from 290 K to the liquefaction temperature of 77 K?
  For diatomic N_2, we have $\gamma = c_P/c_V = 7/5$. From Equation 6.3, we obtain $(290/77) = (V_2/V_1)^{2/5}$, and this gives $V_2/V_1 = 27.5$.

6.7 MULTISTAGE PROCESSES

For combined refrigerator stages, it is possible to reach very low temperatures in bulk samples. For example, a helium dilution refrigerator, which uses a mixture of helium-3 and helium-4 as its working substance, can reach temperatures below 0.01 K (10 mK). This temperature can be used as the initial point for one or more adiabatic demagnetization stages to achieve temperatures in the microkelvin range. Cooling of nuclear subsystems to temperatures well below 1 μK has been achieved in a number of laboratories around the world. Figure 6.7 schematically illustrates multiple adiabatic–isothermal processes on a T–S diagram. The processes shown are not easily achieved because a series of low-temperature baths are required to achieve the consecutive isothermal magnetization steps. In practice, multiple stages with multiple heat switches are used. Figure 6.7 is useful in obtaining a general understanding of why it becomes increasingly difficult to reach lower and lower temperatures.

From Figure 6.7, it can be seen that as the temperature tends toward absolute zero, the effectiveness of each temperature reduction process diminishes. This is because the entropy tends to zero regardless of the conditions under which the system is held. The entropy curves converge as shown, and as the temperature is lowered, it becomes harder and harder to squeeze out entropy from the system. Phase transitions may occur in the system at low T and further complicate the procedure. Research at low temperatures is generally concerned with *how* the entropy tends to zero and phase transitions to new states of matter are of particular interest.

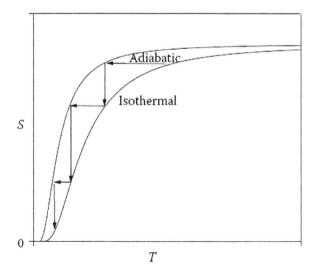

FIGURE 6.7 T–S diagram that illustrates multiple isothermal and adiabatic processes for a paramagnetic system. The entropy curves shown correspond to different applied magnetic fields.

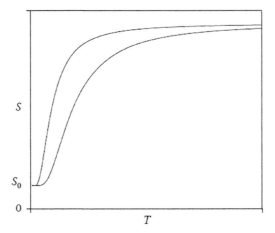

FIGURE 6.8 T–S diagram for a "glassy" spin system in two different applied magnetic fields showing the low-temperature limit entropy S_0 at $T = 0$ K.

6.8   THE THIRD LAW

The third law of thermodynamics may be stated as follows: It is impossible to reach the absolute zero of temperature in a finite number of cooling processes. An equivalent statement is that the entropy S tends to zero as the temperature T tends to zero, that is, $S \to 0$ as $T \to 0$ K.

In common with the other laws of thermodynamics, the third law is based on observation. In our discussion leading up to the statement of the third law, a large body of experimental and theoretical work has been distilled into a few pages. All of the evidence shows that the third law cannot be violated.

As noted above, a question of interest is how the entropy approaches zero for various systems. Phenomena such as ferromagnetism and antiferromagnetism may occur in magnetic materials. Superconductivity occurs in many metallic systems, and superfluidity or frictionless flow occurs in helium-3 and helium-4, which remain liquid down to the lowest temperatures reached and which would still be liquid at 0 K at standard pressure. Other interesting phenomena, such as heavy fermion behavior, occur in certain alloys at low temperatures. All of these transitions are manifestations of small interactions between particles that produce new ordered states at sufficiently low temperatures.

The statement $S \to 0$ as $T \to 0$ may be an idealization as there can be some frozen disorder in a system at low temperatures. This occurs, for example, in noncrystalline glassy materials. To include this possibility, an alternative statement of the third law is given as follows: The entropy tends to a lower limit value as the temperature tends to zero, that is, $S \to S_0$ as $T \to 0$ K. This situation is depicted in Figure 6.8.

The shift in the low T limit entropy value from 0 to S_0 does not alter the statement that it is impossible to reach the absolute zero of temperature in a finite number of processes.

6.9   SUMMARY OF THE LAWS OF THERMODYNAMICS

The laws of thermodynamics, which are based on experience and which form the foundation for the subject, are grouped together here for convenience. They are stated in simple, general ways. We have not previously stated the zeroth law but include it for completeness. As discussed in Chapter 3, alternative equivalent statements for some of the thermodynamic laws can be given, but these are not presented here.

Law 0: Two systems each in thermal equilibrium with a third system are in thermal equilibrium with each other (law of thermometry).

Law 1: The energy of the local universe is constant or $\Delta E_u = 0$.

Law 2: The entropy of the local universe tends toward a maximum or $\Delta S_u \geq 0$.

Law 3: The entropy of a system tends to zero or, more generally, to a constant value as the absolute temperature tends to zero, or $S \to S_0$ as $T \to 0$ K.

In these statements of the laws of thermodynamics, it is clear that the entropy S is of fundamental importance as emphasized previously.

Both macroscopic and microscopic definitions of entropy have been introduced. Microscopic considerations strengthen our grasp of the thermodynamic laws. From a macroscopic point of view, the entropy change because of the transfer of heat energy is given by Equation 3.14 $\Delta S = \int_{T_i}^{T_f} dQ/T$. If the entropy is defined to be zero at $T = 0$ K, then in principle, the absolute entropy at some finite temperature can be obtained by integration, as shown in Equation 3.14. Actual calculations involve knowledge of the heat capacity C as a function of temperature from 0 K to some temperature of interest T, with $dQ = C(T)\, dT$. Microscopically, the entropy can be obtained from the relation as given by Equation 5.12 $S = k_B \ln \Omega$, once the number of states Ω that are accessible to a particular system has been determined. As we have seen, the calculation of Ω for both an ideal gas system and an ideal spin system has been carried out, and specific expressions for the entropy of these two systems are given in Chapter 5. Alternative and more powerful methods for obtaining the entropy of various systems are given in Chapter 10. These methods involve calculation of the partition function Z or the grand partition function $\mathbb{Z}$ for the system of interest corresponding to the canonical and grand canonical ensembles.

Exercise 6.3

The specific heat of liquid helium-3 in the normal fluid range has an approximately linear dependence on temperature between 3 and 70 mK, with the specific heat given by $c = \gamma_{He3} T$. Experiment shows that in this temperature range, $\gamma_{He3} \sim 2.6$ R. What is the entropy change of liquid helium-3 when heated from 10 to 50 mK?

The entropy change is given by $\Delta s = \int_{0.01}^{0.05} \gamma_{He3}\, dT = 2.6R(0.05 - 0.01) = 0.86l\ J/K$.

We have now introduced the basic concepts and all of the laws of thermodynamics. In addition, we have established many of the relationships that are required for applications of thermal physics to processes in bulk matter. In Chapters 7 and 8, the thermodynamic approach is used to describe processes in classical gases and condensed matter, respectively. Chapter 9 deals with phase transitions and critical phenomena. In subsequent chapters (Chapters 10–14), the microscopic approach is developed and applied to systems in the low-temperature, high-density quantum limit.

Before concluding this chapter, we give a brief introduction to an important topic in astrophysics which is concerned with the thermal properties of black holes. In Section 6.10, it is shown that the laws of thermodynamics can, with some adaptation, be applied to processes involving these massive objects.

6.10 BLACK HOLE THERMODYNAMICS

Black holes are extremely cold astronomical objects as indicated for a representative case in Figure 6.1. These massive entities result from the gravitational collapse of large stars, which occurs when nuclear fusion reactions in their cores can no longer generate energy because the hydrogen fuel is exhausted. While black holes are not directly observable, a great deal of information on their remarkable properties has been obtained by studying their gravitational interaction with neighboring astronomical objects and with radiation reaching Earth from more distant sources. The unique features of black hole behavior present challenges to our physical understanding of these objects and the processes they undergo. Progress in developing a microscopically based theory to describe black holes has been hampered because the incorporation of gravity into quantum mechanics has not been accomplished. Thermodynamic concepts together with results from general relativity and quantum field theory have been shown to provide a useful description of black hole physics at the macro level. Pioneering theoretical work done by Stephen Hawking and Jacob Bekenstein in the 1970s stimulated activity in this field.

The properties of a black hole depend on its mass, angular momentum, and electric charge, which together determine the form of what is known as its event horizon. The horizon concept followed from the work of Schwarzschild who solved Einstein's general relativity field equations to show that no matter or light could escape from within the event horizon that surrounds a massive object. For a stationary black hole of mass M, zero angular momentum, and negligible electric charge, the event horizon is spherical with area A given by

$$A = 4\pi R_S^2 \tag{6.4}$$

with $R_S = 2GM/c^2$ the Schwarzschild radius, G the gravitational constant, and c the speed of light. Black holes of this type are called Schwarzschild black holes. Figure 6.9 shows a log-scale plot of R_S versus M measured in solar mass units in the range of 1–1000.

Bekenstein suggested that black holes have entropy and that this entropy is associated with the area of the event horizon, which increases with black hole mass. The need for black holes to possess entropy is a consequence of their ongoing accretion of matter. Once a captured entropy-bearing object crosses the event horizon of a black hole, it is no longer observable. To avoid violating the second law of thermodynamic, which states that the entropy of the universe cannot be decreased in any process, it follows that the entropy of the black hole must increase. An interesting but unanswered question concerns the nature of entropy associated with the event horizon surface. The accretion of matter by a black hole can thus be viewed as the transfer of information from the observable universe to the black hole event horizon surface where it is stored. The general relationship between entropy and information is discussed in Section 10.11.

By making use of general relativity theory together with quantum mechanics, Hawking discovered that black holes emit a low-intensity radiation, now called Hawking radiation. The processes which Hawking considered involve a pair production of photons or particles close to the event horizon. For the particle case, quantum fluctuations allow particle–antiparticle pairs to be created. Theory predicts that

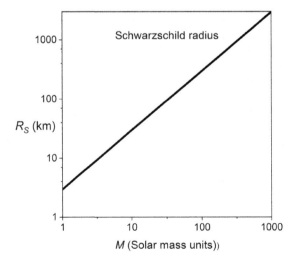

FIGURE 6.9 Log-scale plot of the Schwarzschild radius of an astronomical object as a function of its mass M in solar mass units.

one of the particles escapes (with positive energy), while the other (with negative energy) falls into the black hole. For photons, the detailed theory involves extreme gravitational red shift effects. The emitted radiation is found to obey conventional black body laws that are discussed in Chapter 15. Hawking's finding changed the view that nothing, including light, could escape from a black hole.

The Hawking temperature is given by

$$T_H = \frac{\hbar^2}{8\pi k_B \lambda_P^2 M},$$ (6.5)

where $\lambda_P = \sqrt{\hbar G/c^3} = 1.616 \times 10^{-35}$, m is the Planck length, and k_B is Boltzmann's constant. A log-scale plot of T_H versus M in solar mass units is given in Figure 6.10, which shows that black holes are very cold bodies.

The vertical dashed line in Figure 6.10 indicates the low mass limit for a Schwarzschild black hole to exist, which is taken as seven solar masses. In this brief discussion, we shall consider only Schwarzschild black holes.

Using thermodynamics, together with his result for T_H, Hawking obtained an expression for the entropy of a black hole in terms of the event horizon area consistent with the form obtained by Bekenstein but with improved precision. The Bekenstein–Hawking entropy in dimensionless form involves the event horizon area over Planck's length squared

$$S'_{BH} = \frac{A}{4\lambda_P^2}.$$ (6.6)

Based on these findings, the fundamental relation of thermodynamics $TdS = dE + dW$, as given in Equation 3.18, can be applied to black holes. Changes in the energy E of a black hole are produced by changes in the mass, as given by the Einstein mass–energy relationship, and by changes in the angular momentum. Work W results from the motion of an approaching massive, and possibly electrically charged, object. In addition, the predicted decrease in entropy of a black hole as a result of the Hawking radiation should be taken into account. This is done by allowing for an increase in entropy in the region called the local universe, which lies outside the event horizon.

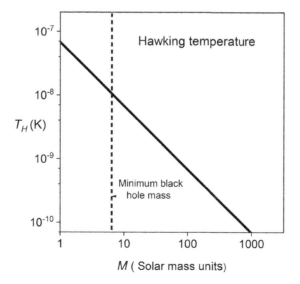

FIGURE 6.10   Log-scale plot of the Hawking temperature as a function of black hole in solar mass units. The dashed line indicates the low mass limit for a black hole to exist.

For a relatively small black hole of 10 solar masses, we obtain the Schwarzschild radius as $R_S = 30$ km and the dimensionless Bekenstein–Hawking entropy is $S_{BH} = 1.1 \times 10^{79}$. Note that the Bekenstein–Hawking entropy corresponds to a very large number of Planck area elements on the event horizon surface. The entropy can be given in SI units by introducing Boltzmann's constant. For a 10 solar mass black hole, the Hawking temperature is 6.8 nK. This low temperature gives rise to the emission of extremely low-intensity, undetectable, Hawking radiation. Using the Stefan–Boltzmann radiation law, it is possible to estimate the time taken for a black hole to evaporate. Values obtained in this way are in general very much larger than age of the universe. Enormous black holes, located at the centers of galaxies, have masses of millions of solar masses, with correspondingly large event horizon areas and extremely low Hawking temperatures. These giant black holes are expected to have lifetimes of the order of 100 billion years.

The applicability of the fundamental relation of thermodynamics to black hole processes is consistent with these objects obeying the laws of thermodynamics, which are summarized in Section 6.9. As mentioned above, allowance must be made for Hawking radiation in dealing with changes in energy and entropy in thermodynamic processes by enlarging the black hole system to include the local universe. In considering the application of the zeroth law and the third law to black holes, it is instructive to introduce the surface gravity, which is the gravitational acceleration of a mass at the event horizon as measured by a distant observer. The surface gravity is related to the Hawking temperature, and both quantities are constant over the event horizon surface consistent with the zeroth law. Because the entropy of a black hole does not tend to zero or to a constant value with decreasing temperature, it is necessary to modify the statement of the third law when applied to black holes. It has been suggested that the surface gravity, which does tend towards zero with decreasing temperature, could serve as an acceptable alternative to the entropy. Calculations for what are called extremal black holes with nonzero angular momentum, no electric charge, and mass, which is close to the lower limit for a black hole to exist, show that κ cannot reach zero. Available theoretical results are consistent with this modified unattainability statement of the third law being applicable to black holes.

In order to develop a more detailed description of black hole properties, it will be necessary to consider the microstates of these objects. This goal presents a major scientific challenge. Interestingly, string theory has opened up interesting and promising research directions related to this objective. Black hole physics is clearly a field in which much remains to be done.

PROBLEMS CHAPTER 6

6.1 A nuclear spin system ($I = 1/2$) in an applied magnetic field B is prepared at a negative temperature by subjecting the system to a short radiofrequency pulse. Sketch the time evolution of the nuclear magnetization as the spin system returns to equilibrium with the surrounding lattice. Show the behavior of the spin temperature with time in a separate plot. Give an expression for the total entropy change of the spin system in returning to equilibrium with the lattice.

6.2 Liquid helium-3 behaves as a Fermi liquid with specific heat $c = \gamma_{He3} T$ below 100 mK before undergoing a transition to a superfluid phase around 2 mK. Sketch an entropy versus a temperature diagram for liquid helium-3. Obtain an expression for the internal energy change in heating 1 mol of liquid helium-3 from 5 to 100 mK. Assume that the volume change is negligible ($\gamma_{He3} \sim 2.6\,R$).

6.3 The low-temperature molar-specific heat of a metal such as copper may be written as the sum of the conduction electron and lattice vibration (phonon) contributions in the form $c = \gamma T + \beta T^3$, with $\gamma = (3/2)\,R(\pi^2/3)1/T_F$ and $\beta = (3/2)R\left(8\pi^2/5\right)\left(1/\theta_D^3\right)$. T_F and θ_D are the Fermi temperature and the Debye temperature of the metal, respectively. Taking, for copper, $T_F = 8.1 \times 10^4$ K and $\theta_D = 343$ K, compare the electron and phonon specific heat contributions at 10, 1, and 0.1 K. Find the entropy change in terms of R for 1 mol of copper in cooling from 1 to 0.01 K. Compare these values with the entropy change of a paramagnetic salt obeying Curie's law, taking $g = 2$ and $S = 1/2$ in a low magnetic field of 0.01 T. Show that $\mu B/k_B T \ll 1$ over the temperature range of interest and use the approximation $\text{sech}^2 x \sim 1$ for $x \ll 1$.

6.4 In an adiabatic demagnetization process, it is clearly advantageous for the paramagnetic material to have a large heat capacity when the cooling process is completed and the material is at final temperature T_f. As shown in Section 5.4(b), the heat capacity of a paramagnetic system that obeys Curie's law is given by $C_H = Nk_B x^2 \, \text{sech}^2(x)$, where $x = \mu B/k_B T$. Find the optimum B_f/T_f ratio for a demagnetization stage in terms of the initial field and temperature. If $B_f = 0.01$ T and $T_i = 1.3$ K (pumped liquid helium-4), what should B_f be set as? What is the final temperature after demagnetization? Take $\mu = \mu_B = 9.27 \times 10^{-24}$ J T^{-1}.

6.5 A number of metals undergo a phase change to a superconducting state at a transition temperature T_C. The phase transition is found to be continuous (see Chapter 9), with no entropy change at T_C. In the normal state, the specific heat is given by $c_H = \gamma T$ and in the superconducting state by $c_s = \beta T^3$, where γ and β are approximately temperature independent. With the use of the third law, compare expressions for the entropy of the normal and superconducting states at T_C and establish a relationship between γ and β. Obtain an expression for the energy difference between the normal and the superconducting states at 0 K.

6.6 In an adiabatic magnetization process carried out on a paramagnetic solid of volume V that obeys Curie's law, the applied field is increased from H_1 to H_2. The H/T ratio remains constant along the adiabatic. Sketch an M–H diagram for the process. How much magnetic work is done in the process? What is the change in internal energy of the system? Obtain an expression for the entropy change in the process.

Section IC

Applications of Thermodynamics to
Gases and Condensed Matter, Phase
Transitions, and Critical Phenomena

7 Application of Thermodynamics to Gases
The Maxwell Relations

7.1 INTRODUCTION

In previous chapters, it has been shown that thermodynamics provides a phenomenological description of processes for systems made up of many particles. The methods of thermodynamics allow us to establish relationships between various properties of a system so that the measurement of one property permits other properties to be deduced without further measurement. Use is often made of the fundamental relation that combines the first and second laws of thermodynamics. For fluids, this has the form $T\,dS = dE + P\,dV$ (Equation 3.18). Together with the equation of state and the heat capacity C_V as a function of temperature, the fundamental relation permits a wide range of processes to be analyzed. For solids and liquids, it is often convenient to use the isothermal compressibility κ, defined in Equation 2.53, and the isobaric thermal expansion coefficient β, defined in Equation 2.52, in place of a formal equation of state. From time to time, it may be convenient to use the adiabatic equation that provides a relationship between two thermodynamic variables for an adiabatic process, such as $PV^r = $ constant for an ideal gas. For a fluid, the equation of state provides a relationship between the three thermodynamic variables, P, V, and T. The state of a system may be specified with any two of these variables. Because the internal energy E and the entropy S are state functions, these two quantities may also be used to specify the state of a system.

In Section 3.13 of Chapter 3, we introduced three thermodynamic potentials that involve combinations of the five variables E, S, P, V, and T. In these special functions, T and S occur together, as do P and V. The conjugate pairs PV and TS, which are products of intensive and extensive variables, have dimensions of energy. The thermodynamic potentials for a single component system such as a pure gas are defined as follows:

$$\text{Enthalpy} \quad H = E + PV, \tag{7.1}$$

$$\text{Helmholtz potential} \quad F = E - TS, \tag{7.2}$$

$$\text{Gibbs potential} \; G = E - TS + PV. \tag{7.3}$$

These forms for the thermodynamic potentials were obtained in the energy representation with the use of Legendre transforms as described in Appendix D and Section 3.13. The potentials are discussed in detail below, and their usefulness in the development of the subject will become apparent in this chapter. The Helmholtz and Gibbs potentials are particularly important and are involved in bridge relationships between macroscopic and microscopic descriptions of systems.

7.2 ENTHALPY

a. *Useful Relationships.* From the definition of the enthalpy H given in Equation 7.1, the differential of H may be written as

$$dH = dE + P\,dV + V\,dP. \tag{7.4}$$

The first law, as given in Equation 2.7, together with Equation 7.4, leads to

$$dH = dQ + V \ dP. \tag{7.5}$$

For isobaric processes where the pressure P is held constant, $dH = dQ$, and the heat capacity at constant pressure may be written as follows:

$$C_P = \left(\frac{dQ}{dT}\right)_P = \left(\frac{\partial H}{\partial T}\right)_P. \tag{7.6}$$

Many processes are carried out at constant pressure, which is often simply atmospheric pressure, and Equation 7.6, which expresses C_P in terms of the partial derivative of H, is useful in such cases. For an infinitesimal reversible process, $dQ = T \ dS$, and with Equation 7.5, this gives

$$dH = TdS + V \ dP. \tag{7.7}$$

For isobaric processes, we note that $dH = T \ dS$, which shows that the infinitesimal change in enthalpy is proportional to the change in entropy if P is held constant. Enthalpy plays an important role in the treatment of isobaric processes, particularly in fields such as chemistry and chemical engineering. An instructive application of the enthalpy function is in the analysis of throttle processes where a gas passes through a constriction or set of constrictions in the form of a porous plug or partition. Such processes are used in the liquefaction of gases.

b. *Adiabatic Throttle Processes.* Consider a quantity of gas that is forced through a permeable barrier, made up of a set of constrictions, by means of a pair of coupled pistons, as shown in Figure 7.1.

The two pistons are arranged to move in such a way that the pressures on the two sides of the porous partition remain constant during the process, with $P_i > P_f$. The process is carried out adiabatically for a thermally insulated system so that $\Delta Q = 0$. Application of the first law gives $\Delta E = E_f - E_i = -\int_i^f P \ dV$. The work done consists of two parts, which correspond to the two regions on either side of the piston, and is easily seen to be $-\int P \ dV = -P_f \int_0^{V_f} dV - P_i \int_{V_i}^0 dV$. It follows that $E_f - E_i = -P_f V_f + P_i V_i$, and the rearrangement leads to $E_i + P_i V_i = E_f + P_f V_f$. This shows that the enthalpy is the same for the initial and final states or

$$H_i = H_f. \tag{7.8}$$

The fact that the process may be considered isenthalpic is important and will be used in the detailed discussion of throttle processes given in Section 7.8. Our equilibrium thermodynamic discussion

FIGURE 7.1   Schematic depiction of a quantity of gas being forced through a porous partition from initial pressure P_i and volume V_i to final pressure P_f and volume V_f.

does not consider intermediate nonequilibrium states for the gas that flows through the porous partition under the effect of the steady pressure gradient across the partition. We simply consider the gas in equilibrium on either side of the barrier.

7.3   HELMHOLTZ POTENTIAL *F*

a. *Relationships Involving F.* The definition of the Helmholtz potential F is given in Equation 7.2, and the differential is

$$dF = dE - TdS - SdT. \tag{7.9}$$

The fundamental relation (Equation 3.18) $TdS = dE + PdV$ together with Equation 7.9 gives

$$dF = -SdT - PdV. \tag{7.10}$$

For isothermal processes, it follows that

$$dF = -PdV. \tag{7.11}$$

For processes that are both isothermal and isochoric, $dF = 0$, which shows that F remains constant in such processes. This is a useful and important result. From Equation 7.10, it follows immediately that the entropy and the pressure are given by partial derivatives of F

$$S = -\left(\frac{\partial F}{\partial T}\right)_V \tag{7.12}$$

and

$$P = -\left(\frac{\partial F}{\partial V}\right)_T. \tag{7.13}$$

Equations 7.12 and 7.13 require knowledge of F as a function of V and T. For a system in equilibrium, the energy may be fixed within certain limits, and the entropy will tend to a maximum value, as required by the second law. This suggests that the Helmholtz potential will exhibit a minimum value in equilibrium. The maximum work output from a heat engine is given by the change in the Helmholtz potential in the operation of the engine. In the past, the Helmholtz potential was named the Helmholtz free energy because of this connection to the work output of an engine.

Exercise 7.1

A heat engine in contact with a heat bath produces an amount of work W. Show that the maximum work output of the engine, in a process where heat Q is absorbed from the heat bath, is given by $-\Delta F_{Eng}$, the change in the Helmholtz potential or *free energy* of the engine in this process. As a corollary to the work output analysis, the Helmholtz potential F_{Eng} of the engine system is a minimum in equilibrium.

The change in internal energy ΔE_{eng} of the engine in a process in which heat Q is absorbed from the heat bath and work W is done by the engine is from the first law $\Delta E_{Eng} = Q - W$. The total change in entropy for the combined system of heat bath plus engine is, from the second law, either positive or zero. If we denote the heat bath by subscript B, we have for the total entropy change $\Delta S = \Delta S_{Eng} + \Delta S_B = \Delta S_{Eng} - Q/T \geq 0$. Using the first law relationship gives $Q = \Delta E_{Eng} + W$, and this leads to $(T\Delta S_{Eng} - \Delta E_{Eng} - W)/T \geq 0$ or, as required, $-\Delta F_{Eng} \geq W$. The maximum work output is obtained when the equality holds, and we see that the term Helmholtz *free energy* was used because of the relationship to the work output as mentioned above. If no work is done ($W = 0$), it follows that $\Delta F_{Eng} \leq 0$, and in equilibrium, the Helmholtz potential tends to a minimum. To prove this result more formally, we can expand the Helmholtz potential, expressed as a function of two variables T and V, in a Taylor series about the extremum point and examine the coefficients of the second-order terms.

As an exercise, show that $(\partial^2 F/\partial T^2)_V \geq 0$, consistent with the extremum in F being a minimum.

b. *Connection to the Partition Function.* The partition function, or sum over states, Z will be introduced in Chapter 10. For a system whose energy eigenstates are known, it is in general simpler to calculate Z rather than $\Omega(E)$, the sum over accessible states introduced in Chapter 4, because Z involves an unrestricted sum over states. The following relationship is established in Chapter 10:

$$F = -k_B T \ln Z. \tag{7.14}$$

This is an important bridge relationship between the microscopic and the macroscopic formulations. Once Z is calculated, the Helmholtz potential F is obtained from Equation 7.14 and then the entropy S from Equation 7.12. From S, other thermodynamic quantities of interest may be determined with the use of the fundamental relation as described in Section 3.12 of Chapter 3.

c. *Changes of State.* When a substance changes from one state to another, for example, from liquid to gas, at constant temperature and at constant volume, the Helmholtz potential remains constant, as shown by Equation 7.10. Very often, however, T and P are held constant rather than T and V. The Gibbs potential is useful in such situations.

7.4   GIBBS POTENTIAL G

a. *Relationships Involving G.* From Equation 7.3, which defines G, the differential dG is immediately obtained:

$$dG = dE - TdS - SdT + PdV + VdP \tag{7.15}$$

Making use of the fundamental relation, Equation 3.18 gives

$$dG = -SdT + VdP. \tag{7.16}$$

Expressions for S and V, as partial derivatives of G, follow from Equation 7.16:

$$S = -\left(\frac{\partial G}{\partial T}\right)_P \tag{7.17}$$

and

$$V = -\left(\frac{\partial G}{\partial P}\right)_T. \tag{7.18}$$

The second-order partial derivatives are $-C_P/T = (\partial^2 G/\partial T^2)_P$ and, with the use of Equation 2.53, $-\kappa V = (\partial^2 G/\partial P^2)_T$. Arguments similar to those made for the Helmholtz potential, with allowance for small changes in both T and V in the fundamental relation, show that the Gibbs potential will be a minimum for a system in equilibrium.

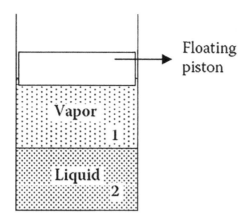

FIGURE 7.2   A two-phase system, such as liquid and vapor, in equilibrium at constant T and P.

b. *Changes of State.* When a change of state occurs at constant *pressure* and constant *temperature*, the molar Gibbs potentials for the two different states, or phases, in equilibrium are the same. This statement is justified below. An example of two phases in equilibrium is a liquid in contact with its vapor at some temperature and under a pressure of, say, 1 atm, as shown in Figure 7.2.

The Gibbs potential for the two-phase system may be written as

$$G = n_1 g_1 + n_2 g_2, \tag{7.19}$$

where g_1 and g_2 are the molar Gibbs potentials for the two phases and n_1 and n_2 are the equilibrium molar quantities for the two-phase system at the particular instant considered. Particles will traverse the interface between the liquid and the vapor phases such that the net rate of transfer is zero in equilibrium. The total number of moles of substance is fixed so that $n = n_1 + n_2$, giving $dn_1 = -dn_2$. In equilibrium, G is at a minimum, and Equation 7.19 leads to

$$dG = g_1 dn_1 + g_2\, dn_2 = 0, \tag{7.20}$$

from which it follows that

$$g_1 = g_2. \tag{7.21}$$

This proves the statement made above that the molar Gibbs potentials for the two states are the same. Use will be made of the result given in Equation 7.21 when phase transitions are discussed in Chapter 9.

7.5   THE GIBBS POTENTIAL, THE HELMHOLTZ POTENTIAL, AND THE CHEMICAL POTENTIAL

We are now able to express changes in F and G in terms of the chemical potentials μ_i introduced in Section 3.12 for a multicomponent system. Applications to phase equilibria and chemical reactions show the importance of μ in such processes. The general form of the fundamental relation for a multicomponent system is given in Equation 3.26, $TdS = dE + PdV - \Sigma_i \mu_i\, dN_i$, where μ_i is the chemical potential for the ith particle species and dN_i is the change in the number of molecules of this type. If Equation 3.26 is used together with the expressions for the differentials dF and dG given in Equations 7.9 and 7.15, we obtain

$$dF = -SdT - PdV + \sum_i \mu_i \, dN_i \qquad (7.22)$$

and

$$dG = -SdT + VdP + \sum_i \mu_i \, dN_i. \qquad (7.23)$$

It follows from Equation 7.22 that

$$\mu_i = \left(\frac{\partial F}{\partial N_i}\right)_{T,V,N_j} \qquad (7.24)$$

and similarly from Equation 7.23

$$\mu_i = \left(\frac{\partial G}{\partial N_i}\right)_{T,P,N_j}. \qquad (7.25)$$

We see that the chemical potential is given by the partial derivative of F with respect to N_i, with T, V, and the other particle numbers held fixed. Similarly, it is given by the partial derivative of G with respect to N_i for constant T, P, and $N_j \neq N_i$. For processes carried out at constant T and V,

$$dF = \sum_i \mu_i \, dN_i, \qquad (7.26)$$

whereas for constant T and P,

$$dG = \sum_i \mu_i \, dN_i. \qquad (7.27)$$

Equations 7.26 and 7.27 are useful and important results for multicomponent systems, such as systems in which chemical reactions can occur or systems that undergo phase transitions. As an example, consider a system in which a phase transition occurs at some temperature T for constant pressure P. For phases that coexist, Equation 7.20 gives

$$dG = \sum_i g_i \, dn_i = \left(\frac{1}{N_A}\right) \sum_i g_i \, dN_i, \qquad (7.28)$$

for the composite system in equilibrium. Comparison of Equation 7.28 with Equation 7.27 leads to the following identification:

$$\mu_i = \frac{g_i}{N_A}. \qquad (7.29)$$

This shows that the chemical potential is equal to the molar Gibbs potential divided by Avogadro's number, or the Gibbs potential per particle, which is a very useful relationship. As noted previously, the chemical potential is an important quantity in thermal and statistical physics. Gradients in μ drive diffusion processes in systems in which there is a nonuniform particle concentration. We shall see that μ plays a crucial role in the quantum distribution functions that are discussed in Chapter 12.

7.6 CHEMICAL EQUILIBRIUM

Consider a gaseous mixture made up of different kinds of molecules i, with numbers N_i, which undergo chemical reactions. It is of interest to determine the equilibrium condition for a multicomponent system of this kind. We denote different types of molecules by the symbol X_i. A chemical reaction may then be written in the form

$$\sum_i x_i X_i = 0, \qquad (7.30)$$

where x_i denotes the number of molecules that participate in a single reaction process. We take the sign of x_i to be positive for products and negative for reactants.

As a simple example, consider the reaction

$$H_2 + Cl_2 \rightleftarrows 2HCl. \qquad (7.31)$$

This reaction occurs when hydrogen and chlorine gases are mixed. Equation 7.31 is rewritten in the form of Equation 7.30 as follows:

$$-1H_2 - 1Cl_2 + 2HCl = 0. \qquad (7.32)$$

In general, any change in the Gibbs potential for a multicomponent gaseous system is given by Equation 7.23. If the pressure and temperature are held constant, Equation 7.23 takes the form given in Equation 7.27. Changes in the numbers of molecules of type i are proportional to the x_i in Equation 7.30. It follows that in equilibrium, when G is a minimum, Equation 7.27 may be written as

$$\sum_i x_i \mu_i = 0. \qquad (7.33)$$

Equation 7.33 is a useful relation, and for the specific example considered above, we obtain

$$2\mu_{HCl} - \mu_{H_2} - \mu_{Cl_2} = 0, \qquad (7.34)$$

which establishes a relationship between the chemical potentials of the different species for the system in equilibrium. Using Equation 7.24 together with Equation 7.14 permits the chemical potential to be obtained from the partition function Z. We have previously given an expression in Equation 5.17 for the chemical potential of a classical *monatomic* ideal gas, which, with the use of the equipartition theorem, may be written as

$$\mu = -T \left(\frac{\partial S}{\partial N} \right)_{E,V} = k_B T \left[\ln \left(\frac{N}{V} \right) + \ln \left(\frac{h^2}{2\pi m k_B T} \right)^{3/2} - \frac{3}{2} \right] = k_B T \left[\ln \left(\frac{V_Q}{V_A} \right) - C \right],$$

with $C = \frac{3}{2} \left(\ln(2\pi/3) + 1 \right)$, $V_A = V/N$ the atomic volume, and $V_Q = (h/\langle p \rangle)^3$ the quantum volume, as discussed in Section 4.3 of Chapter 4. Monatomic gases are, of course, generally nonreactive. Expressions for the chemical potential of polyatomic molecules are similar to that given above, but allowance must be made for internal degrees of freedom as well as the three translational degrees of freedom. In Chapters 16 and 19, it is shown how explicit expressions for the μ_i may be obtained for polyatomic molecules. This approach allows a *quantitative* discussion of chemical equilibrium to be given in terms of the important *law of mass action*, which has the form $\prod_i N_i^{x_i} = K_N(T,V)$, with N_i as the number of molecules of species i. K_N is the equilibrium constant that may be experimentally measured, for various reaction conditions, and which may also be theoretically calculated for a given reaction using molecular partition functions, as is shown in detail in Chapter 19. For the above reaction, under particular conditions of T and V, the law of mass action provides a simple relationship between the numbers of molecules of each species. For the present, we regard K_N as an experimentally determined constant for a particular reaction, under given conditions. The law of mass action is of great usefulness in optimizing chemical reaction processes.v

Exercise 7.2

For the reaction described by Equation 7.32, obtain an expression for the equilibrium number of molecules of HCl for the case $N_{H_2} = N_{Cl_2}$.
From the law of mass action $N_{HCl}^2 N_{H_2}^{-1} N_{Cl_2}^{-1} = K_N$, we find $N_{HCl} = \sqrt{K_N N_{H_2}^2}$.

7.7   MAXWELL'S THERMODYNAMIC RELATIONS

The four Maxwell relations that are derived in this section are of great use in thermodynamics because they relate various partial derivatives of thermodynamic functions to each other. This permits substitution of one partial derivative by another in deriving thermodynamic expressions. In obtaining the Maxwell relations, use is made of a number of identities from multivariable calculus. Consider a function $f = f(x, y)$ of two variables x and y, for which the following useful identities involving the partial derivatives hold:

$$\left(\frac{\partial^2 f}{\partial x \partial y}\right) = \left(\frac{\partial^2 f}{\partial y \partial x}\right), \tag{7.35}$$

$$\left(\frac{\partial x}{\partial y}\right)_f = -\frac{(\partial f/\partial y)_x}{(\partial f/\partial x)_y}, \tag{7.36}$$

$$\left(\frac{\partial x}{\partial f}\right)_y = -\frac{1}{(\partial f/\partial x)_y}. \tag{7.37}$$

We make use of these identities in our derivation of the Maxwell relations and consider, in turn, the differentials of the energy E, the entropy S, the Helmholtz potential F, and the Gibbs potential G, each of which leads to a different Maxwell relation.

a. *The First Maxwell Relation for a Single-Component System.* The fundamental relation Equation 3.18 is $dE = TdS - PdV$. Consider E to be a function of S and V or $E = E(S, V)$, so that the total differential is

$$dE = \left(\frac{\partial E}{\partial S}\right)_V dS + \left(\frac{\partial E}{\partial V}\right)_S dV. \tag{7.38}$$

Comparison of Equations 3.18 and 7.38 permits the following identifications to be made:

$$\left(\frac{\partial E}{\partial S}\right)_V = T \tag{7.39}$$

and

$$\left(\frac{\partial E}{\partial V}\right)_S = -P. \tag{7.40}$$

From the identity given in Equation 7.35, we have $\partial^2 E/\partial V \partial S = \partial^2 E/\partial S \partial V$ or, equivalently, $(\partial/\partial V)_S(\partial E/\partial S)_V = (\partial/\partial S)_V(\partial E/\partial V)_S$. With Equations 7.39 and 7.40, it follows that

$$\left(\frac{\partial T}{\partial V}\right)_S = -\left(\frac{\partial P}{\partial S}\right)_V. \tag{7.41}$$

This is the first Maxwell relation, and it expresses the fact that T, V, P, and S cannot be chosen independently; any two specify the state of the system. Useful connections between the partial derivatives follow as a consequence. Derivations of the other Maxwell relations are similar to those given above.

b. *The Second Maxwell Relation.* The differential for the enthalpy is given by $dH = TdS + VdP$ (Equation 7.7). If we let $H = H(S, P)$, we obtain $dH = (\partial H/\partial S)_P dS + (\partial H/\partial P)_S dP$ and identify the partial derivatives as

$$\left(\frac{\partial H}{\partial S}\right)_P = T \tag{7.42}$$

and similarly

$$\left(\frac{\partial H}{\partial P}\right)_S = V. \tag{7.43}$$

From Equation 7.35, with $f = H$ and $x = P$, $y = S$, it is easily shown that

$$\left(\frac{\partial T}{\partial P}\right)_S = \left(\frac{\partial V}{\partial S}\right)_P. \tag{7.44}$$

This is the second Maxwell relation.

c. *The Third Maxwell Relation.* Consider the differential of the Helmholtz potential given in Equation 7.10, $dF = -SdT - PdV$. With $F = F(T, V)$, we obtain $dF = (\partial F/\partial T)_V dT + (\partial F/\partial V)_T dV$, and this permits the following identifications for the entropy,

$$-S = \left(\frac{\partial F}{\partial T}\right)_V, \tag{7.45}$$

and the pressure,

$$-P = \left(\frac{\partial F}{\partial V}\right)_T. \tag{7.46}$$

The use of Equation 7.35, with $f = F$, leads to

$$\left(\frac{\partial S}{\partial V}\right)_T = \left(\frac{\partial P}{\partial T}\right)_V, \tag{7.47}$$

the third Maxwell relation.

d. *The Fourth Maxwell Relation.* From Equation 7.16, the differential of G is given by $dG = -SdT + VdP$, and with $G = G(T, P)$, this gives $dG = (\partial G/\partial T)_P dT + (\partial G/\partial P)_T dP$. This expression permits the following identifications: the entropy is

$$-S = \left(\frac{\partial G}{\partial T} \right)_P \qquad (7.48)$$

and the volume

$$V = \left(\frac{\partial G}{\partial P} \right)_T . \qquad (7.49)$$

Finally, Equation 7.35, with $f = G$, results in the fourth Maxwell relation:

$$\left(\frac{\partial S}{\partial P} \right)_T = -\left(\frac{\partial V}{\partial T} \right)_P . \qquad (7.50)$$

e. *Summary of the Maxwell Relations.* The four Maxwell relations are summarized as follows:

from Equation 7.41, M1, $\left(\partial T / \partial V \right)_S = -\left(\partial P / \partial S \right)_V$;

from Equation 7.44, M2, $\left(\partial T / \partial P \right)_S = \left(\partial V / \partial S \right)_P$;

from Equation 7.47, M3, $\left(\partial S / \partial V \right)_T = \left(\partial P / \partial T \right)_V$; and

from Equation 7.50, M4, $\left(\partial S / \partial P \right)_T = -\left(\partial V / \partial T \right)_P$.

As mentioned above, the Maxwell relations are a consequence of the fact that T, S, P, and V are not independent. The Maxwell relations may be written down immediately with the aid of the fundamental relation (Equation 3.18) in the form $dE = T \, dS - P \, dV$, with the following rule that involves the variables T, S, P, and V, which occur on the right-hand side of the relations. If in M1–M4 the variables, with respect to which one differentiates, occur as differentials in the fundamental relation, a minus sign must be used in the corresponding Maxwell relation, while any single permutation away from this results in a change in sign. The physical significance of the relations may be understood from an examination of the partial derivatives. In M1 (Equation 7.41), $(\partial T/\partial V)_S$ gives the infinitesimal change in temperature with volume in an isentropic (quasistatic, adiabatic) process, whereas $(\partial P/\partial S)_V$ gives the infinitesimal change in pressure when a gas is heated at constant volume. The negative sign signifies that the two infinitesimal changes have opposite signs. Similar consideration may be used in interpreting the other relations. The great usefulness of the Maxwell relations is that different properties can be connected, and this permits a complete description of a system on the basis of a few of its experimentally measured properties. We make use of the Maxwell relations in Sections 7.8 and 7.9.

7.8   APPLICATIONS OF THE MAXWELL RELATIONS

The Maxwell relations together with the equation of state for a system permit the thermodynamic analysis of a variety of physical processes to which the system may be subjected. As mentioned in Section 7.1, it is convenient to use the isothermal compressibility κ and the thermal expansion coefficient β for liquids and solids in lieu of an equation of state. Results may often be expressed in terms of the heat capacities C_P and C_V. When adiabatic processes are involved, the adiabatic equation for the system is useful. Armed with the Maxwell relations, the thermodynamic potentials, and the fundamental relation, we are now in a position to deal with a large variety of situations. Illustrative examples for expansion processes in gases are given below. Chapter 8 deals with processes in condensed matter.

a. *The Joule Effect and the Joule–Kelvin Effect.* In the mid-nineteenth century, Joule considered the free expansion of a gas and attempted to measure what is now called the *Joule coefficient.* Figure 7.3 gives a sketch illustrating Joule's experiment.

   Initially, the right-hand reservoir in Figure 7.3 contains gas, whereas the left-hand reservoir is evacuated. When the tap between the two reservoirs is opened, gas flows through the tube that connects the vessels until the pressures are the same. No pistons are involved and no work is done by the system, that is, $\Delta W = 0$. Furthermore, no heat energy is transferred during the sudden and irreversible expansion process for which $\Delta Q = 0$ but, as we have seen in Section 3.6, $\Delta S > 0$. Because ΔQ and ΔW are both zero, it follows from the first law that $\Delta E = 0$ in the expansion. Joule was interested in possible temperature changes of the gas in a free expansion process. The Joule coefficient is defined as

$$\Gamma_{\text{J}} = \left(\frac{\partial T}{\partial V} \right)_E. \tag{7.51}$$

Joule attempted to determine the coefficient Γ_{j} for a gas from measurements of the temperature of the external thermal bath before and after the valve was opened but found no detectable effect. An analysis of the experiment is given below, but we first consider a somewhat different expansion process, known as the Joule–Kelvin effect, which involves a porous plug or throttle valve between two reservoirs, as shown in Figure 7.4.

   Gas from the left-hand container is forced through the throttle valve by the pressure gradient across the valve. The pistons are moved so that the pressures in the two containers are maintained at constant values P_i and P, respectively. As discussed previously in Section 7.2(b), the enthalpy H remains constant provided the process is carried out adiabatically.

   The Joule–Kelvin coefficient is defined as

$$\Gamma_{\text{JK}} = \left(\frac{\partial T}{\partial P} \right)_H. \tag{7.52}$$

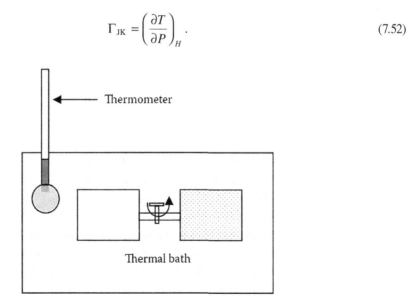

FIGURE 7.3   Schematic representation of Joule's free expansion experiment in which he attempted to detect a change in temperature of a gas that is allowed to expand into an evacuated space by opening a valve while the temperature of the surrounding heat bath is monitored.

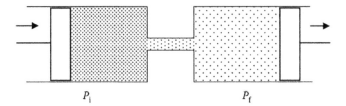

P_i P_f

FIGURE 7.4 The Joule–Kelvin controlled expansion process that involves the flow of gas through a throttle valve. The tube that connects the two containers has a small diameter, and this permits a pressure difference to be maintained between the two large volumes.

and the temperature change in a throttled expansion process is

$$\Delta T = \int_{P_i}^{P_f} \left(\frac{\partial T}{\partial P}\right)_H dP = \int_{P_i}^{P_f} (\Gamma_{JK}) dP. \tag{7.53}$$

The Joule–Kelvin effect can lead to a significant cooling of gases and is used, for example, in the liquefaction of helium.

It is straightforward to obtain expressions for Γ_J and Γ_{JK} from the fundamental relation, the definition of the enthalpy, and certain of the Maxwell relations.

Consider the Joule coefficient Γ_J, defined in Equation 7.51, $\Gamma_J = (\partial T/\partial V)_E$. We make use of the identity given in Equation 7.36 in the form $(\partial T / \partial V)_e = -(\partial E / \partial V)_t / (\partial E / \partial T)_v$.

From the fundamental relation $dE = T\, dS - P\, dV$, we obtain $(\partial E/\partial V)_T = T(\partial S/\partial V)_T - P$, and if we put $(\partial E/\partial T)_v = C_v$, we get $\Gamma_J = -(1/C_v)[T(\partial S / \partial V)_t - P]$. Using Maxwell M3 (Equation 7.47), we obtain the Joule coefficient as

$$\Gamma_J = \frac{1}{C_V}\left[P - T\left(\frac{\partial P}{\partial T}\right)_V\right]. \tag{7.54}$$

If the equation of state for the gas is known, the Joule coefficient can immediately be determined. Note how useful the Maxwell relations are in obtaining an expression for a quantity of interest in a form that is convenient to apply.

A similar procedure on the basis of the identity Equation 7.36 is given Γ_{JK}:

$$\Gamma_{JK} = (\partial T / \partial P)_H = -(\partial H / \partial P)_T / (\partial H / \partial T)_P.$$

From Equation 7.7, $(\partial H/\partial P)_T = [T(\partial S/\partial P)_T + V]$, and with Equation 7.6, we have $C_P = (\partial H/\partial T)_P$. These expressions lead to $\Gamma_{JK} = -(1/C_P)[T(\partial S/\partial P)_T + V]$. Inserting M4 yields

$$\Gamma_{JK} = \frac{1}{C_P}\left[T\left(\frac{\partial V}{\partial T}\right)_P - V\right], \tag{7.55}$$

and knowledge of the equation of state allows an explicit expression for the J–K coefficient to be obtained.

It is a simple matter to show that both Γ_J and Γ_{JK} are zero for an ideal gas so that no cooling occurs in either process for this system. The ideal gas equation of state $PV = nRT$ gives $(\partial P/\partial T)_V = nR/V = P/T$, which means that the right-hand side of Equation 7.54 vanishes and $(\Gamma_J)_{ideal} = 0$. Similarly, $(\partial V/\partial T)_P = nR/P = V/T$ and insertion in Equation 7.55 shows that $(\Gamma_{JK})_{ideal} = 0$. The physical reason for the vanishing of the Joule and Joule–Kelvin coefficients for an ideal gas is that there are no interparticle forces present and therefore no potential energy contributions to the internal energy of such a gas. The internal energy is a function of temperature only ($E = E(T)$), as noted previously, and does not depend on the volume V.

b. *Joule and Joule–Kelvin Coefficients for a van der Waals Gas.* It is of interest to examine both Γ_J and Γ_{JK} for a *real* gas. The van der Waals equation of state provides a useful model for this calculation. Equation 1.2 gives the van der Waals equation of state as $(P + a/V^2)(V - b) = nRT$, and it is convenient to choose $n = 1$ mol. To obtain the coefficients, we require $(\partial P/\partial T)_V$ and $(\partial V/\partial T)_P$.

From Equation 1.2, we get $(\partial P/\partial T)_V = R/(v - b)$, with v as the molar volume. (It is straightforward to allow for $n \neq 1$.) Substitution in Equation 7.54 gives

$$\Gamma_J = \frac{1}{C_V}\left(\frac{a}{v^2}\right), \tag{7.56}$$

with c_V as the specific heat per mole. Similarly, for the Joule–Kelvin case, we obtain

$$\left(\frac{\partial V}{\partial T}\right)_P = \frac{1}{(\partial T/\partial V)_P} = \frac{R}{\left[P - \left(a/v^2\right) + 2ab/v^3\right]}. \tag{7.57}$$

With the approximation that small terms ab/v^2 and $2ab/v^3$ may be neglected and use of Equation 7.57 in Equation 7.55, we get

$$\Gamma_{JK} \simeq -\frac{1}{c_P}\frac{\left[bP - \left(2a/v\right)\right]}{\left[P - \left(a/v^2\right)\right]}. \tag{7.58}$$

We can now calculate the temperature change in the Joule-free expansion of a van der Waals gas if we assume that the specific heat is approximately constant,

$$\Delta T = \int_i^f \Gamma_J \, dV = -\frac{a}{c_V}\int_{v_i}^{v_f}\frac{dV}{V^2} = \frac{a}{c_V}\left[\frac{1}{v_f} - \frac{1}{v_i}\right]. \tag{7.59}$$

Inspection of Equation 7.59 shows that a thermally insulated van der Waals gas cools down in a free expansion process. Physically, this is because the molecules or particles are further apart, on average, after the expansion process than they were before. For a van der Waals gas, the intermolecular potential has the form shown in Figure 7.5.

The short-range repulsive core with radius $r = b$ corresponds to a hard sphere-type potential. The long-range attractive potential is produced by weak fluctuating electric dipole–dipole interactions between molecules and falls off approximately as $1/r^6$. The forces responsible for the attractive potential are generally referred to as the *van der Waals forces*. After an expansion process, the total potential energy U of the system of molecules has *increased* because

$$U = \sum_{i<j} u\left(r_{ij}\right) \tag{7.60}$$

will be made up of contributions $u(r_{ij})$, which are on average less negative than before. To keep the total energy fixed, the total kinetic energy $K = \sum_i p_i^2/2m$ of the system must *decrease* as required by energy conservation. The total energy is given by

$$E = K + U, \tag{7.61}$$

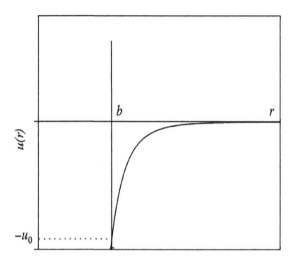

FIGURE 7.5 Intermolecular potential for a van der Waals gas with a short-range infinite repulsive core at *r = b* and a long-range attractive potential that reaches a minimum value of $-U_0$ as shown.

with K and U as defined above. A decrease in kinetic energy corresponds to a drop in temperature of the gas.

We have previously shown that in an irreversible free expansion process of the kind used by Joule, the entropy of a gaseous system increases. The increase in entropy reflects the increase in the number of accessible states available to the system as a result of the expansion. An entropy increase occurs whether or not the gas cools.

The change in temperature of a van der Waals gas caused by a throttled or Joule–Kelvin expansion can be obtained, at least for small changes in pressure, using

$$\Delta T = \int \Gamma_{JK} dP, \tag{7.62}$$

with Γ_{JK} given by Equation 7.58. Because of the presence of terms involving the volume V, which is a function of the pressure, the integral is not simple and must be carried out numerically.

Measurements and calculations show that, depending on the conditions, the temperature of a real gas may increase or decrease in an isenthalpic Joule–Kelvin expansion. Figure 7.6 shows representative schematic plots of final temperature T_f versus final pressure P_f for the isenthalpic expansion of a real gas for various initial pressure P_i and temperature T_i conditions.

A number of features are apparent from the curves based on a series of points, which correspond to particular initial and final conditions. For given P_i and T_i, the gas may either warm or cool, depending on the final pressure P_f. In general, the isenthalpic curves show a maximum at what is called the *inversion point*. The slope $(\partial T/\partial P)_H$ changes from positive to negative at the inversion point. From Equation 7.55, it can be seen that $\Gamma_{JK} = 0$ when $(\partial V/\partial T)_P = V/T$. It is important to remember that the curves plotted in Figure 7.6 correspond to finite changes in T and P as the gas passes through the throttle valve. This process involves nonequilibrium intermediate states, as mentioned in Section 7.2(b). Γ_{JK} is given by the slopes of the fitted curves at constant enthalpy as shown in Figure 7.6. Clearly, Γ_{JK} may be positive or negative, with the sign dependent on the conditions which apply. At a sufficiently high initial temperature, it can be seen that the inversion point occurs at zero pressure. This is the maximum inversion temperature. No cooling of the gas can be achieved by means of a Joule–Kelvin expansion at this or higher temperatures. To cool a particular gas in a Joule–Kelvin expansion, the gas temperature must first be lowered below the maximum

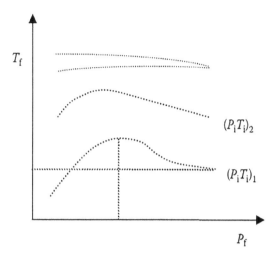

FIGURE 7.6 Representative plots of final temperature T_f versus final pressure P_f of a gas in an isenthalpic throttled expansion from a series of fixed initial pressure P_i and temperature T_i conditions. The maximum, or inversion point, shifts to lower final pressures with changes in the initial conditions as shown. No cooling of the gas occurs when T_i exceeds the maximum inversion temperature.

inversion temperature and the starting conditions chosen to ensure that cooling occurs. Table 7.1 gives the maximum inversion temperature for important gases that can be liquefied in expansion processes.

The physical reason for the change from cooling to heating in a throttled expansion, with different starting temperatures and pressures, is linked to the relative importance of repulsive and attractive interactions between particles. At sufficiently high starting temperatures, repulsive interactions become very important and make a major contribution to the potential energy of the gas. Figure 7.7 shows a representative intermolecular potential commonly referred to as the *Lennard–Jones* or the *6–12 potential*.

The pair potential is described by the empirical expression

$$u(r) = u_0 \left[\left(\frac{r_0}{r} \right)^{12} - 2 \left(\frac{r_0}{r} \right)^6 \right]. \tag{7.63}$$

The form of the potential is similar, in some respects, to that shown in Figure 7.5 for a van der Waals gas. Overlap of the molecular wave functions at small molecular separations gives rise to

TABLE 7.1

Maximum Joule–Kelvin Inversion Temperatures for a Number of Gases

Gas	Maximum Inversion Temperature (K)
Carbon dioxide	1500
Argon	723
Nitrogen	621
Air	603
Hydrogen	202
Helium	40

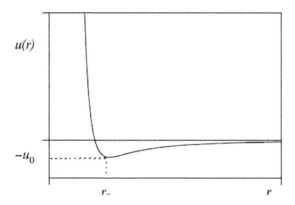

FIGURE 7.7   The Lennard–Jones 6–12 intermolecular potential for a real gas, which reaches a minimum value of $-u_0$ at the molecular separation R_0.

the repulsive part of the potential represented by the first term in Equation 7.63. (The repulsion is linked to the Pauli exclusion principle for fermions, which is given in Chapter 12.) It can be seen that when the repulsive interactions make a dominant contribution to the total potential energy U of the gas, the drop in pressure in a throttled expansion will lead to a decrease in U and a corresponding increase in the kinetic energy K. An increase in K is associated with an increase in the temperature of the gas. Conversely, if the attractive interaction is dominant in the potential energy, the throttled expansion will lead to cooling. The kinetic energy will decrease as the potential energy increases (becoming less negative) using an argument similar to that which predicts cooling in the Joule expansion of a van der Waals gas.

7.9   THE ENTROPY EQUATIONS

It is useful for a number of applications to establish general expressions for entropy changes in terms of changes in state variables. The entropy is a state function and may therefore be written as a function of the state variables T, V, and P. As noted previously, because the state variables are related by an equation of state, two variables are sufficient to specify the state completely.

Generally, we choose

$$S = S(T,V) \tag{7.64}$$

or

$$S = S(T,P) \tag{7.65}$$

although we could also take $S = (P, V)$. Expressions for the total differential of S may be obtained immediately.

    a. *First* T dS *Equation.* From Equation 7.64, we obtain the differential $dS = (\partial S/\partial T)_V$ $dT + (\partial S/\partial V)_T\, dV$. This may be written as

$$T\ dS = C_V\ dT + T\left(\frac{\partial P}{\partial T}\right)_V dV, \tag{7.66}$$

where the use has been made of M3 given in Equation 7.47. Integration of Equation 7.66 results in

$$\Delta S = S(T_f, V_f) - S(T_i, V_i) = \int_i^f \left(\frac{C_V}{T}\right) dT + \int_i^f \left(\frac{\partial P}{\partial T}\right)_V dV. \tag{7.67}$$

This shows that the entropy change in a process specified by initial and final temperatures and volumes may be calculated from a knowledge of the heat capacity C_V and the equation of state.

b. *Second TdS Equation.* It follows from Equation 7.65 that $dS = (\partial S/\partial T)_P \, dT + (\partial S/\partial P)_T \, dP$ or, using M4 Equation 7.50,

$$T dS = C_P \, dT + T\left(\frac{\partial V}{\partial T}\right)_P dP. \tag{7.68}$$

The entropy change in a process in which the initial and final temperatures and pressures are known may be calculated from C_P and the equation of state,

$$\Delta S = \int_i^f \left(\frac{C_P}{T}\right) dT - \int_i^f \left(\frac{\partial V}{\partial T}\right)_P dP. \tag{7.69}$$

Application of Equation 7.69 to ideal gases, where C_P is known and the equation of state has a particularly simple form, is straightforward. In performing the integrals in Equations 7.67 and 7.69, reversible paths are considered with the system passing through a succession of equilibrium states to which the equation of state can be applied. Actual processes in which the system goes from i to f may be irreversible. In calculating entropy changes, the irreversible process is replaced by a reversible process for integration purposes. Because the entropy is a state function, it does not matter which reversible path is followed in the calculation of ΔS. The temperature–entropy or TdS equations are useful and may be applied in a variety of situations, which include processes in condensed matter as described in Chapter 8.

c. *The Entropy Equations for an Ideal Gas.* The equation of state for a molar quantity ($n = 1$) of an ideal gas is $Pv = RT$, with v as the molar volume. If the gas is monatomic, then the specific heat per mole at constant volume is $c_V = \frac{3}{2}R$ and at constant pressure $c_P = \frac{5}{2}R$. For a monatomic ideal gas, Equation 7.67 becomes

$$\Delta S = \frac{3}{2} R \ln\left(\frac{T_f}{T_i}\right) + R \ln\left(\frac{v_f}{v_i}\right) \tag{7.70}$$

and similarly, the second entropy Equation 7.69 may be written as

$$\Delta S = \frac{5}{2} R \ln\left(\frac{T_f}{T_i}\right) - R \ln\left(\frac{P_f}{P_i}\right) \tag{7.71}$$

It is instructive to compare the above results with those obtained from our definition of entropy in terms of microscopic accessible states $S = k_B \ln \Omega$. In Section 5.4 of Chapter 5, where we used the fixed total energy microcanonical ensemble approach, the Sackur–Tetrode equation (Equation 5.15) for the entropy was obtained:

$$S = Nk_B \ln\left[\left(\frac{V}{Nh^3}\right)\left(\frac{4\pi mE}{3N}\right)\right]^{3/2} + \frac{5}{2} Nk_B.$$

For fixed $N = N_A$, the entropy change corresponding to a change in V and E is

$$\Delta S = R \ln\left(\frac{V_f}{V_i}\right) + \frac{3}{2} R \ln\left(\frac{E_f}{E_i}\right) \tag{7.72}$$

For a mole of monatomic ideal gas, the equipartition theorem gives $E - \frac{3}{2} N_A k_B T$, and the insertion of this expression for E in Equation 7.72 results in $\Delta S = \frac{3}{2} R \ln (T_f/T_i) + R \ln(V_f/V_i)$, in agreement with Equation 7.70. The use of the ideal gas equation of state to replace v_f by (RT_f/P_f) and v_i by (RT_i/P_i) in Equation 7.72 immediately leads to Equation 7.71. It is clear that a complete consistency is found in calculating entropy changes for an ideal gas using the various expressions we have for S. If the equation of state is known, similar results may be obtained for other more complicated systems. Condensed matter systems are discussed in Chapter 8. The results derived in this chapter provide a comprehensive and useful description of the thermodynamics of gases. As illustrations, the ideal gas equation and the van der Waals equation have been used to obtain explicit expressions for certain properties of gases with the use of the general formalism.

PROBLEMS CHAPTER 7

7.1   By writing E and S as functions of T and P, obtain the following expressions for the differentials dE and dF: $dE = (C_P - PV\beta)dT + V(\kappa\beta - \beta T)dP$ and $dF = -(PV\beta + S)$ $dT + PV\kappa dP$. β and κ are the isobaric thermal expansion coefficient and the isothermal compressibility, respectively.

7.2   Show that the Gibbs potential for a system is a minimum in equilibrium.

7.3   The equation of state for a nonideal gas is written in the virial expansion form as $P = (N/V)$ $k_B T[1 + B_2(N/V) + ...]$, where $B_2(T)$, the second virial coefficient, is an increasing function of T. Obtain an expression for $(\partial E/\partial V)_T$ for the gas, and determine whether the energy of the gas increases or decreases with volume in a Joule expansion experiment.

7.4   Dieterici's equation of state is written in terms of constants a and b as $P(V - b) = RT$ $\exp(-a/RTV)$. Determine the Joule and Joule–Kelvin coefficients for a gas obeying this equation.

7.5   If a gas obeys the van der Waals equation of state $(P + a/V^2)(V - b) = RT$ (with $N = N_A$), obtain $(\partial E/\partial V)_T = [T(\partial E/\partial V)_T - P]$ for the gas and determine whether the energy of the gas increases or decreases with volume in a Joule-free expansion experiment.

7.6   Use the first TdS equation to obtain an expression for the entropy change ΔS of a van der Waals gas in a process in which the volume and temperature increase. Assume as an approximation that C_V is almost constant in the process and express your result in terms of the initial and final temperatures and the initial and final volumes. Compare your result for ΔS with that for an ideal gas undergoing the same changes in volume and temperature.

7.7   Use the Maxwell relations to obtain an expression for the dependence of the specific heat c_P on pressure P at constant temperature in terms of the volume expansion coefficient β at constant P.

7.8   Consider a gas-phase chemical reaction of the form $\Sigma_i x_i X_i = 0$, where x_i may be positive or negative, which is carried out under constant volume conditions. Show that the law of mass action may be written in the form $\prod_i (P_i)^{x_i} = K_P(T)$, $K_P(T)$ where is the temperature-dependent equilibrium constant and P_i is the partial pressure of constituent i. Use the result to rewrite the law of mass action in terms of the concentrations $c_i = N_i/N$, where N_i is the number of molecules of chemical species i in equilibrium and N is the total number of molecules. First show that the pressure dependence of a reaction may be determined with the relationship $(\partial \ln(K_x (T, P)/\partial P)_T = -\Delta V/RT$.

7.9 A gaseous dissociative reaction of a diatomic molecule X_2 into two monatomic components X is carried out under constant volume conditions. Use the law of mass action to relate the partial pressures and concentrations of the diatomic and monatomic components in equilibrium at some high temperature T. Discuss the effect that an increase in pressure would have on the concentrations of the components.

8 Applications of Thermodynamics to Condensed Matter

8.1 INTRODUCTION

For thermodynamic processes that involve solid or liquid systems, it is convenient, as noted in Section 7.1 (Chapter 7), to use the isothermal compressibility κ and the isobaric thermal expansion coefficient β in place of an equation of state. Condensed matter is typically rather incompressible, and volume changes produced by pressures available in laboratory equipment are generally quite small. Thermal expansion effects are also generally small, and taking $\Delta V \ll V$ permits helpful approximations to be made. When subjected to changes in temperature and/or pressure, solids and liquids may undergo a change of state referred to as a *phase change*. Phase diagrams that show boundaries between solid, liquid, and vapor phases are available for many materials. A discussion of phase changes and critical phenomena is given in Chapter 9, and this topic will not be dealt with in this chapter. In addition to κ and β, it is generally necessary to have knowledge of the specific heat of a substance, and if it is magnetic, the magnetization properties. We shall find, as might be expected, that the specific heats c_P and c_V are almost equal for processes in which the volume changes are small. Tables of the measured physical properties are available for many materials.

Magnetic solids form a special class of materials, and in this chapter, we again confine our attention to the thermodynamics of ideal paramagnetic materials that obey Curie's law. The value of the Curie constant obtained in Chapter 5 and the expression derived there for the specific heat of a paramagnet are used in the discussion. The Maxwell relations, derived for gases in Chapter 7, are extended to paramagnets to facilitate the analysis of magnetic processes in these systems. Magnetic order that is found in nonideal magnetic systems, where interactions between spins are important, is dealt with in Chapter 17.

8.2 SPECIFIC HEATS OF SOLIDS: THE LAW OF DULONG AND PETIT

During the nineteenth century, Dulong and Petit observed that the molar specific heats of many solids, measured at room temperature and above, at a pressure of around 1 atm (101 kPa), were approximately equal and given by

$$c_P = 3R, \tag{8.1}$$

with $R = 8.314\,\mathrm{J\,mol^{-1}\,K^{-1}}$ the gas constant. This result is called the law of Dulong and Petit, although it is not a strict law but rather a convenient rule. When the temperature is lowered, the specific heat is found to decrease steadily to zero. Figure 8.1 shows, schematically, the representative behavior found for solids.

Two important features of the specific heat behavior with temperature are worth noting. First, the temperature at which the Dulong–Petit plateau value is attained depends on the elastic properties of the solid. For diamond, which is an extremely hard material, it is necessary to increase the temperature to well above 1000 K in order for c_P to tend to $3R$. For ductile materials, the plateau value is achieved at much lower temperatures. Second, the decrease in the specific heat at low temperatures

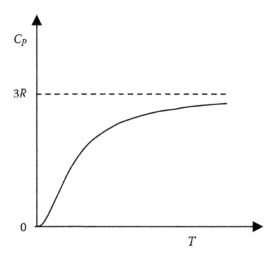

FIGURE 8.1 The specific heat c_P of a typical solid as a function of temperature. The Dulong–Petit value $3R$ is shown as the dashed line.

cannot be explained by classical physics. Following the introduction of quantum physics ideas at the beginning of the twentieth century, new theories for the specific heats of solids were proposed first by Einstein and later by Debye. The Debye model, in particular, satisfactorily explains the low-temperature behavior of c_P. The Einstein and the Debye models are discussed in Chapter 15 and essentially treat the solid as a set of coupled oscillators in which the atoms vibrate about their average positions with vibrational amplitudes dependent on the lattice temperature. Both models correctly predict the high-temperature behavior of the specific heat that can, as shown below, be explained using the equipartition of energy theorem for temperatures at which the classical approximation holds. Differences in the elastic properties of solids are taken into account in a straightforward way in the two models through the introduction of fitting parameters with dimensions of temperature, termed the Einstein temperature and the Debye temperature, respectively. Hard materials have higher Einstein/Debye temperatures than more ductile materials.

For an ideal gas, we have seen that $c_V = \dfrac{3}{2}R$. In the case of solids, there is little difference between c_P and c_V because the solid scarcely changes its volume as the temperature changes and the mechanical work done in the constant pressure case is negligible. A quantitative expression for c_P-c_V, expressed in terms of β and κ as given previously in Equation 2.54, is obtained in Section 8.4. The expression is used to show that the difference in specific heats is indeed very small for typical solids. A comparison of the specific heat of a monatomic ideal gas with that given by the Dulong–Petit law for a solid at high temperature shows that they differ by a factor 2. This comparison suggests that the equipartition theorem can be used to explain the high temperature or *classical* specific heat of solids.

In Equation 4.2, the classical Hamiltonian for a gas of N particles is given as $\mathcal{H} = \sum_{i=1}^{N}\left(P_i^2/2m\right) + \sum_{i<j=1}^{N} u\left(r_{ij}\right)$, where the first term gives the total kinetic energy contribution and the second term the total potential energy because of interactions between pairs of particles. The restriction $i < j$ is inserted to avoid the inclusion of the same pair twice. For a solid of N atoms, the classical Hamiltonian may formally be written in the same way. Figure 8.2 depicts a simple model for a solid in which atoms are joined together by springs. A two-dimensional slice through a three-dimensional structure is shown for clarity.

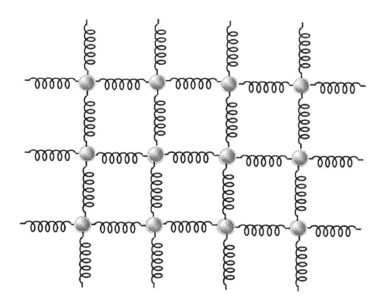

FIGURE 8.2 A two-dimensional representation of a simple model for a solid in which atoms are connected by springs. The temperature-dependent energy of the solid is made up of both kinetic energy and potential energy contributions.

This is essentially the model used by Einstein, who made the additional simplification that all the atoms vibrate about their equilibrium positions with the same frequency ω_E or the model shown in Figure 8.2, each atom has both kinetic energy and potential energy associated with vibrations about an equilibrium position. Because the lattice is three dimensional, each atom has six degrees of freedom (along x, y, and z), corresponding to three kinetic energy degrees of freedom and three potential energy degrees of freedom. Each of these may be represented in quadratic form by $\varepsilon_\kappa = p_{iv}^2/2m$ and $\varepsilon_P = \frac{1}{2}kq_{iv}^2$, where m is the mass of each particle. p_{iv} and q_{iv} are, respectively, the momentum and displacement of particle i along direction v, while k is the effective spring constant of each spring. Use of the equipartition theorem, introduced in Section 1.6 (Chapter 1), predicts that the total energy of a solid of N_A atoms is

$$E = 6N_A\left(\frac{1}{2}k_BT\right) = 3RT. \tag{8.2}$$

From the definition of the molar specific heat $c_V = 1/n\,(dQ/dT)_V = 1/n\,(\partial E/\partial T)_V$ (where n is the number of moles) with the use of Equation 8.2, we obtain $c_V = 3R$, in agreement with the Dulong and Petit law in Equation 8.1. The above derivation applies to monatomic systems that are nonmetallic. In metallic systems, the conduction electrons make a contribution to the specific heat that becomes very important at low temperatures where the lattice vibration contributions tend to zero. The electronic contribution to the specific heat of conduction electrons is dealt with later in the quantum statistics section of this book, specifically Chapter 13.

Each atom or ion in a solid contributes an energy $3k_BT$ to the internal energy E. For an ionic solid such as KCl, there are $2N_A$ ions (N_A of K^+ and N_A of Cl^-) in 1 mol of the substance, giving a specific heat $c_V = 6R$ in this case. Similar considerations may be applied to more complicated systems in which the structural units consist of several atoms. The specific heat contribution per atom is $3k_B$.

8.3   HEAT CAPACITIES OF LIQUIDS

Similar arguments to those applied to solids in Section 8.2 may be applied to liquids in the high-temperature classical regime. Although for a liquid the atoms or molecules are not fixed in a lattice, they nevertheless experience strong interactions with neighboring atoms or molecules. For a monatomic liquid, such as liquid helium-4 (^{4}He) just below its boiling point of 4.2 K, the specific heat is found to be close to $3R$, showing that, to a good approximation, the equipartition theorem applies, with six degrees of freedom for each helium atom.

At lower temperatures in liquid helium, a specific heat anomaly occurs which is linked to the superfluid transition at 2.17 K. Quantum statistics is needed to explain the transition, and further details are given in Chapter 14. For other simple liquids such as liquid nitrogen (N_2), the molar specific heat just below the boiling point of 77 K is again approximately given by $3R$. For more complicated systems such as organic liquids, where molecules may contain several different types of atom, it is necessary to allow for this increase in estimating the molar specific heat.

8.4   THE SPECIFIC HEAT DIFFERENCE $C_P - C_V$

From the first $T\,dS$ equation (Equation 7.66) in Chapter 7, we have $TdS = C_V dT + T(\partial P/\partial T)_V\, dV$, and this gives the heat capacity C_P as

$$C_P = T\left(\frac{\partial S}{\partial T}\right)_P = C_V + T\left(\frac{\partial P}{\partial T}\right)_V\left(\frac{\partial V}{\partial T}\right)_P \text{ or } C_P - C_V = T\left(\frac{\partial P}{\partial T}\right)_V\left(\frac{\partial V}{\partial T}\right)_P. \tag{8.3}$$

Equation 8.3 may be applied to any system for which the equation of state is known. For 1 mol of an ideal gas, we obtain for the difference in specific heats $c_P - c_V = R$, as shown previously. More generally, it is convenient to manipulate Equation 8.3 into a form that involves the thermal expansion coefficient β (Equation 2.52) and the compressibility κ (Equation 2.53). The use of Equations 7.36 and 7.37 for the partial derivatives permits us to write

$$\left(\frac{\partial P}{\partial T}\right)_V = -\frac{\left(\frac{\partial V}{\partial T}\right)_P}{\left(\frac{\partial V}{\partial P}\right)_T} = \left(\frac{\partial V}{\partial T}\right)_P\left(\frac{\partial P}{\partial V}\right)_T$$

and substitution into Equation 8.3 results in the relation $C_P - C_V = -T\left(\partial V/\partial T\right)_P^2\left(\partial P/\partial V\right)_T$ or, using the definitions of β and κ,

$$C_P - C_V = \frac{VT\beta^2}{\kappa} \tag{8.4}$$

Because κ is chosen positive and β^2 is positive, it follows that $C_P > C_V$ in all instances, as expected when allowance is made for the mechanical work done in the constant P case. The ratio of the specific heats γ is given by

$$\gamma = \frac{c_P}{c_V} = 1 + \frac{vT\beta^2}{c_V\kappa}, \tag{8.5}$$

with v as the molar volume. For solids and liquids, it is found that $\gamma \approx 1$ with $c_P \approx c_V$.

Exercise 8.1

Make an estimate of $c_P - c_V$ for metallic copper at ambient temperature.

For copper at 295 K, $\beta = 5 \times 10^{-5}$ K^{-1} and $\kappa = 4.5 \times 10^{-12}$ m^{2}N^{-1} (Pa^{-1}). The molar volume of copper is $v = 7 \times 10^{-6}$ m^3 mol^{-1} ($\rho = 8.92 \times 10^3$ kg m^{-3}) and $c_P = 24.5$ J mol^{-1} K^{-1}. Substitution of values for v, β, and κ into Equation 8.4 gives $c_P - c_V = 1.15$ J mol^{-1} K^{-1}. From the quoted value for c_P, we obtain $\gamma = c_P/c_V = 1.05$.

It is clear that $c_P \approx c_V$ (within 5%) for copper at room temperature. Furthermore, it is seen that $c_P \approx 3R$, as predicted by Dulong and Petit's law.

8.5 APPLICATION OF THE ENTROPY EQUATIONS TO SOLIDS AND LIQUIDS

In Chapter 7, the two $T\,dS$ equations were derived. The first $T\,dS$ equation (Equation 7.66) is given as $TdS = C_V dT + T(\partial P/\partial T)_V\, dV$, whereas the second $T\,dS$ equation (Equation 7.68) is $T\,dS = C_P dT - T(\partial V/\partial T)_P\, dP$. For condensed matter, it is convenient to rewrite these equations in terms of β and κ. From the definitions given in Equations 2.52 and 2.53, we see that Equation 7.66 can be expressed as

$$TdS = C_V\, dT + T\left(\frac{\beta}{\kappa}\right)dV. \tag{8.6}$$

Similarly, Equation 7.68 becomes

$$TdS = C_P dT - TV\beta dP. \tag{8.7}$$

We consider molar quantities of solid or liquid that undergo processes in which the pressure is changed. The molar volume v, the isothermal compressibility κ, and the thermal expansion coefficient β are relatively insensitive to changes in pressure. This permits Equation 8.6 or Equation 8.7 to be integrated, with v, κ, and β taken outside the integral and replaced by average values $\bar{v}$, $\bar{\kappa}$, and $\bar{\beta}$ over the range of integration. Considerable simplification, compared with the case of gases, results from this approximation. Two important cases are considered based on Equation 8.7.

a. *Adiabatic Compression of a Solid.* For a reversible adiabatic compression process, $dQ = T\, dS = 0$, and Equation 8.7 becomes $0 = C_P\, dT - TV\beta dP$. Integration gives $\int_i^f dT/T = \int_i^f (V\beta/C_P)dP$, and hence $\ln\left(T_f/T_i\right) \simeq \left(\bar{V}\bar{\beta}/c_P\right)\left(P_f - P_i\right)$. If it is assumed that $T_f = T_i + \Delta T$, with $\Delta T \ll T_i$, the log function may be expanded to yield for molar quantities of the solid material

$$\frac{\Delta T}{T} \simeq \left(\frac{\bar{v}\bar{\beta}}{c_P}\right)\left(P_f - P_i\right). \tag{8.8}$$

Exercise 8.2

Consider a compression process involving 1 mol of solid copper at 295 K. If the pressure is increased from zero to 10^8 Pa (~10^3 atm), find the rise in temperature of the copper assuming that the metal is thermally isolated from its surroundings.

Substitution of values for $\bar{v}$, $\bar{\beta}$, and c_P from Section 8.4 into Equation 8.8 gives $\Delta T/T_i \approx 10^{-3}$, and we obtain $\Delta T \simeq 0.4\text{K}$. It is seen that, in this case for a metallic system, $\Delta T \ll T_i$.

b. *Isothermal Compression of a Solid.* In the case of a reversible isothermal process, Equation 8.7 becomes $dQ = T\,dS = -TV\beta\,dP$. Integration of this equation gives

$$\Delta Q = -T\bar{V}\bar{\beta}\left(P_f - P_i\right). \tag{8.9}$$

The minus sign shows that heat energy is transferred from the solid to the surroundings.

Exercise 8.3

Find the heat given out when 1 mol of copper is subjected to a pressure increase from zero to 10^8 Pa while kept at a constant temperature of 295 K.

Substitution for $\bar{T}$, $\bar{v}$, and $\bar{\beta}$ in Equation 8.9, with the same values as used above, results in $\Delta Q = -10\,\text{J}$. This relatively small amount of heat energy is transferred to the constant temperature heat bath with which the copper is in contact.

8.6 MAXWELL RELATIONS FOR A MAGNETIC SYSTEM

In Chapter 2, the first law for a magnetic system was written in the two alternative forms $dE^* = dQ + \mu_0 VH\,dM$ (Equation 2.19) and $dE^+ = dQ - \mu_0 VM\,dH$ (Equation 2.21), where H is the applied field strength and M the magnetization. The two forms correspond to different views of what constitutes the system, as explained in Chapter 2. The energy E^* includes the vacuum field energy, whereas E^+ includes the vacuum field energy and the mutual field energy. In the latter case, the work term corresponds to work done on the sample alone. For the present discussion, we shall use the first law in the form of Equation 2.21, dropping the superscript once more for simplicity. The fundamental relation for a magnetic system is therefore

$$T\,dS = dE + \mu_0 VM\,dH. \tag{8.10}$$

In the derivation of the Maxwell relation for fluids in Chapter 7, use was made of the fundamental relation in the form $dE = T\,dS - P\,dV$ (Equation 3.18). Comparison of Equation 8.10 with Equation 3.18 shows a correspondence between the intensive variables $M \leftrightarrow P$ and the extensive variables $V \leftrightarrow \mu_0 VH$. The Maxwell relations are easily written down for the magnetic case simply by the replacement of P with M and V by $\mu_0 VH$ in the set of equations M 1 to M4 given in Equations 7.41, 7.44, 7.47, and 7.50. We obtain the relations:

$$\text{M1,}\left(\frac{\partial T}{\partial H}\right)_S = -\mu_0 V\left(\frac{\partial M}{\partial S}\right)_H, \tag{8.11}$$

$$\text{M2,}\left(\frac{\partial T}{\partial M}\right)_S = \mu_0 V\left(\frac{\partial M}{\partial S}\right)_M, \tag{8.12}$$

$$\text{M3,}\left(\frac{\partial S}{\partial H}\right)_T = \mu_0 V\left(\frac{\partial M}{\partial T}\right)_H, \tag{8.13}$$

$$\text{M4,}\left(\frac{\partial S}{\partial M}\right)_T = -\mu_0 V\left(\frac{\partial H}{\partial T}\right)_H. \tag{8.14}$$

We note that the same set of equations is obtained whether we consider the system to consist of the sample plus mutual field or the sample alone. Either Equation 2.19 or Equation 2.21 may be used.

8.7 APPLICATIONS OF THE MAXWELL RELATIONS TO IDEAL PARAMAGNETIC SYSTEMS

a. *Magnetic Susceptibility and Specific Heat of an Ideal Paramagnet.* For an ideal paramagnetic system, the magnetic susceptibility obeys Curie's law, with the magnetization given by Equation 1.4 $M = \mathcal{M}/V = CH/T$, where C is Curie's constant and V is the volume of the magnetic material. This is the equation of state for this system. The inverse magnetic susceptibility $1/\chi = T/C$ depends linearly on T. For a system of N spins $\frac{1}{2}$, we have from Equation 5.20 $C = N\mu^2\mu_0/Vk_B$, where μ is the magnetic dipole moment associated with each spin. As a check on whether a given material obeys Curie's law or not, it is convenient to plot $1/\chi$ versus T. If magnetic ordering occurs at some temperature, the form $1/\chi = (T - T_C)/C$ on the basis of the Curie–Weiss law (Equation 1.5) is generally found to hold, and this form is depicted in Figure 8.3. For a transition to a ferromagnetic phase, a positive T_C called the Curie temperature is obtained from a plot of this kind, whereas for an antiferromagnet, a negative T_N called the Neél temperature is obtained corresponding to the form $1/\chi = (T+T_N)/C$. Further discussion of nonideal spin systems in which interactions between spins are important is given in Chapter 17. For the present, we confine our discussion to paramagnets well above any transition temperature.

The heat capacity of an ideal paramagnet at the constant applied magnetic field C_H is given by Equation 5.21:

$$C_H = \left(\frac{dQ}{dT}\right)_H = T\left(\frac{\partial S}{\partial T}\right)_H = Nk_B\left(\frac{\mu B}{k_B T}\right)^2 \operatorname{sech}^2\left(\frac{\mu B}{k_B T}\right).$$

The specific heat per mole is obtained immediately for $N = N_A$. In fields that are not too large and at temperatures that are not too low, $\mu B \ll \kappa_B T$ and $\operatorname{sech}^2(\mu B/k_B T) \to 1$, giving

$$c_H \simeq N_A k_B\left(\frac{\mu B}{k_B T}\right)^2. \tag{8.15}$$

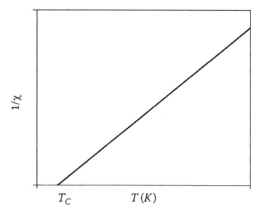

FIGURE 8.3 Illustrative plot of the inverse magnetic susceptibility $1/\chi$ versus T for a material that obeys the Curie–Weiss law. The transition temperature T_C to an ordered phase is estimated by extrapolation from measurements made at high T.

This shows that the specific heat decreases with temperature as $1/T^2$ on the high-temperature side of the Schottky peak in the specific heat.

Earlier in this chapter, an expression for $c_P - c_V$ was obtained in terms of T, β, and κ. Similar procedures may be applied to obtain $c_H - c_M$.

Exercise 8.4

Obtain an expression for $c_H - c_M$ for a paramagnetic system of N spins with magnetic moment μ in a field B. Discuss the form of the heat capacity c_M.

We choose the entropy to be a function of T and H. The total differential of $S = S(T, H)$ is $dS = (\partial S/\partial T)_h dT + (\partial S/\partial H)_t dH$, and this gives $T(\partial S/\partial T)_m = T(\partial S/\partial T)_h + T(\partial S/\partial H)_t(\partial H/\partial T)_M$, from which we obtain $C_H - C_M = -T(\partial S/\partial H)_t (\partial H/\partial T)_M$. With M3 from Equation 8.13 plus application of the partial derivative rules given in Equations 7.36 and 7.37 to $(\partial H/\partial T)_M$, it follows that

$$C_H - C_M = \mu_0 VT\left(\frac{\partial M}{\partial T}\right)_H^2 \left(\frac{\partial H}{\partial M}\right)_T. \tag{8.16}$$

$(\partial M/\partial T)_H$ and $(\partial H/\partial M)_T$ are determined from the equation of state (Equation 1.4), with $B = \mu_0 H$, and substitution in Equation 8.16 together with our expression for the Curie constant gives for 1 mol of the material

$$c_H - c_M = \frac{\mu_0 C_V H^2}{T^2} = N_A k_B \left(\frac{\mu B}{k_B T}\right)^2. \tag{8.17}$$

In Equation 8.17, v is the molar volume of the system. Comparison with Equation 8.15 shows that $C_M = 0$. This result for C_M may appear strange but is a consequence of the special form of the equation of state given in Curie's law, as shown in Equation 1.4. To keep M constant, the H/T ratio must be kept constant. Quasistatic processes with constant H/T are isentropic with no heat energy transfer. These processes are discussed in greater detail below. The entropy is obtained from Equation 5.18, with the use of the relation $E = -MB$. Because M is held constant by keeping H/T constant, the entropy remains fixed and such processes are therefore isentropic.

b. *Adiabatic Demagnetization.* Adiabatic demagnetization of a paramagnetic system has been dealt with previously in Sections 2.7 and 6.6(a). The temperature–entropy equations together with the Maxwell relations provide a concise means for dealing with such processes. Let the entropy S be a function of T and H. The total differential of S is given by $dS = (\partial S/\partial T)_H dT + (\partial S/\partial H)_T dH$. It follows that

$$TdS = C_H\, dT + \mu_0 VT\left(\frac{\partial M}{\partial T}\right)_H dH, \tag{8.18}$$

where the use has been made of M3 given in Equation 8.13.

In an isothermal magnetization process, the heat energy transferred is

$$\Delta Q = \int_i^f T\, dS = -\left(\frac{\mu_0 CV}{T}\right)\int_{H_i}^{H_f} H\, dH = -\frac{1}{2}\left(\frac{\mu_0 CV}{T}\right)\left(H_f^2 - H_i^2\right), \tag{8.19}$$

where C is Curie's constant. The minus sign in Equation 8.19 shows that heat energy is given out to the heat bath during the magnetization process. For an adiabatic demagnetization process from H_1 to H_2, Equation 8.18 gives $0 = C_H \, dT - (\mu_0 CV/T)H \, dH$. Substituting for C_H from Equation 8.15 and for Curie's constant C from Equation 5.20 results in the first-order differential equation, $\left(Nk_B\mu^2\mu_0^2H^2/k_B^2T^2\right)dT = \left(N\mu^2\mu_0^2VH/Vk_BT\right)dH$ or $dT/T = dH/H$. Integration followed by antilogs gives $H_i/T_i = H_f/T_f = \text{constant}$, and hence $T_f = (H_f/T_i)T_i$, in agreement with Equation 2.32. This confirms that H/T remains constant during the adiabatic demagnetization process, and it follows that M remains constant according to Curie's law, as shown in Equation 1.4.

In this chapter, we have considered the properties of condensed matter in the classical limit. Chapter 9 is concerned with the thermodynamic description of phase transitions and critical phenomena again in the classical limit. Many interesting phenomena in condensed matter occur at low temperatures. Examples are superconductivity in metals and superfluidity in the helium liquids. It is necessary to use a quantum mechanical approach in discussing phenomena of this kind. In particular, it is important to introduce quantum statistics for systems of particles that may be either fermions or bosons, and this is done in later chapters of this book.

PROBLEMS CHAPTER 8

8.1   A 200-g cylinder of metallic copper is compressed isothermally and quasistatically at 290 K in a high-pressure cell. Find the change in internal energy of the copper when the pressure is increased from 0 to 12 kbar. How much heat is exchanged with the surrounding fluid? If the process is instead carried out adiabatically, find the temperature increase of the copper. For copper, $c_P = 16$ J(mol K)$^{-1}$, $\beta = 32 \times 10^{-6}$ K^{-1}, $\kappa = 0.73 \times 10^{-6}$ atm^{-1}, and $v = 7\,\text{cm}^3\text{mol}^{-1}$.

8.2   Use the Maxwell relations to obtain an expression for the dependence of the specific heat of a solid c_P on pressure P at a given temperature in terms of the volume expansion coefficient β at constant P.

8.3   Determine the specific heat at constant volume for liquid mercury at 273 K, and at a pressure of 1 atm, from the information provided below. What is the ratio γ of the specific heats? Molar volume = 14.5 cm^3, $\beta = 1.8 \times 10^{-4}$ K^{-1}, $\kappa = 4.0 \times 10^{-11}$ m^{2}N^{-1}, and $c_P = 28.0$ J mol^{-1} K^{-1}.

8.4   The Einstein expression for the specific heat of an insulating solid is $c_V = 3Rx^2[e^x/(e^x - 1)^2]$, where $x = \hbar\omega_E/k_BT = \theta_E/T$ and θ_E is the Einstein temperature. Obtain the low-temperature limit of the Einstein expression and use this form to find the change in entropy of a solid when heated from T_1 to T_2 at a constant volume. Assume that both T_1 and T_2 are much less than θ_E.

8.5   Some solids such as graphite have a layered structure with strong in-plane bonding and much weaker interplanar bonding. The Einstein frequency for atomic in-plane vibrations is therefore much higher than for interplanar vibrations. Give a qualitative prediction of the behavior of the specific heat c_V with temperature for a solid of this type.

8.6   The magnetic susceptibility of a solid of volume V follows the Curie–Weiss law $\chi = C/(T - \theta)$, where C and θ are constants for the material. Obtain an expression for the heat capacity of the solid at a constant applied field. By how much would the temperature of the solid decrease in an adiabatic demagnetization process from H_i to H_f? Sketch the processes on a T–S diagram.

8.7   The work done on a system of electric dipoles in increasing an applied electric field $\boldsymbol{E}_e$ by an infinitesimal amount is given by $dW = V\boldsymbol{P}_e \cdot d\boldsymbol{E}_e$, where $\boldsymbol{P}_e$ is the electric polarization and V is the volume of the material considered. This result is analogous to that for the magnetic case $dW = -V\boldsymbol{M} \cdot d\boldsymbol{H}$ given in Chapter 2, and similar considerations apply as to what constitutes the system in the electric polarization case as discussed in detail for the magnetic case. (Ignore dipole moments induced by the applied electric field.) Obtain the form of the fundamental relation for a system of electric dipoles, and use this expression to write down the Maxwell relations for this case by analogy with the magnetic case. Find the difference in the specific heats $c_E - c_{P_e}$.

9 Phase Transitions and Critical Phenomena

9.1 INTRODUCTION

Phase transitions occur widely in nature and are part of everyday experience. For example, water freezes into ice at 0°C, or vaporizes into steam at 100°C at sea level. Over the past several decades, considerable insight into the nature of phase transitions has been gained using the methods of thermal and statistical physics. As discussed in Chapter 6, the third law states that the entropy S tends to zero as the temperature tends to zero. As the temperature is lowered, interactions between particles become increasingly important and may lead to the onset of some type of long-range order accompanied by symmetry breaking in the system. For example, a magnetic material, such as iron or nickel, undergoes a change at its Curie temperature T_C from disorder and high symmetry in the paramagnetic phase to order and lower symmetry in the ferromagnetic phase. The symmetry change is associated with the orientation of the magnetic dipoles associated with the ions in the material; in zero magnetic field, there is no preferred direction for the dipole moments in the paramagnetic phase, while, within magnetic domains, the dipoles are aligned with each other in the ferromagnetic phase. We shall see that it is convenient to introduce an order parameter η to describe order–disorder transitions. As an example, for a ferromagnet, the order parameter is defined as $\eta = M/M_0$, with M as the bulk magnetization at the temperature of interest and M_0 the low-temperature saturated magnetization.

Phase transitions may be classified into several different types on the basis of the behavior with temperature of the thermodynamic potentials and the order parameter. The order parameter may change discontinuously or continuously at the transition point. In the thermodynamic description of phase transitions, the Helmholtz potential F (constant V) and the Gibbs potential G (constant P) play important roles. A scheme on the basis of the behavior of F or G in the vicinity of the transition point has been developed for the classification of phase transitions as continuous or discontinuous and is discussed in Section 9.3. Figure 9.1 shows a representative phase diagram in the P–T plane for a single-component system.

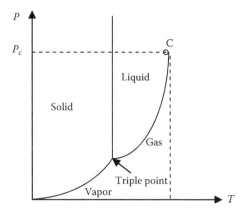

FIGURE 9.1 Phase diagram for a single-component system in the P–T plane with phase boundaries shown between gas, liquid, and solid phases. In general, two phases coexist along the phase boundaries. At the triple point, the three phases coexist. The liquid–gas coexistence curve ends at the critical point C.

The curves in Figure 9.1 show the boundaries between solid, liquid, and vapor phases. At the triple point, which corresponds to particular values of P and T, all three phases coexist. Note that the gas–liquid coexistence curve terminates at the critical point given by P_C and T_C. For $T > T_C$, the system does not liquefy at any pressure. For multicomponent systems, phase diagrams can become much more complicated than for single-component systems. This chapter will focus on the single-component systems.

9.2 NONIDEAL SYSTEMS

a. *van der Waals' Fluid.* The empirical van der Waals equation of state for nonideal gases involves attractive interactions between particles that have a repulsive hard core. The form of the intermolecular potential is shown in Figure 7.5. If values for the van der Waals constants a and b are chosen, it is possible to generate P–V isotherms for various temperatures, as shown in Figure 9.2.

Examination of Figure 9.2 shows that for T lower than a critical temperature T_C, the isotherms exhibit a sigmoid shape over a region where the volume changes significantly, whereas the pressure remains fairly constant. For $T < T_C$, three different values of the volume correspond to the same pressure for each isotherm. For the stability of a fluid system, we expect the pressure to decrease with an increase in volume or $(\partial P/\partial V)_T < 0$, which corresponds to a negative slope of an isotherm shown in a P–V diagram. Those portions of each isotherm that, for $T < T_C$, have positive slopes are unphysical, which shows that the van der Waals equation of state has limitations at high densities. Nevertheless, the isotherms in Figure 8.2 indicate that the van der Waals fluid undergoes a transition from the gaseous to the liquid state, provided that $T < T_C$ and the pressure is increased sufficiently. For $T > T_C$, the van der Waals system remains a gas for all pressures. Point C, at pressure P_C and volume V_C, is the critical point where the fluid behaves in a special way, as illustrated in Figure 9.3 plot of the isobaric density of the fluid ρ as a function of temperature T.

Figure 9.3 shows that, for $P < P_C$, a discontinuity in density occurs in a transition from a gaseous to a liquid phase. For $P = P_C$, however, the transition from high to low density occurs *continuously* as the temperature is increased through T_C. Finally, for $P > P_C$, the density varies smoothly, with no major change in value, as a function of temperature. The critical density P_C is the density of the fluid at the critical point (P_C, T_C). A phenomenon,

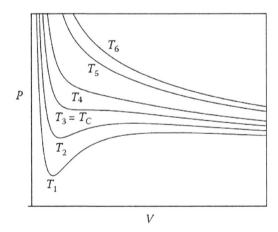

FIGURE 9.2 P–V isotherms for a van der Waals gas as a function of increasing temperature T. The critical isotherm, which shows a point of inflection, corresponds to T_C.

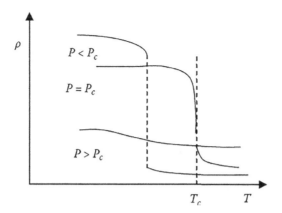

FIGURE 9.3 Schematic plot of the isobaric variation of the density of a van der Waals fluid with temperature at and near the critical point.

called *critical opalescence*, occurs in fluids under conditions close to the critical point. It is found that light shone through the fluid is strongly scattered because of critical fluctuations in the local order parameter that occur in this region. We discuss this phenomenon in slightly greater detail later in this chapter.

b. *Nonideal Spin System.* The Curie–Weiss equation of state, which applies in the paramagnetic phase for a magnetic system, is given by $M = CVH/(T - T_C)$ (Equation 1.5), with T_C as the Curie–Weiss empirical fit temperature, which is generally found to be somewhat higher than the actual transition temperature. Equation 1.5 predicts that the magnetization M and the susceptibility $X = (\partial M/\partial H)_T$ will tend to diverge as $T \to T_C$. As shown in Figure 8.3, a plot of $1/\chi$ versus T for Curie–Weiss systems gives a straight-line form, from which T_C can be estimated. Experiment shows that below the critical temperature T_C, in a low applied magnetic field, the magnetization increases steadily, reaching a limiting value M_0 for $T \to 0$ K. This behavior is shown in Figure 9.4 for fields $H = 0$ and $H > 0$.

For $H = 0$, the magnetization is zero for $T > T_C$ but increases rapidly for $T < T_C$ because of spontaneous ordering and the onset of ferromagnetism. For the present discussion, we ignore the formation of magnetic domains separated by domain walls that occur in ferromagnetic crystals. In effect, we are focusing on the magnetization in a single domain.

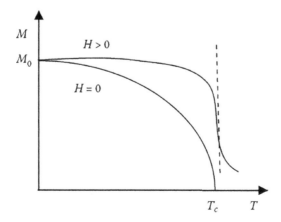

FIGURE 9.4 The magnetization M versus T behavior for a nonideal magnetic system in zero and nonzero applied magnetic fields illustrating changes that occur particularly near T_C.

9.3 CLASSIFICATION OF PHASE TRANSITIONS

In Section 7.4(b), it is shown that, for a system in which two phases coexist in equilibrium at constant pressure, the molar Gibbs potentials are equal for the two phases. Away from coexistence, the system is found in the phase that has the lower Gibbs potential under the given conditions. This topic is discussed in detail in Section 9.4. Historically, Ehrenfest introduced a classification system for phase transitions on the basis of the behavior of the Gibbs potential, and its derivatives, in the vicinity of the transition under constant pressure conditions. A similar classification can be given in terms of the Helmholtz potential when the volume is kept fixed. From Equation 7.16, we have $dG = -S\,dT + V\,dP$. The entropy and the volume are given by the first-order derivatives, $S = -(\partial G/\partial T)_P$ (Equation 7.17) and $V = (\partial G/\partial P)_T$ (Equation 7.18), whereas the second-order derivatives give the heat capacity, $C_P/T = (\partial S/\partial T)_P = -(\partial^2 G/\partial T^2)_P$, and compressibility, $\kappa = -1/V(\partial V/\partial P)_T = -1/V(\partial^2 G/\partial P^2)_T$. Higher-order derivatives may be formed but the above are sufficient for our purposes.

In the Ehrenfest scheme, a transition is classified as the first order if there are finite discontinuities in the first derivatives of G, that is, in the entropy S and in the volume V. A transition is classified as the second order if the first derivatives of G are continuous but there is a finite discontinuity in the second derivatives, C_P and κ. Transitions of still higher order are classified according to which order derivative first shows a discontinuity. Figure 9.5 graphically illustrates the Ehrenfest classification scheme for the first- and second-order transitions.

Many phase transitions are found to be of first order, with a finite change in molar volume and with a latent heat of transition. A well-known example of a first-order transition is the melting of ice. For second- and higher-order transitions, the entropy and the volume are continuous through the transition with no latent heat. These phase changes are therefore referred to as *continuous transitions*, and this terminology is nowadays preferred to that used in the Ehrenfest scheme. Although the Ehrenfest classification does provide useful guidance in considering phase transitions, the subtleness of the changes that accompany continuous transitions makes it increasingly difficult to

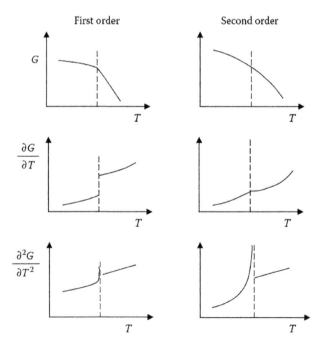

FIGURE 9.5 The behavior of the Gibbs potential G and its derivatives with respect to temperature for first- and second-order phase transitions in the Ehrenfest scheme. Note that G changes slope at T_C for a first-order transition but does not change slope at T_C for a second-order continuous transition.

distinguish between transitions of order higher than two. Systems in which continuous phase transitions are observed include the following:

Metal-superconductor ($H = 0$)
Paramagnet–ferromagnet (or antiferromagnet)
Liquid ^{4}He normal–super fluid transition
α to β brass.

Careful measurements of the specific heat have been carried out for these systems and reveal discontinuities of the type shown in Figure 9.5 for Ehrenfest second-order transitions. Continuous phase transitions are considered in some detail later in this chapter.

9.4 THE CLAUSIUS–CLAPEYRON AND THE EHRENFEST EQUATIONS

a. *First-Order Transitions: The Clausius–Clapeyron Equation.* Consider a system that undergoes a discontinuous or first-order transition from phase 1 to phase 2. Figure 9.6 shows the equilibrium curve that separates the two phases on a P–T diagram.

If the equilibrium curve is crossed at any point (P, T) from phase 2 to phase 1, there is a molar volume change Δv and a molar latent heat l is involved in the transition. The associated molar entropy change is Δs. (Lowercase symbols denote the molar quantities.) The Clausius–Clapeyron equation relates the slope of the equilibrium curve at the point at which it is traversed to the ratio of the entropy change and the volume change that accompanies the first-order transition. The derivation is straightforward and is presented below.

If Equation 7.21 is applied to points a and b in Figure 9.6, $g_1(T, P) = g_2(T, P)$ and $g_1(T+dT, P+dP) = g_2(T+dT, P+dP)$. If we expand both $g_1(T+dT, P+dP)$ and $g_2(T+dT, P+dP)$ in Taylor series in two variables about the points T and P and retain only the first-order terms as a sufficiently good approximation, we obtain

$$g_1(T,P) + \left(\frac{\partial g_1}{\partial T}\right)_P dT + \left(\frac{\partial g_1}{\partial P}\right)_T dP = g_2(T,P) + \left(\frac{\partial g_2}{\partial T}\right)_P dT + \left(\frac{\partial g_2}{\partial P}\right)_T dP.$$

With the use of Equation 7.21 followed by rearrangement, the above expression leads to

$$\left[\left(\frac{\partial g_1}{\partial T}\right)_P - \left(\frac{\partial g_2}{\partial T}\right)_P\right] dT = -\left[\left(\frac{\partial g_1}{\partial P}\right)_T - \left(\frac{\partial g_2}{\partial P}\right)_T\right] dP. \tag{9.1}$$

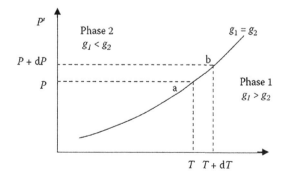

FIGURE 9.6 P–T phase diagram for a system that undergoes a phase transition. The diagram shows the equilibrium curve along which two phases 1 and 2 coexist. Points a and b lie close to each other on the coexistence curve.

Equation 7.16 may be written for molar quantities as $dg = -s\,dT + v\,dP$. It follows that the molar entropy is given by

$$s = -\left(\frac{\partial g}{\partial T}\right)_P,\qquad(9.2)$$

whereas the molar volume is

$$v = \left(\frac{\partial g}{\partial P}\right)_T.\qquad(9.3)$$

Substituting Equations 9.2 and 9.3 into Equation 9.1 gives $dP/dT = (s_1 - s_2)/(v_1 - v_2)$ or

$$\frac{dP}{dT} = \frac{\Delta s}{\Delta v},\qquad(9.4)$$

which is the Clausius–Clapeyron equation. An alternative form is

$$\frac{dP}{dT} = \frac{l}{T\Delta v},\qquad(9.5)$$

where the molar entropy change is

$$\Delta s = \frac{l}{T}.\qquad(9.6)$$

Exercise 9.1

Use the Clausius–Clapeyron equation to obtain an expression for the vapor pressure above a liquid in a sealed container as a function of temperature. The situation is illustrated in Figure 9.7.

Well away from the critical point, the vapor pressure will, to a good approximation, obey the ideal gas equation of state, which, in terms of the molar volume, may be expressed as $Pv = RT$. Furthermore, $\Delta v = v_1, -v_2 \simeq v_1$ because the molar volume of the vapor is very much larger than that of the liquid. Substitution into Equation 9.5 gives $dP/dT \simeq l/Tv_1 = lP/RT^2$. Integration of this equation leads to $\int dP/P = l/R \int dT/T^2$, and hence

$$P = P_0 e^{(-l/RT)}.\qquad(9.7)$$

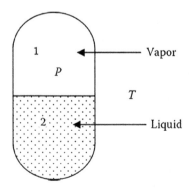

FIGURE 9.7   Liquid and vapor in equilibrium at temperature T in a sealed container. The vapor pressure at temperature T is P.

The constant P_0 is a constant of integration and gives the vapor pressure in the high T limit. Equation 9.7 shows that the vapor pressure depends exponentially on $1/T$ and the slope of a plot of $\ln P$ versus $1/T$ is $-l/R$. Experimental results for the vapor pressures of many liquids, measured as a function of T, are described very well by Equation 9.7.

b. *Higher-Order Transitions: The Ehrenfest Equations.* The Clausius–Clapeyron equation cannot be applied to second- or higher-order transitions because Δs and Δv are both zero in these cases. For continuous (second-order) transitions, the entropy does not change through the transition. This implies that $s_1(T, P) = s_2(T, P)$ and $s_1(T+dT, P+dP) = s_2(T+dT, P+dP)$. Expansion of s_1 and s_2 about T and P in a Taylor series in two variables up to the first order leads to

$$\frac{dP}{dT} = \frac{(\partial s_1/\partial T)_P - (\partial s_2/\partial T)_P}{(\partial s_2/\partial P)_T - (\partial s_1/\partial P)_T} \tag{9.8}$$

From the Maxwell relation M4 given in Equation 7.50 and the definitions of the specific heat c_P and the thermal expansion coefficient β, we obtain

$$\frac{dP}{dT} = \frac{1}{vT}\left(\frac{c_{P_1} - c_{P_2}}{\beta_1 - \beta_2}\right). \tag{9.9}$$

Equation 9.9 relates the slope at a chosen point on the P–T equilibrium curve for the two phases to the specific heat difference divided by the expansion coefficient difference between the phases at the chosen point.

A similar procedure in terms of volume continuity rather than entropy continuity gives

$$\frac{dP}{dT} = \left(\frac{\beta_1 - \beta_2}{\kappa_1 - \kappa_2}\right), \tag{9.10}$$

where κ_1 and κ_2 are the isothermal compressibilities of the two phases. Equations 9.9 and 9.10 are the Ehrenfest equations for a second-order transition. In the case of a third-order transition, where C_p, β, and κ are continuous, three Ehrenfest equations apply.

9.5 CRITICAL EXPONENTS FOR CONTINUOUS PHASE TRANSITIONS

The modern discussion and development of the subject of critical phenomena, which is concerned with the behavior of physical properties of systems that undergo continuous phase transitions, has, to a large extent, involved critical exponents. Before critical exponents are introduced, we consider the experimental results for fluids and magnets. Figure 9.8 shows a schematic plot of the liquid–vapor coexistence curve for a number of pure fluid substances plotted in reduced temperature and density coordinates.

Figure 9.8 illustrates that the experimental data (not shown but which cluster along the curve) for eight substances, when plotted in reduced coordinates, can be well fit by a single curve. T_C is the critical temperature and ρ_c the critical density. The use of reduced coordinates is based on what is called the *law of corresponding states*, which implies that in reduced coordinates, all fluids of this kind have the same equation of state. As an example, the van der Waals equation of state may be rewritten in terms of reduced coordinates.

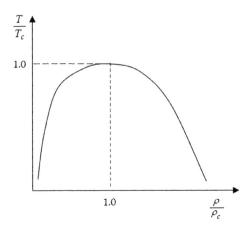

FIGURE 9.8 Representation of the liquid–vapor coexistence curve for eight substances (Ne, Ar, Kr, Xe, N_2, O_2, CO, and CH_4) plotted in reduced temperature and density coordinates. Experimental results which are not shown lie very close to the curve.

Exercise 9.2

Express the van der Waals equation of state in terms of reduced coordinates T/T_C, P/P_C, and v/v_C instead of the van der Waals constants.

The critical point is characterized as a point of inflection on the critical isotherm plotted on a PV diagram, that is, $(\partial P/\partial V)_{T_C} = 0$ and $(\partial^2 P/\partial V^2)_{T_C} = 0$. For molar quantities, the van der Waals equation is $(P+a/v^2)(v-b) = RT$. We obtain the first- and second-order partial derivatives of P with respect to v at $T = T_C$ and place the results equal to zero to get

$$\left(\frac{\partial P}{\partial v}\right)_{T_C} = -\frac{RT}{\left(v_C - b\right)^2} + \frac{2a}{v_C^3} = 0 \tag{9.11}$$

and

$$\left(\frac{\partial^2 P}{\partial v^2}\right)_{T_C} = \frac{2RT}{\left(v_C - b\right)^3} - \frac{6a}{v_C^4} = 0. \tag{9.12}$$

If we then equate the right-hand part of Equations 9.11 and 9.12 and solve for the critical volume v_C, we find

$$v_C = 3b. \tag{9.13}$$

It follows that

$$T_C = \frac{8a}{27bR}, \tag{9.14}$$

with

$$P_C = \frac{a}{27b^2}. \tag{9.15}$$

Introduction of reduced coordinates v/v_C, T/T_C, and P/P_C gives

$$\left[\left(\frac{P}{P_C}\right) + 3\left(\frac{v_C}{v}\right)^2\right]\left[3\left(\frac{v}{v_C}\right) - 1\right] = 8\left(\frac{T}{T_C}\right). \tag{9.16}$$

The constants a and b no longer appear. In terms of the reduced pressure, temperature, and volume, Equation 9.16 leads to a law of corresponding states for all substances that obey the van der Waals equation of state. The law of corresponding states permits the collapse of a family of P–V curves into a single curve by scaling P and V with the values of P_C and V_C, respectively.

We turn now to critical exponents. Near the maximum in Figure 9.8, the universal curve has the form

$$\frac{|\rho_l - \rho_g|}{\rho_c} \propto \left[\frac{|T - T_C|}{T_C}\right]^{1/3} \tag{9.17}$$

At T_C, the liquid density ρ_l and the gas phase density ρ_g are equal. For magnetic systems, the magnetization in zero applied field, as depicted in Figure 9.4, can be fit close to T_C with an expression of the form

$$\frac{M}{M_0} \propto \left[\frac{|T - T_C|}{T_C}\right]^{1/3} \tag{9.18}$$

The identical exponents in Equations 9.17 and 9.18, which apply to quite different systems, suggest the importance of underlying universal features in critical phenomena.

It is convenient to define the order parameter η for each system and to introduce the reduced temperature $\varepsilon = (T - T_C)/T_C$. For a magnetic system, we take $\eta = M/M_0$, whereas for a fluid $\eta = (\rho_l - \rho_g)/\rho_c$. For $T \leq T_C$, Equations 9.17 and 9.18 may immediately be rewritten as

$$\eta \sim \varepsilon^\beta \tag{9.19}$$

Equation 9.19 applies to continuous phase transitions in general, with β as the order parameter critical exponent. Order parameters for certain low-temperature continuous phase transitions, such as the superconducting and superfluid transitions, are defined with the use of the quantum mechanical wave functions for those systems. A number of other critical exponents may be defined for fluids and magnetic systems. These include the heat capacity exponent α, the isothermal compressibility/susceptibility exponent γ, and the critical isotherm ($T = T_C$) exponent δ. Considerable effort has been devoted to the measurement and theoretical prediction of the critical exponents for a variety of systems that undergo continuous phase transitions. Using scaling ideas, it is possible to show that, for a given system, the determination of two of the exponents fixes the values of all the other exponents.

Exercise 9.3

Obtain the order parameter critical exponent β for a system obeying the van der Waals equation of state.

In terms of reduced variables $\tilde{p} = (P - P_C)/P_C$, $\tilde{v} = (V - V_C)/V_C$, and $\varepsilon = (T - T_C)/T_C$, Equation 9.16 becomes

$$\left(\frac{(1 + \tilde{p}) + 3}{(1 + \tilde{v})^2}\right)\left[3(1 + \tilde{v}) - 1\right] = 8(1 + \varepsilon).$$

This gives

$$2\tilde{p}\left[1 + \frac{7}{2}\tilde{v} + 4^2\tilde{v} + \frac{3}{2}\tilde{v}^3 + 8\varepsilon\left(1 + 2\tilde{v} + \tilde{v}^2\right)\right].$$

With the use of the binomial theorem and retention of leading terms as an approximation, we obtain

$$\tilde{p} = -\frac{3}{2}\tilde{v}^3 - 6 \in \tilde{v} + 4\varepsilon + \cdots. \tag{9.20}$$

Consider the $\tilde{p} - \tilde{v}$ isotherms for ε at or close to the critical point, as shown in Figure 9.9.

For $\varepsilon \leq 0$, points a and b in Figure 9.9 represent the limits of coexistence of liquid and vapor. The pressure at points a and b is the same, and in terms of liquid and gas volumes at the points a and b, Equation 9.20 becomes

$$\tilde{p}_a = \frac{3}{2}\tilde{v}_l^{\,3} + 6\varepsilon\tilde{v}_l + 4\varepsilon + \cdots, \tag{9.21}$$

$$\tilde{p}_b = -\frac{3}{2}\tilde{v}_g^{\,3} - 6\varepsilon\tilde{v}_g + 4\varepsilon. \tag{9.22}$$

The negative sign for $|\tilde{v}_l|$, which takes into account our choice of origin, has been used in Equation 9.21. Subtraction of Equation 9.22 from Equation 9.21 gives

$$0 = \left(\tilde{v}_g^{\,3} + \tilde{v}_e^{\,3}\right) + 4\varepsilon\left(\tilde{v}_g + \tilde{v}_l\right). \tag{9.23}$$

For $\varepsilon \to 0$, $|\tilde{v}_g| \to |\tilde{v}_l|$. As an approximation in this limit, $|\tilde{v}_g| = |\tilde{v}_l| = 2|\varepsilon|^{1/2}$ or

$$\frac{\left(v_g - v_l\right)}{v_c} = \tilde{v}_g - \tilde{v}_l - 4|\varepsilon|^{1/2}. \tag{9.24}$$

With

$$\tilde{v}_g - \tilde{v}_l = \frac{v_g - v_l}{v_c} = \frac{\rho_g^{-1} - \rho_l^{-1}}{\rho_c^{-1}} \approx \frac{\rho_l - \rho_g}{\rho_C}$$

to a good approximation, because we need only allow for small differences in ρ_g and ρ_l from ρ_c, we obtain from Equation 9.24 the van der Waals critical exponent value $\beta = \frac{1}{2}$. This contrasts with the experimental value $\beta = \frac{1}{3}$ obtained for pure fluids, as given in Equation 9.17. In view of the simple empirical form chosen for the van der Waals equation of state, it is not a surprise that the predicted order parameter exponent disagrees with the measured value. More elaborate theoretical approaches are required to predict the values of the critical exponents.

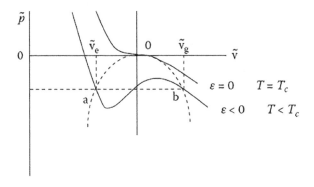

FIGURE 9.9 $\tilde{p} - \tilde{v}$ isotherms for a van der Waals gas. $\tilde{p} = (P - P_C)/P_C$ and $\tilde{v} = (V - V_C)/V_C$ are the reduced pressure and reduced volume, respectively. The isotherm with $\varepsilon = 0$ has a point of inflection at 0 as shown.

9.6 LANDAU THEORY OF CONTINUOUS TRANSITIONS

In the mid-twentieth century, Ginzburg and Landau developed a phenomenological approach that provides considerable insight into continuous phase transitions. The theory involves the expansion of a thermodynamic potential in a power series in the order parameter η. Consider a fluid at constant volume, or a magnetic system in zero field ($H = 0$). In equilibrium, the Helmholtz potential F is a minimum. As independent variables, we choose the absolute temperature T and the order parameter η. Assume that $F(T, \eta)$ may be expanded as follows:

$$F(T,\eta) = F_0(T) + + L_2(T)\eta^2 + L_4(T)\eta^4 + \cdots. \tag{9.25}$$

We treat the order parameter as a vector quantity, with $\eta^2 = \boldsymbol{\eta}...\boldsymbol{\eta}$. For symmetry reasons, only even powers of η, which are scalars and do not depend on the sign of η, are retained in the expansion. The coefficients $L_2(T)$ and $L_4(T)$ are chosen so that $\eta = 0$ for $T > T_C$, whereas $\eta > 0$ for $T < T_C$. $F(T, \eta)$ is required to vary smoothly and continuously through the transition, following the Ehrenfest description. The coefficients $L_2(T)$ and $L_4(T)$ should go to zero at $T = T_C$. Landau suggested the following form for $L_2(T)$:

$$L_2(T) = \alpha(T - T_C) + \cdots. \tag{9.26}$$

This is of the form of an expansion about T_C, with the leading term chosen to be zero in order that $L_2(T) \to 0$ as $T \to T_C$. In the case of $L_4(T)$, it is sufficient to require that $L_4(T) > 0$ in order that the Helmholtz potential increases for large values of η to ensure the stability of the system. Although the choice of $L_2(T)$ in Equation 9.26 appears somewhat arbitrary, and it is furthermore not clear that the expansion of F in terms of η can be justified mathematically, the approach has the virtue of simplicity.

As shown in Chapter 7, the Helmholtz potential is a minimum in equilibrium, and the equilibrium value of η at a given temperature is obtained by differentiation of F with respect to η,

$$\left(\frac{\partial F}{\partial \eta}\right)_T = 0 = 2L_2(T)\eta + 4L_4(T)\eta^3 + \cdots. \tag{9.27}$$

It is obvious that $\eta = 0$ is always a solution. Finite η solutions exist and are given by

$$\eta = \pm\sqrt{-\frac{L_2}{2L_4}}. \tag{9.28}$$

In order for the finite solutions to be real, we require $L_2 < 0$, and with Equation 9.26, this implies $T < T_C$. In this case, we obtain

$$\eta = \left(\frac{\alpha T_C}{2L_4}\right)^{1/2}|\varepsilon|^{1/2}. \tag{9.29}$$

Comparison with Equation 9.19 shows that the Landau theory predicts $\beta = \dfrac{1}{2}$ for the order parameter critical exponent. This agrees with the van der Waals equation of state prediction but is not in agreement with the experimentally determined value $\beta = \dfrac{1}{3}$ for fluids and magnets. It is instructive to plot curves of $F(T)$ versus η as shown in Figure 9.10. Note the form of the minimum in the Helmholtz potential as T passes through T_C.

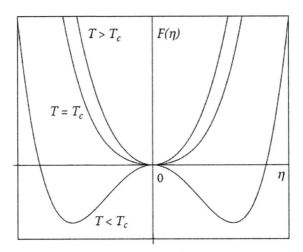

FIGURE 9.10 Landau theory plot of the Helmholtz potential versus the order parameter η for various temperatures in the vicinity of the critical temperature T_C.

For $T < T_C$, nonzero equivalent solutions $\pm\eta$ are found, which correspond to the two minima in the symmetric Helmholtz potential. For $T \to T_C$, the two minima move toward $\eta = 0$ and, at $T = T_C$, a single broad minimum emerges. Finally, for $T > T_C$, a parabolic-shaped minimum centered at $\eta = 0$ is obtained. Note that for $T \simeq T_C$, the broad shallow minimum allows large fluctuations in η with little change in F. These critical fluctuations turn out to be important in more advanced treatments of critical phenomena. Critical opalescence, which is mentioned in Section 9.2, is linked to the large fluctuations in η. For temperatures in the vicinity of the critical point, dynamical clusters or *islands* of order appear and disappear, as a function of time. The large-scale fluctuations in density give rise to enhanced scattering of light.

Exercise 9.4

Use the Landau theory to obtain expressions for the heat capacity C of a magnet or fluid system for T just above and just below the critical temperature T_C. These heat capacity values give the magnitude of the discontinuity in C at the critical point. Assume that $L_4(T)$ varies slowly with temperature.

From Equation 7.12, $S = -(\partial F/\partial T)_V$, and the heat capacity in terms of the Helmholtz potential is $C = T(\partial S/\partial T)_V = T(\partial^2 F/\partial T^2)_V$. For $T \le T_C$, with $\varepsilon = (T - T_C)/T_C$, we get from Equations 9.25 and 9.29

$$C_- = -T\left(\frac{\partial^2 F_0}{\partial T^2} - \frac{\partial}{\partial T}\frac{\alpha^2(T-T_C)}{L_4} + \cdots\right) \approx T_C\left(\frac{\alpha^2}{L_4} - \frac{\partial^2 F_0}{\partial T^2}\right),$$

and for $T \ge T_C$ $C_+ \approx -T_C\,\partial^2 F_0/\partial T^2$. The change in the heat capacity across the transition is

$$\Delta C = C_- C_+ \approx \left(\frac{\alpha^2}{L_4}\right)T_C. \tag{9.30}$$

The expected form of the heat capacity dependence on temperature from the Landau theory is given in Figure 9.11. The discontinuity in C at T_C is based on Equation 9.30. A more complete analysis is required to predict the shapes of the curves on either side of T_C. The form of the curves is determined by the temperature dependence of the Helmholtz potential as shown in the analysis given above.

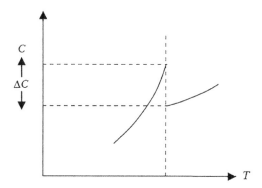

FIGURE 9.11 Heat capacity as a function of temperature for a continuous phase transition as given by the Landau theory. The predicted shape is similar to that observed in systems that exhibit continuous transitions.

For ferromagnetic materials in the vicinity of the Curie temperature, the experiment gives heat capacity curves with shapes similar to those shown in Figure 9.11. Over the past several decades, techniques that are applied to various model systems have been developed to describe continuous phase transitions and to predict critical exponent values. Wilson pioneered the use of what is called the renormalization group method, and the modern theory of continuous phase transitions is often referred to as Ginzburg–Landau–Wilson theory. An introduction to some of the ideas and the models used in the development of the theory is given in Chapter 17.

This chapter has shown how the thermodynamic approach provides a means for classifying phase transitions on the basis of the behavior of the Helmholtz potential in the vicinity of the critical point. For the second-order transitions, phenomenological approaches based first on the van der Waals equation of state for the vapor–liquid transition and second, and more generally, on the Landau expansion of the Helmholtz potential in terms of the order parameter η permit estimates of the order parameter critical exponent β to be made. The value obtained for β is somewhat higher than the values obtained experimentally for a variety of systems.

The following chapters are concerned with the microscopic statistical approach to thermal physics. Several chapters are devoted to quantum statistics and the properties of systems of noninteracting fermions and bosons. Further discussion of phase transitions, specifically in magnetic materials, is given in Chapter 17, which deals with nonideal systems.

PROBLEMS CHAPTER 9

9.1 Obtain an expression for the entropy of a vapor just above the boiling point in terms of the molar entropy of the condensed state, the latent heat of vaporization, and the temperature T. Equate your expression to the Sackur–Tetrode expression for the entropy of a monatomic ideal gas obtained in Chapter 5. With the use of the ideal gas equation of state, find the entropy of the condensed (liquid) state in terms of measured quantities, including the latent heat of vaporization and the pressure of the vapor at its normal boiling point.

9.2 Apply the result of Question 9.1 to find the entropy of liquid helium at its boiling point of 4.2 K. The latent heat of vaporization of helium is $21 \times 10^3 \, \mathrm{J \, kg^{-1}}$.

9.3 Obtain an expression for the entropy of an Einstein solid in the low-temperature limit with the use of results given in Question 8.4. If the solid is in a sealed container in equilibrium with its vapor, find the vapor pressure at temperature T. Assume that the vapor may be treated as an ideal gas, and use the Sackur–Tetrode equation for the entropy of this phase.

9.4 For a single-component system, the solid, liquid, and vapor phases coexist at the triple point as shown in Figure 9.1. The latent heat of sublimation is equal to the sum of the latent heats of melting and vaporization. Use this condition to relate the slopes of the sublimation curve to the slopes of the vaporization and melting curves in the vicinity of the triple point.

9.5 A particular solid undergoes a first-order phase transition from crystallographic phase 1 to another phase 2 at a temperature T_0 under a constant pressure of one atmosphere. By how much would the transition temperature change if the pressure were increased by an amount ΔP? The enthalpy change at the transition is H, and the densities of the solid in the two phases are ρ_1 and ρ_2, respectively.

9.6 At a pressure of 32 atm, the liquid and solid phases of helium-4 can coexist down to absolute zero temperature. Find the slope of the coexistence curve as the temperature tends to 0 K.

9.7 Derive the second Ehrenfest equation (as given in Equation 9.10) which applies to continuous second-order phase transitions in which the volume remains constant through the transition.

9.8 The Dieterici equation of state for a nonideal gas has the form $P = (RT/V-b)\exp(-a/RTV)$. Obtain expressions for the critical temperature, volume, and pressure in terms of a and b. Write the equation of state in terms of reduced dimensionless variables P/P_C, V/V_C, and T/T_C. Determine constants a and b for nitrogen gas for which T_C is 77 K and P_C is 35 atm.

9.9 Use the van der Waals equation in terms of reduced variables $p_r = P/P_C$, $v_r = V/V_C$, and $t_r = T/T_C$ to determine how the slope of the critical isotherm changes in the vicinity of the critical point. Gaseous carbon dioxide has a critical temperature of 304 K and a critical pressure of 74 atm. If it is assumed that the van der Waals equation applies to this gas, determine the values for the van der Waals constants a and b and obtain the critical volume. Compare b with the molar volume of the gas.

9.10 According to the Ehrenfest classification scheme, phase transitions of order higher than two may exist. Describe distinguishing features of a third-order phase transition in the Ehrenfest scheme. Give sketch graphs of the derivatives of the Gibbs potential for this type of transition.

9.11 A magnetic system is situated in a small applied magnetic field H at temperature T. Make use of the Landau theory for continuous phase transitions to predict the behavior of the magnetic susceptibility as $T \rightarrow T_C$. Assume the Gibbs potential has the form $G(H, T) = G_0(H, T) + \alpha_2(T)M^2 + \alpha_4(T)M^4$, with $G_0(T) = F(M, T) - H \cdots M$.

Part II

Quantum Statistical Physics and
Thermal Physics Applications

Section IIA

The Canonical and Grand Canonical
Ensembles and Distributions

10 Ensembles and the Canonical Distribution

10.1 INTRODUCTION

In Chapter 5, the microscopic interpretation of entropy is presented in terms of the logarithm of the number of accessible microstates Ω (E, V, N) for a system such as a gas with the fixed total energy E, volume V, and particle number N. The high point of this discussion is the relationship between the entropy S and Ω given in Equation 5.12, $S = k_B \ln \Omega$ (E, V, N). This relationship shows that the entropy of a system can be determined by counting the number of accessible microstates. As pointed out in Chapter 5, Equation 5.12 is an important bridge that connects the microscopic and macroscopic descriptions of systems with large numbers of particles. Once an expression for the entropy has been obtained, various other thermodynamic quantities may be derived from it. Examples of what is termed the microcanonical ensemble description are given in Chapter 5 for the ideal gas and ideal spin systems.

For systems not in equilibrium, the approach to equilibrium, which is characterized by an increase in entropy, may be understood in terms of the fundamental postulate given in Section 5.1. In equilibrium, all accessible microstates occur with the equal probability. This implies that the probability of finding the system in a particular microstate $\mid r$ is $p_r = 1/\Omega(E, V, N)$. A system not in equilibrium at some instant will not have equal probabilities for all the *accessible* microstates. The system will then evolve with time until the probabilities are equal. These considerations do not, however, provide any information on how long it will take to reach an equilibrium.

Because Ω (E, V, N) is a restricted sum over states, its evaluation may, in some cases, present difficulties. This is seen, for example, in the discussion of the entropy of an ideal gas in Chapter 5, where a number of approximations were made. In this chapter, a more general approach is developed in terms of the partition function Z. This quantity involves an unrestricted sum over the states of the system, which, in most cases, simplifies the summation procedure. For systems with high densities of states, the unrestricted sum may be expressed as an integral over the range of energy values involved, such as 0 to ∞ for a gas. The partition function is extremely useful in the determination of mean values for state variables. In this chapter, we shall focus mainly on systems of localized and, in principle, distinguishable particles such as spins on fixed lattice sites in solids. The evaluation of the partition function for ideal systems of this kind is straightforward as we shall see. For systems of indistinguishable delocalized particles, we have to allow for the particle indistinguishability to avoid overcounting states, as discussed in Chapter 5. Our strategy will be to derive complete expressions for ideal Bose and Fermi gases in the quantum limit in Chapter 12 and then to obtain the proper classical limit form for the partition function. This approach avoids the *ad hoc* introduction of correction factors as used in Chapter 5 in evaluating $\ln \Omega$ for an ideal gas. Before introducing Z, we summarize a number of concepts and results from probability theory.

10.2 STATISTICAL METHODS: INTRODUCTION TO PROBABILITY THEORY

10.2.1 Discrete Variables and Continuous Variables

Probability theory deals with both discrete variables and continuous variables. In the case of discrete variables, there are discrete outcomes, which are observed when an event occurs. Examples of events of this kind are the tossing of sets of coins or dice for which different outcomes are possible. The discrete outcomes involve heads or tails for each of the coins in a set and, similarly, numbers 1–6 for each die in a set. Continuous variables, as the term implies, can take a continuous set of values within some range. For simplicity, discrete statistical variables are considered first. Statistical *ensembles* are important in probability theory and involve a large number of identical systems prepared in similar ways. For example, an ensemble of dice would involve N good dice thrown in a similar fashion.

The number of times n_i that a particular outcome i is found in a large ensemble of N systems gives the probability p_i of that outcome:

$$p_i = \lim_{N \to \infty} \frac{n_i}{N}. \tag{10.1}$$

It is readily seen that $\sum_i p_i = 1$. If p_i is the probability of a particular outcome, then $1 - p_i$ is the probability of not obtaining that outcome. The ensemble average of a variable X that takes discrete values x_i is defined as

$$\langle x \rangle = \sum_i p(x_i) x_i, \tag{10.2}$$

where $p(x_i)$ is the probability of the outcome x_i. It is useful to consider moments of the distribution, with the nth moment of x given by

$$\langle x^n \rangle = \sum_i p(x_i) x_i^n. \tag{10.3}$$

The variance or dispersion is defined as

$$\sigma_x^2 = \langle x^2 \rangle - \langle x^2 \rangle, \tag{10.4}$$

where σ_x is called the standard deviation. The standard deviation provides information on the width of the probability distribution $p(x_i)$.

For continuous variables, the summations in Equations 10.2 and 10.3 are replaced by integrals. The mean value $\langle x \rangle$ of the variable X is defined as

$$\langle x \rangle = \int_a^b p(x) x \, dx, \tag{10.5}$$

where the limits a and b specify the range of the continuous variable. The normalization condition becomes

$$\int_a^b p(x) dx = 1. \tag{10.6}$$

The nth moment of the variable X is

$$\langle x^n \rangle = \int_a^b p(x) x^n \, dx, \tag{10.7}$$

whereas the average of a function $f(x)$ is given by

$$\langle f \rangle = \int_a^b p(x) f(x) dx. \tag{10.8}$$

Exercise 10.1

For a good die with six faces, show that $\langle x \rangle = 3.5$ and $\langle x^2 \rangle = 15.17$. Use these values to obtain σ_x. Discuss the form of the probability distribution.

We have $\langle x \rangle = \sum_i p(x)x_i$ with $p(x_i) = 1/6$ and $x_i = 1$ to 6. Substitution in the expression gives $\langle x \rangle = 3.5$. A value for $\langle x^2 \rangle$ is obtained with the use of Equation 10.3 for $n = 2$ and the dispersion σ_x^2 from Equation 10.4. The probability distribution $p(x_i)$ is constant for x_i in the range of 1–6 and zero outside this range.

Exercise 10.2

The continuous variable θ has a probability distribution given by $p(\theta) \, d\theta = d\theta/\pi$ in the range from 0 to π. Obtain values for $\langle \theta \rangle$ and $\langle \theta^2 \rangle$ and the variance. Repeat the calculation for the function $f(\theta) = \cos^2 \theta$.

From Equation 10.5, we have $\langle \theta \rangle = (1/\pi) \int_0^\pi \theta \, d\theta = \pi/2$. Similarly, $\langle \theta^2 \rangle = \pi^2/3$ is obtained with the use of Equation 10.7, and the variance from Equation 10.4 is $\langle \theta^2 \rangle - \langle \theta^2 \rangle^2 = \pi^2/12$. The similar calculation for $f(\theta) = \cos^2 \theta$ is straightforward.

10.2.2 JOINT PROBABILITIES

Joint probabilities arise when two or more variables are involved. As an example, the statistical ensemble may consist of dice with several different colors, and we could, for example, obtain the probability of blue dice with the number 6 uppermost. The joint probability for discrete variables is specified as $p(x_i, y_j)$, where x_i is the value of variable X and y_j is the value of variable Y. The normalization condition for continuous variables becomes

$$\iint p(x, y) \, dx \, dy = 1. \tag{10.9}$$

In special cases when the two variables are statistically independent, the joint probability is simply the product of the separate probabilities

$$p(x, y) = p(x)p(y). \tag{10.10}$$

For statistically independent variables, it follows that

$$\langle xy \rangle = \langle x \rangle \langle y \rangle, \tag{10.11}$$

or the mean of the product is equal to the product of the means.

Consider functions $f(x, y)$ and $g(x, y)$ of the variables x and y. The mean value of the function f is

$$\langle f \rangle = \int p(x, y) f(x, y) \, dx \, dy \tag{10.12}$$

with a similar expression for $\langle g \rangle$. It is easy to see by carrying out the integrals that

$$\langle f + g \rangle = \langle f \rangle + \langle g \rangle, \tag{10.13}$$

or the mean of the sum of two functions is equal to the sum of the means. The mean of the product of two functions is, in general, *not* given by the product of the means. However, if the variables are statistically independent, then

$$\langle fg \rangle = \langle f \rangle \langle g \rangle. \tag{10.14}$$

The proof is straightforward:

$$\langle fg \rangle = \iint p(x,y) f(x) g(y) \, dx \, dy = \int p(x) f(x) \, dx \int p(y) g(y) \, dy = \langle f \rangle \langle g \rangle.$$

10.2.3 The Binomial Distribution

The binomial distribution is important in many statistical calculations. Consider an ensemble of N similar systems, each of which is described by a variable that has two possible discrete outcomes. Examples are N coins or N spins $\frac{1}{2}$. Let p be the probability of one of the two outcomes and $q = 1 - p$ the probability of the other outcome. The probability of n outcomes of one kind (e.g., heads) and $N - n$ outcomes of the other kind (tails) is

$$P^N(n) = \begin{pmatrix} N \\ n \end{pmatrix} p^n q^{N-n}, \tag{10.15}$$

with $\begin{pmatrix} N \\ n \end{pmatrix} = N! / [n!(N-n)!]$, a degeneracy factor which is simply the number of ways of arranging N objects where n are of one kind and $(N-n)$ of another kind, as discussed for spin systems in Chapter 4. Equation 10.15 is called the binomial distribution. The name derives from the identical form that the right-hand side of Equation 10.15 has with the general term in the binomial expansion

$$(p+q)^n = \sum_n \begin{pmatrix} N \\ n \end{pmatrix} p^n q^{N-n}. \tag{10.16}$$

Because the sum of the probabilities is $(p + q) = 1$, Equation 10.16 may be used immediately to show that the distribution is normalized. The following relations are obtained for the mean value and variance:

$$\langle n \rangle = Np \tag{10.17}$$

and

$$\sigma_N^2 = \langle n^2 \rangle - \langle n^2 \rangle = Npq. \tag{10.18}$$

Details are given in Appendix B. From Equation 10.18, the standard deviation is

$$\sigma_N = \sqrt{Npq}, \tag{10.19}$$

and it follows that the fractional deviation is given by

$$\frac{\sigma_N}{\langle n \rangle} = \left(\frac{q}{p} \right)^{1/2} \frac{1}{\sqrt{N}}. \tag{10.20}$$

Equation 10.20 shows that the fractional deviation decreases as $1/\sqrt{N}$ and, for N very large, will tend to zero. This is consistent with $n/N \to p$ as $N \to \infty$.

For large N, the binomial distribution is well approximated by the Gaussian distribution

$$P_N(n) \simeq \frac{1}{\sqrt{2\pi}\sigma_N} e^{-\langle n - \langle n^2 \rangle \rangle / \sigma_N^2}. \tag{10.21}$$

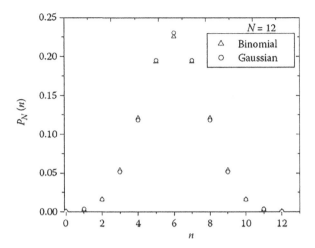

FIGURE 10.1 The binomial distribution (triangles) for $N = 12$ and $p = q = \dfrac{1}{2}$. The Gaussian approximation to the binomial distribution given by Equation 10.21 is also shown (circles). Even for this comparatively small value of N, the agreement between the two distribution functions is good and improves as N is made larger.

Equation 10.21 is obtained using a Taylor expansion of $\ln P_N(n)$ retaining terms up to the second order. Further details are given in Appendix B. Figure 10.1 shows a comparison of the binomial distribution with the Gaussian approximation to it for $N = 12$ and $p = q = 0.5$.

10.3 ENSEMBLES IN STATISTICAL PHYSICS

In Section 10.2, it is pointed out that, in probability theory, statistical ensembles are used in the definition of probabilities of particular outcomes. The various ensembles introduced into statistical physics correspond to particular physical situations of interest. They are known by the names *microcanonical ensemble*, *canonical ensemble*, and *grand canonical ensemble*. Canonical ensemble simply means the standard ensemble. The microcanonical ensemble approach is dealt with in Chapter 5, although this terminology is not emphasized there.

The number of accessible microstates for a system with the fixed total energy E is $\Omega(E, V, N)$. In Section 5.1, the fundamental postulate is introduced and leads to the expression for the probability that a system is found in a particular microstate. In Sections 5.2 and 5.3, equilibrium conditions for two interacting systems are developed on the basis of considerations of $\Omega(E, V, N)$ for the combined system. These conditions lead to the identification $S = k_B \ln \Omega(E, V, N)$, given in Equation 5.12. For the microcanonical ensemble, a large number of identically prepared systems with the fixed total energy E and the fixed particle number N are considered. Members of the ensemble are in the same macrostate but may be found in any one of the accessible microstates. The statistical methods applied in Chapter 5 are based on probabilities determined using this approach.

An important point arises in comparing the ensemble average of some physical quality with the time average, which is measured experimentally for a given system. Over a period of time, a system, even one that is close to ideal, will make transitions between its accessible microstates because of interactions between particles. For systems close to ideal, the interactions between particles are extremely small but nevertheless do allow the exchange of energy between particles, with a simultaneous change in the set of quantum numbers that describe the system. It is asserted that the time average of any quantity is equal to the ensemble average of that quantity. In most systems of interest, the time needed for transitions between microstates to occur is very short and the assertion is well justified. The proof that ensemble and long-time averages are the same is not simple and is linked

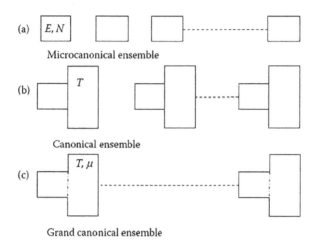

(a) Microcanonical ensemble

(b) Canonical ensemble

(c) Grand canonical ensemble

FIGURE 10.2   (a) Microcanonical ensemble, E and N constant for each member; (b) canonical ensemble, T and N held constant; and (c) grand canonical ensemble, T and μ held constant. In the canonical ensemble, a large heat bath is used to fix the temperature for each member of the ensemble. In the grand canonical ensemble, a large reservoir is used to fix the temperature and the chemical potential for each member of the ensemble.

to the ergodic hypothesis mentioned in Section 4.2. This matter is discussed in somewhat greater detail in Chapter 18.

In the canonical ensemble, the N systems do not have a fixed energy but each is in contact with a heat bath at temperature T, with which heat may be exchanged. The grand canonical ensemble involves N systems, each in thermal and diffusive contact with a reservoir at temperature T and with chemical potential μ. In this case, both heat exchange and particle exchange can occur. Figure 10.2 depicts the three ensembles. For each of the three ensembles, a probability distribution function is obtained. These are called the microcanonical, canonical, and grand canonical distributions, respectively.

10.4   THE CANONICAL DISTRIBUTION

In the derivation of the canonical distribution, it is convenient to consider a small system 1 that has discrete energies E_1. We defer until later specification of the eigenstates $| n$ for a particular system. To avoid complications because of indistinguishability of particles and the statistics they obey, which depends on whether they are fermions or bosons, we shall in this chapter largely confine the application of the formalism that we develop to systems of localized particles. Examples are spins, with associated magnetic dipole moments, at fixed lattice sites in a solid situated in a magnetic field or fixed electric dipoles in an electric field. The expressions we obtain in this section are quite general, and the approach is readily extended to systems of delocalized indistinguishable particles as shown in Chapters 12 and 16.

For each member of the canonical ensemble that we consider, a heat bath at temperature T is in contact with a small system, as shown in Figure 10.2. Each heat bath is effectively infinite in size in comparison with its small system. For convenience in the discussion that follows, we label the small system 1 and the heat bath 2. As a starting point, we use the results obtained for the microcanonical ensemble case in Chapter 5. The total number of accessible microstates for the combined system is given by

$$\Omega = \Omega_1\left(E_1\right)\Omega_2\left(E_0 - E_1\right),\qquad(10.22)$$

where E_0 is the total combined energy of the small system plus heat bath. From the fundamental postulate given in Chapter 5, the probability $P(E_1)$ of finding the small system with energy E_1 and the heat bath with energy $E_0 - E_1$ (with $E_1 \ll E_0$) is proportional to the total number of accessible microstates for the combined system when energy is shared in this way. Therefore,

$$P_1(E_1) \propto \Omega_1(E_1)\Omega_2(E_0 - E_1). \tag{10.23}$$

Now, we choose system 1 to be in a particular microstate, with energy E_1, so that $\Omega_1(E_1) = 1$.
Consequently, we have

$$P_1(E_1) \propto \Omega_2(E_0 - E_1). \tag{10.24}$$

In Chapter 5, it is pointed out that, for a large system, $\Omega(E)$ is a very rapidly varying function of E, which reaches a maximum when $E_2 = E_0$. We expand $\ln \Omega_2(E_0 - E_1)$, which varies more slowly with energy than $\Omega_2(E_0 - E_1)$, in a Taylor series about E_0

$$\ln \Omega_2(E_0 - E_1) = \ln \Omega_2(E_0) + \left(\frac{\partial \ln \Omega_2}{\partial E_2}\right)(-E_1) + \frac{1}{2}\left(\frac{\partial^2 \ln \Omega_2}{\partial E_2^2}\right)E_1^2 + \cdots \tag{10.25}$$

Because $E_1 \ll E_0$, this series may be expected to converge rapidly, and as an approximation, terms of order higher than unity are neglected. With $(\partial \ln \Omega_2 / \partial E_2) = \beta$, as defined in Equation 5.14, we obtain

$$\ln \Omega_2(E_0 - E_1) \simeq \ln \Omega_2(E_0) - \beta E_1 = \ln\left[\Omega_2(E_0)e^{-\beta E_1}\right]. \tag{10.26}$$

After taking antilogs, insertion of the resultant expression into Equation 10.24 leads to

$$P_1(E_1) = Ce^{-\beta E_1}. \tag{10.27}$$

The proportionality constant C contains the quantity $\Omega_2(E_0)$. C is determined by the normalization condition

$$\sum_{E_1} P_1(E_1) = 1, \tag{10.28}$$

and this gives

$$P_1(E_1) = \frac{e^{-\beta E_1}}{Z}, \tag{10.29}$$

with the partition function defined as

$$Z = \sum_{E_1} e^{-\beta E_1}. \tag{10.30}$$

The symbol Z is derived from the German word *zustandsumme* or sum over states. Equation 10.29 is the canonical distribution and is of central importance in statistical physics and a high point in our development of the subject. As pointed out in Chapter 7, the Helmholtz potential is given in terms of the partition function by $F = -k_B T \ln Z$, and this bridge relationship is established in Section 10.6. From F, other thermodynamic quantities can be obtained. A number of approximations have been made in the derivation of the canonical distribution given above. Justification for these approximations can be given in mathematical terms but comes also from the agreement of predictions on the basis of the canonical distribution with experimental results for macroscopic systems.

Mean values may readily be obtained with the canonical distribution. Equations 10.29 and 10.2 give the mean energy of the system as

$$\langle E \rangle = \left(\frac{1}{Z}\right)\sum_{E_1} E_1 e^{-\beta E_1} \tag{10.31}$$

or, in equivalent form in terms of $\ln Z$,

$$\langle E \rangle = -\frac{\partial \ln Z}{\partial \beta}, \tag{10.32}$$

where $Z = \sum_{E_1} e^{-\beta E_1}$. Equation 10.32 is a useful form for calculations on a variety of systems.

In general, we have for the partition function $Z = \sum_{n_1,n_2,n_3,\dots} e^{-\beta(n_1\varepsilon_1 + n_2\varepsilon_2 + n_3\varepsilon_2 + \cdots)}$, where n_j is the number of particles found in state 1 with energy ε_1, n_2 is the number with energy ε_2, and so on subject to the constraint on particle numbers $N = \sum_{i=1,2,\dots} n_i$. The number of particles in a particular state depends on the statistics that the particles obey. In the classical limit, the single particle states are sparsely occupied as pointed out in Chapter 4. However, for a gas, allowance has to be made for the indistinguishability of particles. As mentioned above, we defer consideration of the partition function for delocalized particles in an ideal gas until after the introduction of the quantum distribution functions in Chapter 12. In the classical limit of the quantum distributions for bosons and fermions, we obtain a result that automatically corrects for overcounting of states. A short discussion of the partition function for an ideal classical gas is given in Section 10.10 at the end of this chapter.

In the next section, we consider a system of distinguishable localized spins for which the partition function is easily evaluated. The spin system in contact with a heat bath may consist of a single localized moment or a large number of localized moments. In the latter case, each quantum state will correspond to a set of quantum numbers, and the energy E_1 will be the net energy of system 1 in this state. In such cases, it is necessary to allow for degeneracy effects for the set of microstates with the same energy if we require the probability of finding the system with a particular energy rather than in a particular quantum state labeled by a set of quantum numbers. The degeneracy factor is usually denoted by g, and if we put $\Omega_1(E_1) = g_1$, Equation 10.29 becomes

$$P_1(E_1) = \frac{g_1 e^{-\beta E_1}}{Z}, \tag{10.33}$$

with

$$Z = \sum_{E_1} g_1 e^{-\beta E_1}. \tag{10.34}$$

The distribution in Equation 10.33 gives the probability of finding system 1 in any one of the degenerate states with energy E_1.

10.5 CALCULATION OF THERMODYNAMIC PROPERTIES FOR A SPIN SYSTEM USING THE CANONICAL DISTRIBUTION

In Chapter 5, various results for the properties of an ideal spin system of N localized electron spins $\frac{1}{2}$ with dipole moment $\boldsymbol{\mu} = -g\mu_B \boldsymbol{S}$ in a magnetic field $\boldsymbol{B}$ were obtained from the expression for $\Omega.(E)$. These results may readily be derived with the canonical distribution. It is necessary to evaluate Z for the system with $E_1 = \sum_{i=1}^{N} \varepsilon_i = g\mu_B B \sum_{i=1}^{N} m_i$, and $m_i = \pm\frac{1}{2}$ for a spin $\frac{1}{2}$ system. The properties of the exponential function permit the partition function to be factorized as follows:

$$Z = \sum_{\{m1,m2,\ldots,mN\}} e^{-\beta_g \mu_B B(m_1 + m_2 + \cdots + m_N)} = \sum_{m_1 = \pm\frac{1}{2}} e^{-\beta\mu_B Bm_1} \sum_{m_2 = \pm\frac{1}{2}} e^{-\beta\mu_B m_2} \cdots \sum_{m_N = \pm\frac{1}{2}} -\beta_g \mu_B m_N$$

or $Z = z^N$, where z, defined as

$$z = \sum_{m = \pm\frac{1}{2}} e^{-\beta_g \mu_B Bm}, \tag{10.35}$$

is the partition function for a single spin and is readily evaluated. We see that for the system of localized spins $\ln Z = N \ln z$, with the single spin partition function given by $z = \left(e^{-1/2\beta_g \mu_B B} + e^{1/2\beta_g \mu_B B}\right) = 2\cosh\left[\beta\left(\frac{1}{2}g\mu_B B\right)\right]$. For $g = 2$, this gives $z = 2\cosh(\beta\mu_B B)$ and hence,

$$\ln Z = N \ln\left(2\cosh\beta\mu_B B\right). \tag{10.36}$$

From Equation 10.31, the mean energy of the system is $\langle E \rangle = (1/Z)\sum_1 E_1 e^{-\beta E_1}$ or, in equivalent form (Equation 10.32) in terms of $\ln Z$, $\langle E \rangle = -\partial \ln Z / \partial \beta$. With the use of Equations 10.36 and 10.32, the mean energy is easily calculated to be

$$\langle E \rangle = -N\mu_B B \tanh\left(\beta\mu_B B\right), \tag{10.37}$$

in agreement with the microcanonical ensemble result given in Equation 5.19.

The mean magnetic moment for a single spin is obtained on differentiation of $\ln z$ with respect to B, $\langle \mu \rangle = +(1/\beta)\,\partial \ln z / \partial B = \mu_B \tanh \beta\mu_B B$, so that for N spins

$$\mathcal{M} = N\mu = N\mu_B \tanh \beta\mu_B B. \tag{10.38}$$

This result may be obtained directly using the following relation involving the partition function Z for N spins:

$$\mathcal{M} = \left(\frac{1}{\beta}\right)\frac{\partial \ln Z}{\partial B}. \tag{10.39}$$

Note that $\langle E \rangle = -\langle \mathcal{M} \rangle B$, as expected. Figure 10.3 shows the behavior of $\langle \mathcal{M} \rangle / N \mu_B$ as a function of $\mu_B B / k_B T$.

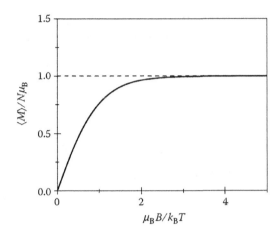

FIGURE 10.3 The reduced magnetic moment $\langle M \rangle / N\mu_B$ plotted as a function of $\mu_B B/k\mu BT$ for an ideal spin system of N spins $\frac{1}{2}$. The Curie law region corresponds to low B/T.

The linear region close to the origin in Figure 10.3 corresponds to the high-temperature Curie law regime. From Equation 10.39 with $B = \mu_0 H$, the magnetic susceptibility per unit volume is given by

$$\chi = \frac{\mathcal{M}}{VH} = \left(\frac{N\mu_B\mu_0}{VB} \right) \tanh\left(\frac{\mu_B B}{k_B T} \right). \tag{10.40}$$

At high temperatures and in fields that are sufficiently small so that $\mu_B B/k_B T \ll 1$, we obtain

$$\chi = \frac{n\mu_B^2\mu_0}{k_B T}, \tag{10.41}$$

where $n = N/V$. Equation 10.41 is Curie's law with an explicit expression for the Curie constant. From Equation 10.37, the heat capacity is as follows:

$$C_H = \left(\frac{\partial E}{\partial T} \right)_H = \left(\frac{\partial E}{\partial \beta} \right)_H \left(\frac{\partial \beta}{\partial T} \right) = Nk_B \left(\frac{\mu_B B}{k_B T} \right)^2 \operatorname{sech}^2\left(\frac{\mu_B B}{k_B T} \right). \tag{10.42}$$

This is the result given previously in Equation 5.21. The behavior of $C_H = N/k_B$ as a function of $k_B T/\mu_B B$ is shown in Figure 5.5 and is discussed in Section 5.4.

10.6 RELATIONSHIP BETWEEN THE PARTITION FUNCTION AND THE HELMHOLTZ POTENTIAL

In Section 10.5, various thermodynamic properties of an ideal spin system have been obtained from Z. We now establish the bridge relationship between F and Z, which is closely connected to the relationship between S and Ω. For an ideal spin system, Equation 10.36 shows that

$$Z = Z(\beta, B). \tag{10.43}$$

The total differential of $\ln Z$ is

$$d \ln Z = \left(\frac{\partial \ln Z}{\partial \beta} \right)_B d B + \left(\frac{\partial \ln Z}{\partial B} \right)_B d\beta = -\langle E \rangle d\beta + \beta \langle \mathcal{M} \rangle dB, \tag{10.44}$$

where we have made use of Equations 10.32 and 10.39. Now $d(\beta\langle E\rangle) = \beta\, d\langle E\rangle + \langle E\rangle\, d\beta$, and with $dB = \mu_0\, dH$, which is a good approximation for a paramagnetic material, this gives $d[\ln Z + \beta\langle E\rangle] = +\beta d\langle E\rangle + \beta\mu_0\langle \mathcal{M}\rangle dH$. From the first law in Equation 2.21, with the work term being the work done on the sample only, which is appropriate here, we obtain $d[\ln Z + \beta\langle E\rangle] = \beta\, dQ = dS/k_B$. Integration and rearrangement leads to the required result

$$F = \langle E\rangle - TS = -k_B T \ln Z. \tag{10.45}$$

Equation 10.45 allows the Helmholtz potential to be obtained from the partition function for the system. Other thermodynamic quantities may be obtained from F with the relationships established in Section 7.3. Determination of the partition function therefore leads to a *complete* thermodynamic description of a system.

Although the expression for Z used in this section refers to an ideal paramagnet of localized moments, Equation 10.45 is a general result applicable to any system. An alternative derivation of Equation 10.45 is given as follows. The partition function is from Equation 10.34 $Z = \sum_{E_1} g_1 e^{-\beta E_1}$, where g_1 is a degeneracy factor for states with energy E_1. For a large system with many degrees of freedom, we can regard the energy levels as forming a quasi-continuum and write Equation 10.34 as

$$Z = \sum_E \Omega(E) e^{-\beta E}, \tag{10.46}$$

where $\Omega(E)$ is the number of accessible states in the small range(E to $E + \delta E$. In Equation 10.46, $\Omega(E)$ increases very rapidly with E, as discussed in Chapter 4. The exponential factor $e^{-\beta E}$ decreases very rapidly with E. The product, which is sharply peaked, has a maximum at $\overline{E}$, as depicted in Figure 10.4. To a good approximation, the partition function written in terms of the peak energy $\overline{E}$ is given by $Z = \Omega(\overline{E}) e^{-\beta \overline{E}} \Delta E/\delta E$, where ΔE is the half-height width of the peak in the function shown in Figure 10.4. Taking logs gives

$$\ln Z = \ln \Omega(\overline{E}) - \beta \overline{E} + \ln\left(\frac{\Delta E}{\delta E}\right). \tag{10.47}$$

For a system with a large number of degrees of freedom, $\ln(\Delta E/\delta E)$ may be neglected in comparison with the other terms. Equation 10.47 may be rewritten as

$$k_B T \ln Z = \left[k_B T \ln \Omega(\overline{E})\right] - \overline{E},$$

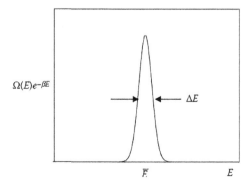

FIGURE 10.4 The product $\Omega(E)e^{-\beta E}$ as a function of E showing a peak of width ΔE at $\overline{E}$. $\Omega(E)$ is a rapidly increasing function of E, whereas $e^{-\beta E}$ is a rapidly decreasing function. The product gives rise to a sharply peaked function with a maximum at $\overline{E}$.

giving

$$F = \bar{E} - TS = -k_{\mathrm{B}}T \ \ln Z, \tag{10.48}$$

with

$$S = k_{\mathrm{B}} \ln \Omega \left(\bar{E} \right).$$

This is the same result as Equation 10.45, provided we put $\bar{E} = \langle E \rangle$. In arriving at Equation 10.48, the summation over all energies has effectively been replaced by the largest term. This is a good approximation because of the very sharply peaked nature of the function depicted schematically in Figure 10.4. It is also clear that the mean energy $\langle E \rangle$ of the system will coincide with the energy $\bar{E}$, at which the function $\Omega(E)e^{-\beta E}$ has its peak value.

10.7   FLUCTUATIONS

In Section 10.6, it is argued that the partition function is sharply peaked at $\langle E \rangle$. This discussion may be placed on a more quantitative basis in terms of the variance in the energy for the ideal spin system. From the definition of the variance in Equation 10.18, we can write

$$\sigma_E^2 = \left\langle E^2 \right\rangle - \left\langle E \right\rangle^2 . \tag{10.49}$$

Equation 10.32 gives $\langle E \rangle = -\partial \ln Z/\partial \beta = 1/Z(\partial Z/\partial \beta)$, and from the definition of an ensemble average Equation 10.2 with Equation 10.29, it follows that

$$\left\langle E^2 \right\rangle = \left(\frac{1}{Z} \right) \sum_{E_1} E_1^2 e^{-\beta E_1} = \left(\frac{1}{Z} \right) \left(\frac{\partial^2 Z}{\partial \beta^2} \right) = \frac{\partial^2 \ln Z}{\partial \beta^2} + \left(\frac{1}{Z^2} \right) \left(\frac{\partial Z}{\partial \beta} \right)^2 .$$

(Use has been made of the identity $\partial^2 \ln Z/\partial \beta^2 = (1/Z)(\partial^2 Z/\partial \beta^2) - (1/Z^2)(\partial Z/\partial \beta)^2$.)

From Equation 10.49 and with the use of Equation 10.32 to obtain $\langle E \rangle^2$, we get

$$\sigma_E^2 = \frac{\partial^2 \ln Z}{\partial \beta^2}. \tag{10.50}$$

$\ln Z$ for an ideal spin system (spin $\dfrac{1}{2}$, $g = 2$) is given by Equation 10.36, and the application of Equation 10.50 leads to

$$\sigma_E^2 = N \left(\mu_{\mathrm{B}}\beta \right)^2 \operatorname{sech}^2 \left(\beta\mu_{\mathrm{B}}B \right). \tag{10.51}$$

The magnitude of the fractional deviation in energy, with $\langle E \rangle$ from Equation 10.37, is

$$\frac{\sigma_E}{\langle E \rangle} = \frac{\sqrt{N} \left(\mu_{\mathrm{B}}B \right) \sqrt{\operatorname{sech}^2 \left(\beta\mu_{\mathrm{B}}B \right)}}{N \left(\mu_{\mathrm{B}}B \right) \tanh \left(\beta\mu_{\mathrm{B}}B \right)}$$

$$\frac{\sigma_E}{\langle E \rangle} \simeq \left(\frac{1}{\sqrt{N}} \right) \operatorname{cosech} \left(\beta\mu_{\mathrm{B}}B \right) \simeq \left(\frac{1}{\sqrt{N}} \right) \text{ for } \beta\mu_{\mathrm{B}}B < 1. \tag{10.52}$$

This is consistent with Equation 10.20 and shows that, for large N comparable with N_A, fluctuations in the energy are of the order of 1 part in 10^{11}. Such tiny fluctuations in energy are not detectable in usual physical measurements. Fluctuations in other quantities, such as the magnetization of the spin system, are also very small.

10.8 CHOICE OF STATISTICAL ENSEMBLE

In Section 10.3, the microcanonical, canonical, and grand canonical ensembles are described. The systems in the microcanonical ensemble have their total energy E fixed to within a small uncertainty δE. The canonical ensemble fixes the temperature T of systems in contact with a heat bath. Although the energy of each member in a canonical ensemble can fluctuate about a mean value, it is shown in Section 10.7 that, for systems containing a large number of particles, with a very large number of microstates, the fluctuations in energy are extremely small and can be neglected for practical purposes. Similarly, it will be shown later that, in the grand canonical ensemble case, fluctuations in both energy and particle number are extremely small.

The above observations have important consequences for large systems because they show that the three ensembles are effectively equivalent. This means that results obtained using the canonical ensemble may, for example, be applied to an isolated system whose total energy is fixed, provided the temperature of the system is known. Flexibility in the choice of the ensemble used for a particular system provides considerable advantages in calculations. In the derivation of the quantum distributions, it will be convenient to make use of the grand canonical ensemble. The results obtained are, nevertheless, applicable to systems with fixed energy and particle number, provided the temperature T and the chemical potential μ are specified for these systems.

10.9 THE BOLTZMANN FACTOR AND ENTROPY

From Equation 10.29, it follows that the probability p_r for a system in contact with a heat bath to be in a particular state r is

$$P_r = \frac{e^{-\beta E_r}}{Z}. \tag{10.53}$$

The factor $e^{-\beta E_r}$ is known as a Boltzmann factor. The ratio of the probabilities for a system to be in two different states r and s is

$$\frac{p_r}{p_s} = e^{-\beta(E_r - E_s)}. \tag{10.54}$$

Equation 10.54 is a useful relation because it only requires knowledge of the energies of the two different states involved.

Exercise 10.3

Obtain the ratio of the probabilities for occupation of the spin states for an ideal paramagnet with spins $\frac{1}{2}$ and $g = 2$ in an applied magnetic field. The energy eigenvalues are $E_\pm = \pm \mu_B B$ as depicted in Figure 4.4.

Use of the Boltzmann factors with the given energies leads to

$$\frac{p_+}{p_-} = e^{-\beta(-\mu_B B - \mu_B B)} = e^{2\beta \mu_B B}.$$

Typically, for electron and nuclear systems at temperatures $T = 4$ K or higher in laboratory fields of a few tesla, $\mu_B B \ll k_B T$, and the exponential may be expanded to give

$$\frac{p_+}{p_-} \simeq 1 + \frac{2\mu_B B}{k_B T}.$$ (10.55)

Equation 10.55 shows that, for the given conditions, the low energy $\left|-\frac{1}{2}\right\rangle$ state has a slightly greater probability of occupation than the high energy $\left|+\frac{1}{2}\right\rangle$ state. At very high temperatures, the probability ratio tends to 1. Note that we have used p_+ to denote the probability to find spins in the low energy state, with magnetic moment parallel to B, and p_- the corresponding probability for the higher energy state. This notation is consistent with that adopted in Chapter 4 for the occupancy of the two states.

It is straightforward to obtain an expression for the entropy of a system in terms of the probabilities p_r using results derived earlier in this chapter. The relationship shown in Equation 10.45 for the Helmholtz potential is $F = \langle E \rangle - TS = -k_B T \ln Z$, and for the entropy, this gives

$$S = \frac{\langle E \rangle}{T} + k_B \ln Z.$$ (10.56)

The mean energy is $\langle E \rangle = (1/Z) \sum_r E_r e^{-\beta E_r}$. Rearrangement of Equation 10.53 gives $E_r = (-1/\beta)\ln Zp_r$, and the insertion of this result into the above equation for the mean energy leads to

$$\langle E \rangle = \sum_r \left(-\frac{1}{\beta}\right) p_r \ln Zp_r.$$ (10.57)

Equations 10.56 and 10.57 with the condition $\sum_r p_r = 1$ give the required result

$$S = -k_B \sum_r p_r \ln p_r,$$ (10.58)

or in alternative form $S/k_B = -\sum_r p_r \ln p_r$.

Once the probabilities p_r for a system to be in states r are determined, the entropy may be obtained immediately. Equation 10.58 is named the Gibbs entropy equation in honor of J. Willard Gibbs. The equation is useful in information theory, as discussed in Section 10.11, and may be applied to other phenomena for which probabilities can be obtained.

Exercise 10.4

An ideal N spin system has two states with occupation probabilities $p_+ = p_- = 0.5$. Find the entropy of this simple system.

The Boltzmann expression gives $S = -2Nk_B(0.5\ln0.5) = Nk_B \ln 2$. This agrees with the result obtained using Equation 5.18. It is a simple matter to obtain S for other values of the occupation probabilities.

10.10 BOLTZMANN AND GIBBS ENTROPY EQUATIONS

The Gibbs equation for the entropy of a thermodynamic system, given in Equation 10.58, provides insight into the entropy concept. Recall that Boltzmann's famous equation $S = k_B \ln \Omega$ given in Equation 5.12, expresses the entropy in terms of the logarithm of the number of states Ω accessible to a system. The Gibbs equation involves the probabilities of a system's various accessible states being occupied. For a system which is not in equilibrium, the probabilities will change with time as the system reaches an equilibrium. Of course, Ω will change as well but in a collective way. The Gibbs entropy equation involves more information than that of Boltzmann.

It is straightforward to obtain the Gibbs entropy equation from the Boltzmann entropy using an expression for $\ln \Omega$ given in terms of the probabilities of microstates of a system being occupied. Consider a large ensemble of N identical systems with each member of the ensemble distributed over the various accessible states labeled r. The probability of a system being in state r is $p_r = n_r/N$, with n_r as the number of systems found in state r over the ensemble. The probabilities are subject to the conditions $\sum_r p_r = 1$ and $\sum_r p_r E_r = \langle E \rangle$, with $\langle E \rangle$ as the mean energy of systems in the ensemble.

The number of accessible states for the *complete ensemble* is

$$\Gamma(n_1, n_2, n_3, \ldots) = \frac{N!}{n_1! \, n_2! , n_3! \ldots} \tag{10.59}$$

Taking natural logs and using Stirling's formula $\ln A! = A \ln A - A$ give

$$\ln \Gamma = N \ln N - N - \sum_r n_r \ln n_r + \sum_r n_r = N \ln N - \sum_r n_r \ln n_r. \tag{10.60}$$

Putting $n_r = N p_r$ in Equation 10.60 leads to

$$\ln \Gamma = N \ln N - \sum_r N p_r \ln N p_r = N \ln N - N \ln N \sum_r p_r - N \sum_r p_r \ln p_r \tag{10.61}$$

Equation 10.61 simplifies to the form

$$\ln \Gamma = -N \sum_r p_r \ln p_r \tag{10.62}$$

In equilibrium, each member of the ensemble has Ω accessible states. From combinatorial considerations, it follows that the number of states for the *entire ensemble* is given by $\Gamma = \Omega^N$ and therefore $\ln \Omega = \ln \Gamma / N$. Inserting this result in Equation 10.62 gives $\ln \Omega = -\sum_r p_r \ln p_r$ Substitution for $\ln \Omega$ in the Boltzmann entropy equation (Equation 5.12) finally leads to the Gibbs equation

$$S = -k_B \sum_r p_r \ln p_r \tag{10.63}$$

Exercise 10.5

Show that the Gibbs entropy equation reduces to the Boltzmann equation for a system in thermodynamic equilibrium.

From the Gibbs equation (Equation 10.63), we have $S = -k_B \sum_r p_r \ln p_r$. For a large system in equilibrium, all Ω accessible states are equally probable as required by the fundamental postulate of statistical physics. It follows that the probabilities are given by $p_r = 1/\Omega$. Substituting for the p_r gives the Boltzmann equation $S = -k_B \sum_r (1/\Omega) \ln (1/\Omega) = k_B \ln \Omega$.

An increase in entropy corresponds to an increase in disorder and, correspondingly, an increase in *uncertainty* in the microscopic probability description of a large system undergoing a process. Information theory is concerned with the uncertainty that can arise in transmitting information as discussed in Section 10.11.

10.11   SHANNON ENTROPY AND INFORMATION THEORY

In addition to thermodynamic systems, the Boltzmann and Gibbs equations for the entropy can be applied to other systems which involve statistical descriptions. A simple and familiar example, introduced previously in this book, is a set of tossed coins, each of which can land with either heads or tails facing upward. If all the coins are fair, with equal probabilities p_h and p_t, respectively, of landing with either heads up or tails up, then by using $p_h + p_t = 1$, it follows that $p_h = p_t = \frac{1}{2}$. It is necessary that the set of coins be sufficiently large to avoid significant statistical fluctuations. Any departure from equal probabilities raises questions about the fairness of the coins and in this way provides us some information on the coin system.

The symbol H is used to denote the distribution entropy in general applications of probability distributions in information theory. This notation is based on Boltzmann's pioneering work on molecular collisions in an ideal gas and in particular his H theorem, in which H is related to thermodynamic entropy.

A major advance in information theory occurred in the mid-twentieth century with the introduction by Claude Shannon of what has come to be called the Shannon entropy H_S. The Shannon equation for H is similar to the Gibbs entropy equation and has the form

$$H_S = -\sum_r p_r \log_b p_r. \tag{10.64}$$

Since $0 \le p_r \le 1$, it follows that $H_S \ge 0$. In contrast to the Gibbs equation, there is no scaling constant with dimensions in Equation 10.64. Note that the *base* of the log function is indicated by subscript b, which is chosen to suit the form in which information is presented. For a binary set of characters (e.g., 0 or 1) or outcomes of events such as tossing a coin (e.g., heads or tails), it is convenient to use $b = 2$.

For the tossed coin event mentioned above, with outcome probabilities $p_h = p_t = \frac{1}{2}$, and using $b = 2$, the Shannon entropy is given by $H_S = -\sum_r p_r \log_b p_r = -\frac{1}{2}\log_2\frac{1}{2} - \frac{1}{2}\log_2\frac{1}{2} = \log_2 2 = 1$. The same result for the Shannon entropy would apply in the case of bits of information represented by 0 or 1 provided the two-bit outputs generated by a device had probabilities $p_0 = p_1 = \frac{1}{2}$. It is of interest to examine the Shannon entropy when the device is programmed to generate the output 0 or 1 with $p_0 \ne p_1$. In the extreme case $p_0 = 1$ and $p_1 = 0$, we obtain $H_S = 0$. This result shows that when the output is totally predictable, with no uncertainty, the Shannon entropy goes to zero. The same result $H_S = 0$ would occur if the probabilities were reversed with $p_0 = 0$ and $p_1 = 1$. These findings show that the Shannon entropy gives a measure of uncertainty in the outcome of an event.

Figure 10.5 displays a plot of the Shannon entropy H_S versus the probability p_0, in the range $0 \le p_0 \le 1$, of the outcome 0 for an event involving a binary bit-generating device of the type described above. It can be seen that H_S exhibits a maximum for $p_0 = 0.5$ and decreases towards zero as p_0 nears its limit value of 0 or 1.

As stated above, in the limit $p_0 = 0$, there is no uncertainty in successive device bit outputs, which are always 1. Similarly for $p_0 = 1$, outputs are always 0. The maximum uncertainty occurs for $p_0 = p_1 = 0.5$.

As noted above, the Shannon entropy plays an important role in information theory, which has applications in communications and the transmission of data. To illustrate the approach, consider a message involving a string of characters q, each of which can take Q values. Familiar examples of character sets in common use are $Q = 26$ for letters of the alphabet and $Q = 128$ for ASCII. In order to allow for uncertainty in transmission of a character q_i, we introduce the normalized probability distribution $(p_1 p_2, ... p_Q)$. For q_i to be transmitted with zero uncertainty, it is necessary that $p_i = 1$ and that the other probabilities to be 0. Less stringent requirements for an acceptable transmission of information can be adopted in practice. Making use of the Shannon entropy $(p_1 p_2, ... p_Q)$, as

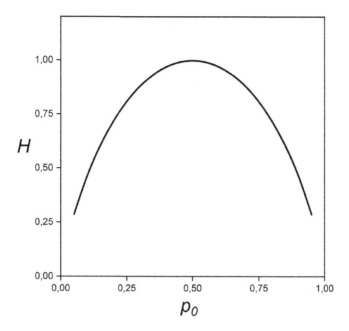

FIGURE 10.5   The Shannon entropy H_S as a function of the probability p_0 for a single bit event occurring with outcome 0. The plot shows that H_S decreases towards zero both above and below the maximum at $p_0 = 0.5$.

defined in Equation 10.64, provides a measure of the uncertainty introduced by relaxing the most stringent probability requirements, thereby assisting in making optimal use of transmission channel capacity.

Exercise 10.6

A device produces a stream of bit pairs made up as follows: [00], [01], [10], and [11]. Determine the Shannon entropy for the set of bit pairs assuming that they occur with equal probabilities. Use logarithms to base 2 in the calculation. $\left(\log_2 x = \log_b x / \log_b 2\right)$.

From Equation.10.54, we obtain $H_S = -\sum_r p_{r_r} \log_b p_r = -4 \times \frac{1}{4} x \log_2 \frac{1}{4} = 2.$

Note that the equal probability case produces a maximum uncertainty in the outcome of a single bit pair event in the stream of bit pairs. The maximum in the Shannon entropy has increased amplitude compared to the single bit case due to the doubling of the number of bits involved in each event.

How would the Shannon entropy change if the probability for a chosen bit pair [11] were increased to 0.7, while the probabilities for the other bit pairs were reduced to 0.1 each?

The Shannon entropy becomes $H_S = -[3 \times 0.1. \times \log_2 0.1] - [1 \times 0.7 \log_2 0.7] = 1.2.$

There is a significant decrease in H_S due to the decrease in uncertainty of a [11] outcome. Information theory is a highly developed subject. The present brief discussion of the Shannon entropy concept simply indicates how uncertainty in the reliability of information transfer can be measured when endeavoring to increase the transmission speed.

10.12   THE PARTITION FUNCTION FOR AN IDEAL GAS

For a system of N localized spins, as considered in Section 10.5, the partition function can from Equation 10.35 be written as $Z = z^N$, where z is the single particle partition function. This result holds in general for distinguishable localized particles. For delocalized, indistinguishable particles, as found in an ideal gas, we have to allow for overcounting of quantum states as discussed in Chapter 5. By taking the classical limit of the quantum distributions, we show in Chapter 12 that to a good approximation $Z = (1/N!)z^N$. It follows that $\ln Z = N \ln z - N \ln N + N$, where the use has been made of Stirling's approximation for $\ln N!$. It is therefore straightforward to obtain $\ln Z$ for an ideal classical monatomic gas once the single particle partition function z has been evaluated. This calculation is carried out in Chapter 16 by making use of the single particle in a box quantum state and conversion of the sum over energy states to a definite integral. The result as given in Equation 16.4 is $z = V (mk_BT/2\pi\hbar^2)^{3/2}$.

From $\ln Z$, the Helmholtz potential follows as $-k_BT \ln Z$, and on differentiation of F with respect to T, the Sackur–Tetrode expression for the entropy, as given in Equation 5.15, is obtained. The thermodynamic properties of the ideal gas are completely determined in this way. Further discussion is given in Chapter 16, where allowance is made for internal degrees of freedom in polyatomic molecules.

Before considering ideal quantum gases, we obtain the results for the grand canonical ensemble and introduce in Chapter 11 the grand partition function or grand sum. The grand canonical ensemble involves baths for which the temperature and chemical potential are specified. In this ensemble, there are no constraints on energy or particle number, and as shown in Chapter 12, this simplifies the derivation of the quantum distribution functions for fermion and boson systems. As pointed out above, the classical ideal gas partition function is obtained in the classical limit of the quantum distributions.

PROBLEMS CHAPTER 10

10.1 A red and a blue die are thrown simultaneously. What is the probability of obtaining a pair of sixes? Find the probability of throwing six on red and three on blue. Compare this probability with the probability of throwing a six and a three if the dice have the same color.

10.2 Ten good coins are tossed simultaneously. Find the probability of obtaining four heads. Use results given in Appendix B to obtain the mean number of heads and the corresponding dispersion in many throws of the coins. Sketch the form of the probability distribution. Construct a table of probabilities $P(n)$ for various numbers of heads n.

10.3 In a one-dimensional random walk with equal step lengths l along a gently inclined plane, the probability of a step down-plane is 20% larger than for a step up-plane. Find the mean displacement from the starting point after 200 steps have been taken. Obtain the corresponding dispersion.

10.4 A monatomic ideal gas at high temperature T in a discharge tube has a triply degenerate excited state at an energy ε above the ground state. Find the ratio of the population of the excited state to that of the ground state. Apply your result to helium in a discharge tube at a temperature of 104 K. The ground state of the helium atom is nondegenerate, whereas the first excited state at 19.8 eV above the ground state is triply degenerate.

10.5 Schottky defects in crystals correspond to the removal of an atom from a lattice site to the surface of the crystal as described in Chapter 5. If the energy of formation of a defect is ε, find the ratio of the probabilities for a given site to be either occupied or unoccupied at a given temperature T. If the number of Schottky defects n is much smaller than the number of lattice sites N, show that $n/N = \exp(-\beta\varepsilon)$. Use the partition function for the system to obtain an approximate value for the total energy associated with defects at a given temperature.

10.6 A system consists of N localized particles each of which has three energy levels at ε, 2ε, and 3ε with degeneracy factors $g = 1, 2,$ and 1, respectively. Use the canonical distribution to obtain the mean energy and the heat capacity of the system.

10.7   Rare earth ions in a crystal have three doubly degenerate energy levels at 0, Δ, and 3Δ. If the host crystal is in contact with a heat bath at temperature T, use the canonical distribution to obtain the mean energy of a rare earth ion. If there are n rare earth ions in the crystal, show how you would obtain their contribution to the low-temperature heat capacity.

10.8   N localized noninteracting $S = 1$ spins with moment μ are situated in a magnetic field B and are in contact with a heat bath at temperature T. Use the canonical distribution to obtain the mean energy of the system and the magnetization as a function of T. Sketch the form of the magnetization curve as a function of $\mu_B/k_B T$, and give a qualitative explanation of its features. Compare your results with those for a spin $S = \dfrac{1}{2}$ system.

10.9   A nonideal spin system consisting of N ions per unit volume with spin $S = \dfrac{1}{2}$ and magnetic moment μ is situated in an applied field B. In the mean field approximation, the energy of each spin is given by $\varepsilon_m = 2m(\mu_B + k_B \alpha M/N_\mu)$, where m takes values $+\dfrac{1}{2}$ and $-\dfrac{1}{2}$. The second term in parentheses is an effective local field with the parameter α as an effective temperature. Write down the partition function for the system and obtain an expression for the energy $E = -Nx \tanh \beta x$, where $x = (\mu_B + k_B \alpha M/\mu)$. By making suitable approximations, obtain the magnetization and the magnetic susceptibility in the high T, low B limit with $M \ll N_\mu$. Compare your expression for the susceptibility with that given by the Curie–Weiss law.

10.10   A system of N localized spins in a solid is in thermal contact with the lattice at temperature T. Each spin has an electric quadrupole moment, and the electric field gradient associated with the crystal field gives rise to two doubly degenerate energy levels with energies $-\varepsilon$ and ε, respectively. Write down the partition function for the system, and use the expression to find the mean energy $\langle E \rangle$ as a function of T. Sketch the behavior of $\langle E \rangle$ with T.

10.11   Consider a system of N noninteracting localized spins $\left(S = \dfrac{1}{2} \right)$ at temperature T in a magnetic field B. Each spin has two quantum states, and the probability of occupation of these states depends on B and T. Use the Boltzmann definition of the entropy to obtain an expression for the entropy of the system as a function of the probability p_r of occupation of the lower energy state for p_r in the range from 0 to 1. Generate a plot of S versus p_r and briefly discuss its form.

11 The Grand Canonical Distribution

11.1 INTRODUCTION

In Chapter 10, the three important ensembles that are used in statistical physics are introduced and discussed. In particular, the canonical ensemble approach is considered in some detail. In Chapter 5, the microcanonical ensemble provides the basis for the microscopic description of an ideal gas and an ideal spin system. The grand canonical ensemble, which completes the ensemble set, is discussed in the present chapter. For this case, both energy and particles can be exchanged with a reservoir. In Section 10.8, it is pointed out that, for systems consisting of macroscopically large numbers of particles, we can choose whichever ensemble is most convenient for a particular calculation. This is because fluctuations in energy and particle number are negligibly small for large systems in contact with a reservoir. In deriving the quantum distribution functions, it is convenient to use the grand canonical ensemble, and this is done in Chapter 12 for systems of fermions and bosons. This approach avoids the constraint on particle number and introduces the chemical potential into the distributions in a straightforward way. Before introducing the grand canonical distribution, we first consider the equilibrium conditions for two systems that interact thermally, mechanically, and via particle exchange.

11.2 GENERAL EQUILIBRIUM CONDITIONS

In Section 5.2, the equilibrium conditions for two systems in thermal contact are obtained. To arrive at the general equilibrium conditions for two systems that interact via thermal, mechanical, and particle exchange processes, we again use accessible state considerations from Chapters 4 and 5 as our starting point. Figure 11.1 shows a container separated into two parts by a movable, thermally conducting, porous piston.

The number of accessible states for the combined system is

$$\Omega(E,V,N) = \Omega_1(E_1,V_1,N_1)\Omega_2(E_2,V_2,N_2) \tag{11.1}$$

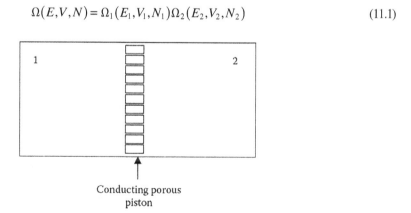

Conducting porous
piston

FIGURE 11.1 Two systems interacting via thermal, mechanical, and particle exchange processes. The movable piston is made of thermally conducting material and is equipped with small channels through which particles can be exchanged. In equilibrium, the temperatures, pressures, and chemical potentials of the two systems are equal.

subject to the constraints

$$E = E_1 + E_2, \quad V = V_1 + V_2, \quad N = N_1 + N_2, \tag{11.2}$$

which correspond to fixed total energy E, total volume V, and total particle number N, all being kept fixed. From the entropy definition, $S = k_B \ln \Omega$, Equation 11.1 may be written in the alternative form

$$S(E,V,N) = S_1(E_1,V_1,N_1) + S_2(E_2,V_2,N_2) \tag{11.3}$$

The entropies of systems 1 and 2 are added to give the entropy for the combined system, as required for an extensive quantity. In equilibrium, according to the second law, the entropy is a maximum. It follows that

$$ds = 0 = \left(\frac{\partial S_1}{\partial E_1}\right)_{V_1,N_1} dE_1 + \left(\frac{\partial S_1}{\partial V_1}\right)_{E_1,N_1} dV_1 + \left(\frac{\partial S_1}{\partial N_1}\right)_{E_1,V_1} dN_1 + \left(\frac{\partial S_1}{\partial E_2}\right)_{V_2,N_2} dN_1$$

$$+ \left(\frac{\partial S_2}{\partial V_2}\right)_{E_2,N_2} dV_2 + \left(\frac{\partial S_1}{\partial N_2}\right)_{E_2,N_2} dN_2. \tag{11.4}$$

The constraints given in Equation 11.2 imply that $dE_1 = -dE_2$, $dV_1 = -dV_2$, and $dN_1 = -dN_2$. Equation 11.4 may therefore be rewritten as

$$\left[\left(\frac{\partial S_1}{\partial E_1}\right) - \left(\frac{\partial S_2}{\partial E_2}\right)\right]dE_1 + \left[\left(\frac{\partial S_1}{\partial V_1}\right) - \left(\frac{\partial S_2}{\partial V_2}\right)\right]dV_1 + \left[\left(\frac{\partial S_1}{\partial N_1}\right) - \left(\frac{\partial S_2}{\partial N_2}\right)\right]dN_1 = 0.$$

With the definitions given in Section 3.12 on the basis of the general form of the fundamental relation, this equation becomes

$$\left(\frac{1}{T_1} - \frac{1}{T_2}\right)dE_1 + \left(\frac{P_1}{T_1} - \frac{P_1}{T_2}\right)dV_1 + \left(\frac{\mu_1}{T_1} - \frac{\mu_1}{T_2}\right)dN_1 = 0. \tag{11.5}$$

Because dN_1, dV_1, and dE_1 are independent and arbitrary in magnitude, all coefficients in Equation 11.5 must be identically zero. The equilibrium conditions therefore become

$$T_1 = T_2, P_1 = P_2, \mu_1 = \mu_2. \tag{11.6}$$

In equilibrium, the temperatures, pressures, and chemical potentials of the two subsystems are separately equal.

11.3 THE GRAND CANONICAL DISTRIBUTION

In Section 10.4, the canonical distribution is derived for a system in thermal contact with a large heat bath at temperature T. We follow a similar approach in obtaining the grand canonical distribution for systems in thermal and diffusive contact with a large reservoir. Consider a small system in both thermal and diffusive contact, through a fixed permeable wall, with a large reservoir at temperature T and with chemical potential μ. Figure 11.2 depicts the situation.

The grand canonical ensemble consists of a large number of replicas of the combined small system 1 plus reservoir 2. As shown in Section 11.2, in equilibrium $T_1 = T$ and $\mu_1 = \mu$, where T and μ refer, respectively, to the temperature and chemical potential of the reservoir. Because of the large size of the reservoir, T and μ may be regarded as constant in any exchange of energy and particles between the reservoir and the small system.

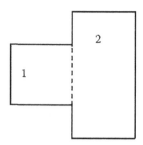

FIGURE 11.2   A small system 1 in thermal and diffusive contact with a large reservoir 2 at temperature T and with chemical potential μ.

Let system 1 be in a particular state with energy E_1 and number of particles N_1. The probability of this state is, from the fundamental postulate, proportional to the total number of accessible states.

$$P(E_1, N_1) \propto \Omega_2(E_0 - E_1, N_0 - N_1), \qquad (11.7)$$

where $E_0 = E_1 + E_2$ is the total energy and $N_0 = N_1 + N_2$, the total number of particles for the combined system plus reservoir. Using the definition of the entropy $S = k_B \ln \Omega$, Equation 11.7 is written in the form

$$P(E_1, N_1,) \propto \exp\left(\frac{1}{k_B} S_2(E_0 - E_1, N_0 - N_1) \right). \qquad (11.8)$$

Because $E_0 \gg E_1$ and $N_0 \gg N_1$, we can to a good approximation expand S_2 in a Taylor series in two variables about E_0 and N_0 and retain only firstorder terms. This gives

$$S_2(E_0 - E_1, N_0 - N_1) = S_2(E_0, N_0,) + \left(\frac{\partial S_2}{\partial E_2}\right)_{N_0} (-E_1) + \left(\frac{\partial S_2}{\partial N_2}\right)_{E_0} (-N_1) + \cdots$$

$$= S_2(E_0, N_0,) - \frac{E_1}{T} + \frac{\mu N_1}{T}. \qquad (11.9)$$

In Equation 11.9, use has been made of the relationships $1/T = (\partial S/\partial E)_N$ and $-\mu/T = (\partial S/\partial N)_E$ that follow from the general form of the fundamental relation given in Chapter 3. Substitution of Equation 11.9 into Equation 11.8 gives $P(E_1, N_1) \propto e^{(1/k_B)S_2(E_0,N_0)} e^{-(1/k_B)[E_1 - \mu N_1]}$.

Because $S_2(E_0, N_0)$ is constant, this quantity may be absorbed into the proportionality constant, which is determined from the normalization condition $\sum_{E_1 \cdot N_1} P(E_1, N_1) = 1$. This leads to the grand canonical distribution

$$P(E_1, N_1) = \frac{e^{\beta[\mu N_1 - E_1]}}{\mathbb{Z}}, \qquad (11.10)$$

where

$$\mathbb{Z} = \sum_{E_1 N_1} e^{\beta[\mu N_1 - E_1]} \qquad (11.11)$$

is the grand partition function or grand sum. If the number of particles N_1 is kept fixed, by blocking particle exchange, the factor $e^{\beta \mu N_1}$ may be taken outside the summation for $\mathbb{Z}$ and cancelled with the same factor in the numerator. In this way, the canonical distribution Equation 10.29 is recovered, that is, $P(E_1) = e^{\beta E_1}/Z$, with $Z = \sum_{E^1} e^{-\beta E_1}$. The exponential function $e^{-\beta E}$ is called Boltzmann factor, whereas $e^{\beta[\mu N_1 - E_1]}$ is called Gibbs factor.

In both the canonical and grand canonical distributions, the probability for the small system to have energy E_1 decreases exponentially as a function of E_1. This reflects the rapid decrease in the number of accessible states, and hence the entropy, for the combined system because energy is transferred from the large reservoir to the small system 1. To understand the dependence of the grand canonical distribution function on particle number N_1, it is necessary to determine the chemical potential μ for the system of interest. In general, regions of high particle density have a higher μ than regions of low particle density. This is determined by the convention adopted in the definition of μ, where a negative sign is introduced, with $\mu = -T(\partial S/\partial N)_{E_1,V}$. From this relation, and the definition of F, it is easily seen that $\mu = (\partial F/\partial N)_{E_1,V}$ gives a convenient alternative expression for determining μ. From Equation 5.17 we obtain, for a classical ideal gas, the result

$$\mu = k_B T \ln \left[(N/V)(4\pi mE/3Nh^2)^{-3/2} \right] = k_B T \left[\ln \left(V_Q/V_A \right) - C \right].$$

Because $V_Q < V_A$, it follows that μ is negative for a classical ideal gas. In Chapter 12, the chemical potential for ideal Bose and Fermi gases is discussed together with the evolution of μ toward the classical form with temperature. For a large classical gas system that is not in equilibrium, particles will transfer from regions of higher μ (less negative corresponding to smaller V_A, i.e., higher particle density) to regions of lower μ until equilibrium is achieved. Further discussion of the chemical potential is given later in this chapter. In the grand canonical distribution, the number of accessible states decreases rapidly as particles are transferred from the reservoir to the small system. To allow for a possible dependence of the energy states of system 1 on the number of particles in the system, it is appropriate to write $E_1 = E_1(N_1)$.

11.4 THE GRAND CANONICAL DISTRIBUTION APPLIED TO AN IDEAL GAS

The expressions given in Section 11.3 for the grand canonical distribution and the grand partition function (grand sum) are quite general, and it is helpful to consider a specific system to see how the summations are carried out. The eigenstates for an ideal gas are those for a particle in a box, as discussed in Section 4.3. Let the states be labeled by a set of quantum numbers denoted r. Because the particles do not interact, all particles have the same set of single particle states. The states may be occupied by zero, one, or more particles, depending on the quantum statistics that the particles obey.

For a noninteracting system of particles, the grand sum $\mathbb{Z}$ given by Equation 11.11 may be written as

$$\mathbb{Z} = \sum_{E_1,N_1} e^{\beta[\mu N_1 - E_1]} = \sum_{n_1,n_2,\cdots} e^{\beta[\mu n_1 - n_1\varepsilon_1 + \mu n_2 - n_2\varepsilon_2 + \cdots]}, \qquad (11.12)$$

where n_1 is the number of particles in state 1 with energy ε_1, n_2 is the number in state 2 with energy ε_2, and so on. The chemical potential μ is assumed constant with the system in equilibrium. Equation 11.12 may be rearranged as follows:

$$\mathbb{Z} = \sum_{n_1} e^{-\beta n_1(\varepsilon_1-\mu)} \sum_{n_2} e^{-\beta n_2(\varepsilon_2-\mu)} \cdots = \prod_r \left[\sum_{n_r} e^{-\beta n_r(\varepsilon_r-\mu)} \right]. \qquad (11.13)$$

Equation 11.13 shows that the grand sum can be written as a product of factors for each single particle state r. The values taken by n_r depend on whether the particles are fermions or bosons.

Evaluation of the summations for these cases is dealt with in Chapter 12. In the next section, we shall see that the mean values of physical quantities may be obtained from $\ln \mathbb{Z}$. For the ideal gas, we have from Equation 11.13 the useful result

$$\ln \mathbb{Z} = \sum_r \left[\ln \sum_{n_r} e^{-\beta n_r (\varepsilon_r - \mu)} \right]. \tag{11.14}$$

The quantum distribution functions are obtained in Chapter 12 with use of Equation 11.14.

11.5 MEAN VALUES

To calculate mean values from $\ln \mathbb{Z}$, we use the general form of Equation 11.11

$$\ln \mathbb{Z} = \ln \left[\sum_{E_1, N_1} e^{\beta[\mu N_1 - E_1]} \right].$$

The mean number of particles in system 1 is

$$N_1 = \sum_{E_1, N_1} N_1 P(E_1, N_1) = \frac{1}{\beta} \left(\frac{\partial \ln \mathbb{Z}}{\partial \mu} \right). \tag{11.15}$$

The form for $\ln \mathbb{Z}$ given in Equation 11.14 for an ideal gas leads to the same result $\langle N \rangle = 1/\beta (\partial \ln \mathbb{Z} / \partial \mu) = \sum_r \langle n_r \rangle$, where we can identify $\langle N \rangle$ with $\langle N_1 \rangle$.

Exercise 11.1

Use the expression for $\mathbb{Z}$ in Equation 11.11 to obtain the difference between the mean energy and the mean particle number multiplied by the chemical potential.

From Equation 11.11, we have $\ln \mathbb{Z} = \ln \left[\sum_{E,N} e^{\beta[\mu N - E]} \right]$, where the subscripts on N and E have been omitted to simplify expressions. On differentiation of $\ln \mathbb{Z}$ with respect to β and with a change in sign, we obtain the required result

$$-\frac{\partial \ln \mathbb{Z}}{\partial \beta} = E - N\mu. \tag{11.16}$$

In Chapter 10, we have shown that the mean energy may be obtained from Z using Equation 10.34 $\langle E \rangle = -\partial \ln Z / \partial \beta$. It is clearly desirable to establish a relationship between $\ln Z$ and $\ln \mathbb{Z}$, and this is done in the following section.

11.6 RELATIONSHIP BETWEEN THE PARTITION FUNCTION AND THE GRAND SUM

In the case of the canonical ensemble, energy fluctuations were shown to be of order $1/\sqrt{N}$. For the grand canonical ensemble, the fractional deviations in E_1 and N_1 for system 1 are expected to be

$$\frac{\sigma_{E_1}}{\langle E_1 \rangle} \simeq \frac{1}{\sqrt{N_1}} \tag{11.17}$$

and

$$\frac{\sigma_{N_1}}{\langle N_1 \rangle} \simeq \frac{1}{\sqrt{N_1}}, \tag{11.18}$$

where σ_{E_1} and σ_{N_1} are the standard deviations from the mean in energy and particle number, respectively. A detailed discussion of fluctuations is given in Chapter 12 for the quantum distribution functions. Because the fluctuations are very small, of the order of parts in 10^{11} for large $N \sim N_A$, the canonical ensemble and the grand canonical ensemble are, to an excellent approximation, equivalent, as pointed out in Section 10.8. This means that, even if the number of particles in a system is kept constant, it is permissible to use the grand canonical ensemble and grand sum in the calculation of quantities of interest. This is of considerable assistance, for example, in the derivation of the quantum distribution functions. Replacement of N_1 by $\bar{N}_1$ in Equation 11.11 with allowance for a sharply peaked distribution in N_1 gives, to a good approximation,

$$\mathbb{Z} \simeq \Delta N_1 e^{\beta \mu \bar{N}_1} \sum_{E_1} e^{-\beta E_1}, \tag{11.19}$$

where ΔN_1 is the width of the distribution. Figure 11.3 depicts the probability distribution in N_1, and the approximation we make in Equation 11.19 replaces the sharply peaked Gaussian distribution by a narrow rectangular distribution.

Taking logarithms of both sides of Equation 11.19 leads to

$$\ln \mathbb{Z} = \beta \mu \bar{N}_1 + \ln Z + \ln \Delta \bar{N}_1. \tag{11.20}$$

The term $\ln \Delta N_1$ is found to be much smaller than the other terms in Equation 11.20 and, as an excellent approximation, may be neglected. Equation 11.20 becomes

$$\ln Z \simeq \ln \mathbb{Z} - \beta \mu N, \tag{11.21}$$

where N_1 has been replaced by N, the fixed number of particles in the system of interest. Equation 11.21 shows that $\ln Z$ may be obtained from $\ln \mathbb{Z}$ by subtraction of $\beta \mu N$. $\mathbb{Z}$ is calculated without any constraint on particle number, which is an advantage that we make use of in the

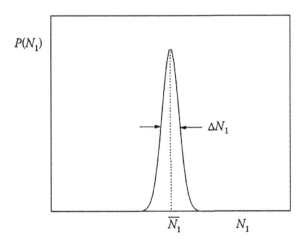

FIGURE 11.3 Probability distribution $P(N_1)$ in the grand canonical ensemble showing the sharp peak at $\bar{N}_1$. The width of the distribution in the figure has been greatly exaggerated for clarity.

derivation of the quantum distribution functions. The mean energy may be obtained from either $\ln Z$ or $\ln \mathbb{Z}$ as follows:

$$\langle E \rangle = -\frac{\partial \ln Z}{\partial \beta} = -\frac{\partial \ln \mathbb{Z}}{\partial \beta} + \mu N. \tag{11.22}$$

11.7 THE GRAND POTENTIAL

Two bridge relations have been established between the microscopic and macroscopic description of many particle systems. For the microcanonical ensemble, we have $S = k_B \ln \Omega$ (Equation 5.12) and, for the canonical ensemble, $F = -k_B T \ln Z$ (Equation 10.48). It is natural to ask whether a similar relation exists between the grand sum $\mathbb{Z}$ and a thermodynamic potential. The following relationship is found to hold:

$$\Omega_G = -k_B T \ln \mathbb{Z}, \tag{11.23}$$

where Ω_G is defined as the grand potential and is not to be confused with the number of accessible states Ω. To establish Equation 11.23, it is convenient to adopt an approach similar to that used in the derivation of Equation 10.48 in Section 10.6. We write $\mathbb{Z}$ in the form $\mathbb{Z} = \sum_{E,N} \Omega(E,N) e^{-\beta(E-\mu N)}$,

where $\Omega(E, N)$ is the number of accessible states for the system in the joint energy-particle number range E to $E + \delta E$ and N to $N + \delta N$. The product $\Omega(E, N) e^{-\beta(E-\mu N)}$ is sharply peaked around the most probable values $\bar{E}$ and $\bar{N}$ as shown in Figure 11.4.

The number of accessible states increases rapidly as a function of both E and N, whereas $e^{-\beta(E-\mu N)}$ decreases rapidly as a function of these quantities. (In examination of the exponential factor, it is helpful to bear in mind that $\mu = -1/\beta \, (\partial \ln \Omega / \partial N)_{E,V}$ and in the classical limit μ is negative.) To a good approximation, the grand sum may be written as

$$\mathbb{Z} \approx \Omega\left(\bar{E},\bar{N}\right) e^{-\beta(\bar{E}-\mu\bar{N})}\left(\frac{\Delta E}{\delta E}\right)\left(\frac{\Delta N}{\delta N}\right), \tag{11.24}$$

where ΔE and ΔN represent the half-height dimensions of the sharply peaked function.

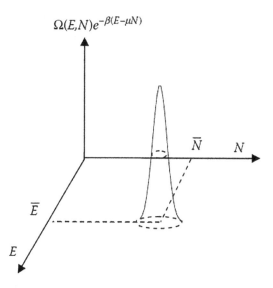

FIGURE 11.4 The product $\Omega(E, N) e^{-\beta(E-\mu N)}$ as a function of E and N shows a very sharp peak at $\bar{E}, \bar{N}$.

Taking logarithms of both sides of Equation 11.24 gives

$$\ln \mathbb{Z} = \ln \Omega\left(\overline{E},\overline{N}\right) - \beta\overline{E} + \beta\mu\overline{N} + \ln\left(\frac{\Delta E}{\delta E}\right) + \ln\left(\frac{\Delta N}{\delta N}\right). \tag{11.25}$$

On the basis of arguments similar to those used in Section 10.6, we neglect the last two terms in Equation 11.25 because they are very small compared with the other terms. For N large, it may be expected that fluctuations in particle number and energy will be very small, as shown in Equations 11.17 and 11.18. The most probable values $\overline{E}$ and $\overline{N}$ are replaced by E and N because these values are effectively fixed. From Equation 11.25, with $S = k_B \ln \Omega$, we get

$$-k_\beta T \ln\mathbb{Z} = E - TS - \mu N. \tag{11.26}$$

The grand potential is defined as

$$\Omega_G = E - TS - \mu N = F - \mu N, \tag{11.27}$$

and this, with Equation 11.26, gives

$$\Omega_G = -k_B T \ln \mathbb{Z}. \tag{11.28}$$

This is the third important bridge relationship between microscopic and macroscopic descriptions. The Gibbs potential is defined in Equation 7.3 as $G = E - TS + PV$, and in terms of the chemical potential for a single component system $G = \mu N$, as shown in Section 7.5. This result, with Equations 11.26 and 7.3, leads to $\Omega_G = F - G = -PV$ and, from Equation 11.28,

$$k_B T \ln \mathbb{Z} = PV. \tag{11.29}$$

Equation 11.29 is particularly useful in the determination of equations of state for single component systems for which $\mathbb{Z}$ can be calculated. Examples are given in later chapters.

Exercise 11.2

Show that the expression $k_B T \ln \mathbb{Z} = PV$ in Equation 11.29 can be derived from the Gibbs definition of entropy, given in Equation 10.56 as $S = -k_B \sum_r p_r \ln p_r$, together with the grand canonical distribution.

The probability of finding the system in state r, with energy E and particle number N, is $p_r = P(E, N) = e^{-\beta(E-\mu N)}/\mathbb{Z}$, and this gives $S = -(k_B/\mathbb{Z})\sum_{E,N}[-\beta(E-\mu N) - \ln\mathbb{Z}]e^{-\beta(E-\mu N)}$. It follows that the entropy becomes

$$S = 1/T(1/\mathbb{Z})\sum_{E,N}(E-\mu N)e^{-\beta(E-\mu N)} + (k_B/\mathbb{Z})(\ln \mathbb{Z})\sum_{E,N}e^{-\beta(E-\mu N)},$$

and this expression can be rewritten as

$$S = \frac{1}{T}(E - \mu N) + k_B \ln \mathbb{Z}, \tag{11.30}$$

where $\langle E \rangle$ and $\langle N \rangle$ are, respectively, the mean energy and the mean particle number for the system. From Equation 11.30, with replacement of $\langle E \rangle$ and $\langle N \rangle$ by fixed values E and N, because fluctuations are extremely small, we obtain

$$-k_B T \ln \mathbb{Z} = E - TS - \mu N = \Omega_G, \tag{11.31}$$

as required. Note that Ω_G, like the other thermodynamic potentials, is an extensive quantity.

Exercise 11.3

Show that the entropy S, the pressure P, and the particle number N may be obtained as partial derivatives of Ω_G.

The differential of Ω_G is

$$d\Omega_G = dE - T\ dS - S\ dT - \mu\ dN - N\ d\mu \tag{11.32}$$

The fundamental relation in the form given by Equation 3.26 for a single component system is $T\ dS = dE + P\ dV - \mu\ dN$ that, when combined with Equation 11.32, gives

$$d\Omega_G = -P\ dV - S\ dT - N\ d\mu. \tag{11.33}$$

It follows that

$$P = -\left(\frac{\partial \Omega_G}{\partial V}\right)_{T,\mu}, N = -\left(\frac{\partial \Omega_G}{\partial \mu}\right)_{V,T}, \text{ and } S = -\left(\frac{\partial \Omega_G}{\partial T}\right)_{V,\mu}.$$

These relationships show that thermodynamic quantities of interest may be obtained once an explicit form for $\Omega_G(V, T, \mu)$ has been obtained from the grand sum $\mathbb{Z}$. The procedure is analogous to obtaining thermodynamic quantities from $F(T, V)$ in the canonical ensemble approach, as mentioned in Chapter 10.

The discussion of the grand canonical distribution and the grand sum given in this chapter is rather formal, but the results obtained will prove extremely useful, for example, in deriving the quantum distributions. As pointed out previously in the application of these expressions to systems of delocalized particles, care must be taken to allow for the quantum statistics that the indistinguishable particles obey. In the next chapter, we allow for the fermion or boson nature of the particles.

We now have the basic expressions of equilibrium statistical mechanics that will permit us to obtain useful descriptions for a wide variety of systems. The microcanonical ensemble (Ω, S) and canonical ensemble (Z, F) approaches have allowed us to obtain results for an ideal spin system of localized, distinguishable spins. In Chapter 5, the microcanonical ensemble approach was used to treat the case of an ideal gas. In the calculation of $\Omega(S)$, and hence S, for an ideal gas of indistinguishable particles, we introduced a factor $1/N!$ to correct for overcounting of states. In Section 10.10, it is pointed out that the partition function obtained in the classical limit of both the Fermi–Dirac and the Bose–Einstein quantum distributions leads to the correct expression for the Helmholtz potential and hence the entropy without any need for a correction factor, as we shall find in Chapter 12.

PROBLEMS CHAPTER 11

11.1   Consider a system in thermal and diffusive contact with a reservoir at temperature T and with chemical potential μ. Show that the mean energy $\langle E \rangle$ and mean particle number $\langle N \rangle$ are related to the grand partition function $\mathbb{Z}$ by the expression $\langle E \rangle - \langle N \rangle \mu = -\partial \ln \mathbb{Z}/\partial \beta$. If particle exchange with the reservoir were prevented, how would this expression change?

11.2   Use the Boltzmann definition of the entropy S to show that $S = (1/T)\ [\langle E \rangle - \mu \langle E \rangle] + k_B \ln Z$.

11.3   Obtain an expression for the Helmholtz potential and hence the chemical potential of a localized ideal spin system at temperature T in a magnetic field of induction B. Obtain the form of the chemical potential in the high B/T and low B/T limits. Sketch the behavior of the chemical potential for the system as a function of B/T.

11.4   In Chapter 16, the following expression is obtained for the partition function for a classical monatomic ideal gas $Z = z^N/N!$ with the single particle partition function, z given by $z = V(mk_BT/2\pi\hbar^2) \approx (V/V_Q)$.

In Chapter 5, it is shown that the chemical potential for a monatomic ideal gas in the classical limit is $\mu = k_B T \left[\ln(N/V) - \frac{3}{2}\ln\left(4\pi mE/3Nh^2\right) \right] \approx k_B T \, \ln\left(V_Q/V_A\right)$. Use the approximate relationship between $\ln \mathbb{Z}$ and $\ln Z$ to show that $\ln \mathbb{Z} = N$ in the classical limit.

11.5   Use the result for the grand partition function obtained in Question 11.4 together with the grand potential $\Omega_G = -k_B T \ln \mathbb{Z}$ to obtain the ideal gas equation of state.

11.6   A classical gas of N particles is contained in a volume V. Show that the probability of n particles being in a small subvolume v of the gas is given by the Poisson distribution $P(n) = \bar{n}^n e^{-\bar{n}}/n!$, where $\bar{n} = Np = N\,(v/V)$ is the mean number of particles in the subvolume v.

11.7   For the situation in Question 11.6, give an expression for the fractional deviation in the number of particles in the subvolume v from the mean value. If the volume $V = 2\,\mathrm{L}$ and $N = 6 \times 10^{22}$ molecules, give an estimate of the volume ratio v/V and the linear dimensions of the subvolume for the fluctuations to be of the order of one part per million.

11.8   Two equal volumes designated 1 and 2, each of which contains a quantity of a classical monatomic gas at the same temperature, are connected by a valve that is closed. If the pressures are initially unequal with $P_1 = 2\,P_2$, establish a relationship between the chemical potentials μ_1 and μ_2. Give a qualitative description of the approach to equilibrium, in terms of changes in pressure and chemical potential, when the valve is opened.

11.9   For the situation described in Question 11.8, obtain expressions for the grand sum of the composite system before and after the valve is opened.

Section IIB

Quantum Distribution Functions, Fermi–Dirac and Bose–Einstein Statistics, Photons, and Phonons

12 The Quantum Distribution Functions

12.1 INTRODUCTION: FERMIONS AND BOSONS

In nature, two different classes of particles are found, called fermi particles, or fermions, and bose particles, or bosons, respectively. Fermions have half-integral spin angular momentum $\left(\frac{1}{2}\hbar, \frac{3}{2}\hbar, \cdots\right)$ and obey the Pauli exclusion principle. Examples are electrons, protons, neutrons, and ^{3}He atoms. Bosons have integral spin angular momentum $(0, \hbar, 2\hbar, \ldots)$ and do not obey the Pauli exclusion principle. Examples are photons, pions, deuterons, and ^{4}He atoms. It is extremely important in the development of theoretical expressions for many particle delocalized systems that the indistinguishability of particles, be they fermions or bosons, is properly taken into account.

Consider a pair of particles that may be either two fermions or two bosons. The Schrödinger equation for the system is written in terms of the Hamiltonian operator for the system as

$$\mathcal{H}(1,2)\chi(1,2) = E\chi(1,2),\tag{12.1}$$

where 1 represents the coordinates of particle 1 and 2 the coordinates of particle 2. The Hamiltonian (1,2) must be symmetric under particle interchange because of the indistinguishability of the particles, giving

$$\mathcal{H}(1,2) = \mathcal{H}(2,1)\tag{12.2}$$

The pair wave function $\chi(1, 2)$ may be either symmetric or antisymmetric. This can be seen by introducing the permutation operator P_{12}, which permutes the coordinates of the particles.

Application of P_{12} in succession to the pair wave function gives $P_{12}\chi(1, 2) = \lambda\chi(1, 2)$ and $P_{12}^2\chi(1, 2) = \lambda^2\chi(1, 2) = \chi(1, 2)$.

It follows that $\lambda^2 = 1$ and P_{12} has eigenvalues $\lambda = \pm 1$. Properly normalized symmetric and antisymmetric pair wave functions are

$$\chi_s = \frac{1}{\sqrt{2}}\left[\chi(1,2) + \chi(2,1)\right]\tag{12.3}$$

and

$$\chi_a = \frac{1}{\sqrt{2}}\left[\chi(1,2) - \chi(2,1)\right].\tag{12.4}$$

For N particles, the generalized symmetric and antisymmetric wave functions are given in terms of the permutation operator P_v, which permutes the coordinator of v particles simultaneously as $\chi_s = \left(1/\sqrt{N!}\right)\left[\sum_{v=1}^{N} P_v\chi(1,2,\cdots,N)\right]$ and $\chi_a = \left(1/\sqrt{N!}\right)\left[\sum_{v=1}^{N}(-1)^v P_v\chi(1,2,\cdots,N)\right]$. These general results are not required in our treatment of quantum statistics and are given here for completeness.

The Hamiltonian for two noninteracting particles may be separated as follows: $\mathcal{H}(1, 2) = \mathcal{H}(1) + \mathcal{H}(2)$. The particles have energy eigenvalues given by $\mathcal{H}(1)\,\phi_m(1) = E_m\,\phi_m(1)$ and $\mathcal{H}(2)\,\phi_n(2) = E_n\,\phi_n(2)$, where $\phi_m(1)$ and $\phi_n(2)$ are the single-particle wave functions. The symmetric and antisymmetric pair wave function may be written in terms of the products of single-particle functions as

$$\chi_{n,m} = \frac{1}{\sqrt{2}}\left[\phi_n(1)\phi_m(2) \pm \phi_m(1)\phi_n(2)\right]. \tag{12.5}$$

Operation on $\chi_{n,m}$ with the Hamiltonian $\mathcal{H}(1, 2)$ gives the eigenvalues $E_n + E_m$. In Equation 12.5, the plus sign applies to bosons and the minus sign to fermions. For fermions, $\chi_{n,m}$ vanishes for $m = n$, which is an expression of the Pauli principle that no two fermions can have identical quantum numbers.

For N noninteracting particles, the symmetric and antisymmetric wave functions are given in terms of single-particle wave functions by $\chi_s = \left[1/\sqrt{N!n_2!n_2!}\right]\sum_{v} P_v\phi_m(1)\phi_n(2)\cdots\phi_P(N)$, where n_1, n_2,... are the numbers of particles associated with a particular eigenvalue and $\chi_a = \left[1/\sqrt{N!}\right]\sum (-1)^v P_v\phi_m(1)\phi_n(2)\cdots\phi_P(N)$.

Before the development of quantum mechanics, the question of indistinguishability of particles was not properly taken into account. Classical ideas were used in what is called the Maxwell–Boltzmann (MB) statistics. In contrast to the symmetry requirements for systems of fermions and bosons, as outlined above, MB statistics ignores these requirements. At high temperatures and low densities, classical statistics works well because of the sparse occupation of accessible states, as discussed in Section 4.3.

The MB statistics is therefore an approximation to the true particle statistics and is applicable only in situations when most states are empty with some states occupied by a single particle and very few states, in the case of bosons, occupied by more than one particle. It follows that for systems in the classical limit, the statistics that the particles obey are no longer of crucial importance in our calculations.

Exercise 12.1

Two noninteracting particles have four accessible quantum states in which they may be found. Give the wave functions, assuming the particles obey (a) Fermi–Dirac (FD) statistics, (b) Bose–Einstein (BE) statistics, and (c) MB statistics.

  a. For fermions obeying FD statistics, the wave functions are antisymmetric. $\chi_a = \left(1/\sqrt{2}\right)\left[\phi_m(1)\phi_n(2) - \phi_n(1)\phi_m(2)\right]$ with $m \neq n$. Indistinguishable states are counted only once. Distinct states are $(m = 1; n = 2, 3, 4)$, $(m = 2; n = 3, 4)$, and $(m = 3; n = 4)$, giving a total of six wave functions for the six distinct states.
  b. For bosons obeying BE statistics, the wave functions are symmetric. $\chi_s = \left(1/\sqrt{2}\right)\left[\phi_m(1)\phi_n(2) + \phi_n(1)\phi_m(2)\right]$. The distinct states are $(m = 1; n = 1, 2, 3, 4)$, $(m = 2; n = 2, 3, 4)$, $(m = 3; n = 3, 4)$, and $(m = 4; n = 4)$, giving a total of 10 wave functions.
  c. For particles obeying MB statistics, the wave functions are $\chi = \phi_m(1)\phi_n(2)$, with $m = 1, 2, 3, 4$ and $n = 1, 2, 3, 4$, giving a total of 16 wave functions. The indistinguishability of particles is not taken into account.

It is of interest to consider the following ratio for the three cases, $r = $ (number of states with $m = n$)/(number of states with $m \neq n$). Calculation of the ratios gives (a) $r_{FD} = 0/6 = 0$, (b) $r_{BE} = 4/6 = 0.66$, and (c) $r_{MB} = 4/12 = 0.33$.

The ratio for BE statistics is enhanced compared with the MB ratio, whereas for the FD statistics, the ratio is zero because of the exclusion principle. The tendency of bosons to clump together in the same state is important particularly when the energy of the system is very low.

12.2   QUANTUM DISTRIBUTIONS

Consider a system of N noninteracting particles, which may be either fermions or bosons, in a container of fixed volume V and in thermal contact with a reservoir at temperature T. The particles possess single-particle states labeled r. To calculate quantities of interest, such as the mean number of particles in a given quantum state, we require the partition function evaluated in the canonical ensemble. From Equation 10.30,

$$Z = \sum_E e^{-\beta E} = \sum_{n_1, n_2, \cdots} e^{-\beta[n_1\varepsilon_1 + n_2\varepsilon_2 + \cdots]} \tag{12.6}$$

where the n_r are subject to the constraint $N = \sum_r n_r$. This constraint introduces a complication in the evaluation of Z. Previously, it has been noted that, for large systems, it does not matter which ensemble is chosen in the calculation of mean values because fluctuations in observable quantities are extremely small.

In the derivation of the quantum distribution functions, it is advantageous to work in the grand canonical ensemble and to calculate the grand sum $\mathbb{Z}$ because this avoids the constraints on N. From Equation 11.14, $\ln \mathbb{Z} = \sum_r \left[\ln \sum_{n_r} e^{-\beta n_r (\varepsilon_r - \mu)} \right]$, where μ is the chemical potential. Evaluation of $\ln \mathbb{Z}$ may be carried out immediately for FD statistics, where $n_r = 0$ or 1 only for all r, giving

$$\ln \mathbb{Z}_{\mathrm{FD}} = \sum_r \ln \left[1 + e^{\beta(\mu - \varepsilon_r)} \right]. \tag{12.7}$$

For BE statistics, $n_r = 0$, 1, ..., ∞ for all r, giving $\ln \mathbb{Z}_{\mathrm{BE}} = \sum_r \left[\ln \sum_{n_r = 0}^{\infty} e^{-\beta n_r (\varepsilon_r - \mu)} \right]$.

The sum may be evaluated as a geometric series, as discussed in Appendix A, with the ratio of successive terms simply $e^{-\beta(\varepsilon_r - \mu)}$, so that $\sum_{n_r = 0}^{\infty} e^{-\beta n_r (\varepsilon_r - \mu)} = 1 + e^{-\beta(\varepsilon_r - \mu)} + e^{-2\beta(\varepsilon_r - \mu)} + \cdots \simeq \left[1/\left(1 - e^{-\beta(\varepsilon_r - \mu)} \right) \right]$.

This gives for the BE case

$$\ln \mathbb{Z}_{\mathrm{BE}} = -\sum_r \ln \left[1 - e^{\beta(\mu - \varepsilon_r)} \right]. \tag{12.8}$$

The grand sum may therefore be written inclusively as

$$\ln \mathbb{Z}_{\mathrm{BE}} = \pm \sum_r \ln \left[1 \pm e^{\beta(\mu - \varepsilon_r)} \right], \tag{12.9}$$

where the plus signs apply for the FD statistics and minus signs for the BE statistics. The mean number of particles in state r is given by $\langle n_r \rangle = (1/\mathbb{Z}) \sum_{n_r} n_r e^{\beta n_r (\mu - \varepsilon_r)}$. With the values of β and μ fixed by the reservoirs with which the systems in the ensemble are in contact, we obtain the convenient form

$$\langle n_r \rangle = -\frac{1}{\beta} \frac{\partial \ln \mathbb{Z}}{\partial \varepsilon_r}. \tag{12.10}$$

Equation 12.10 with $\ln \mathbb{Z}$ from Equation 12.9 gives the important result for fermions and bosons

$$\langle n_r \rangle = -\frac{1}{e^{\beta(\varepsilon_r - \mu)} \pm 1}, \tag{12.11}$$

where the plus sign again applies to FD statistics and the minus sign to BE statistics. The change of sign in the denominator provides a crucial distinction between the two distribution functions. A further very important difference involves the behavior of the chemical potential μ for the different particle statistics.

12.3   THE FD DISTRIBUTION

For the FD statistics, Equation 12.11 gives the FD distribution

$$\langle n_r \rangle = -\frac{1}{e^{\beta(\varepsilon_r - \mu)} + 1}, \tag{12.12}$$

where the chemical potential μ is a function of temperature. At low temperatures, as we shall see, μ is positive and large compared with $1/\beta = k_B T$. From Equation 12.12, a number of special values for $\langle n_r \rangle$ apply, dependent on the value of ε_r in relation to μ. For $\varepsilon_r \ll \mu$, we see that $\langle n_r \rangle \simeq 1$, whereas for $\varepsilon_r = \mu$, $\langle n_r \rangle = 0.5$ and for $\varepsilon_r \gg \mu$, $\langle n_r \rangle = 0$. Figure 12.1 shows the form of the FD distribution for $T = 0$ K and for $T > 0$ K, with $k_B T \ll \mu$.

The distribution shows that at zero temperature, all states up to a maximum energy, called the Fermi energy ε_F, which is the special name given to μ at 0 K, are filled with a single fermion and all higher states are empty. This distribution is required by the Pauli exclusion principle. At finite temperatures, states near ε_F are only partly occupied on average because thermal excitation transfers some fermions with $\varepsilon \leq \varepsilon_F$ to states with $\varepsilon > \varepsilon_F$. Figure 12.2 shows an energy-level diagram depicting the finite temperature situation.

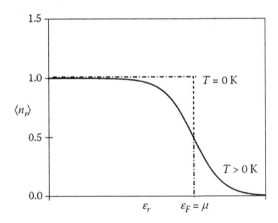

FIGURE 12.1   The FD distribution function for $T = 0$ K (dashed line) and for $T > 0$ K (full line). At 0 K, $\langle n_r \rangle$ drops abruptly from 1 to 0 for $\varepsilon_r = \mu = \varepsilon_F$, whereas for $T > 0$ K, the distribution decreases from 1 to 0 smoothly over a range of energy $\sim k_B T$. The Fermi energy ε_F is the value of μ at $T = 0$ K.

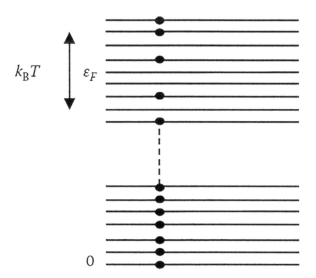

FIGURE 12.2   The diagram shows the energy-level occupation diagram for a system of noninteracting fermions at a finite temperature. Full occupation of the low-energy states occurs with the partial occupation of states for energies close to the Fermi energy. For $T > 0$ K, thermal excitation leads to the transfer of fermions to states above ε_F.

At high temperatures, the chemical potential decreases and the distribution changes its form dramatically. This behavior is discussed in Chapter 13.

12.4   THE BE DISTRIBUTION

The BE distribution obtained from Equation 12.11 has the form $\langle n_r \rangle = 1/\left[e^{\beta(\varepsilon_r - \mu)} - 1 \right]$.

In contrast to the FD case, where μ is large and positive at low temperatures, the chemical potential is always negative or at best close to zero (i.e., $\mu \leq 0$) for boson systems. μ is, in general, a function of T and becomes very small as $T \to 0$ K. This has important consequences that are discussed in Chapter 14. In particular, the BE condensation phenomenon is examined. As discussed above, for bosons, more than one particle may occupy the same single-particle state r, with $\langle n_r \rangle > 1$. Figure 12.3 gives a representation of the BE distribution for some finite temperature.

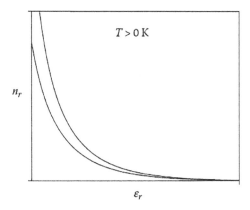

FIGURE 12.3   The BE distribution function representation for two temperatures $T > 0$ K. The ground state ($\varepsilon = 0$) is excluded.

In discussing the quantum distribution functions, we have for convenience chosen the lowest state to have energy zero. Introducing a small finite energy Δ for the ground state simply changes the values taken by the chemical potential slightly. For the BE case, for example, we would have $\mu \to \Delta$ rather than $\mu \to 0$ as the temperature tends to 0 K.

12.5   FLUCTUATIONS

In Section 11.5, it is argued that fluctuations in energy and particle number are extremely small in the grand canonical ensemble. We now examine particle number fluctuations for ideal Fermi and Bose gases. For a noninteracting gas of fermions or bosons, it is sufficient to obtain an expression for the fluctuations in the mean particle number $\langle n_r \rangle$ because $\langle E \rangle = \sum_r \langle n_r \rangle \varepsilon_r$ and energy fluctuations are linked to particle fluctuations. As discussed in Chapter 10, the dispersion in n_r is given by $\sigma_{n_r^2} = \langle n_r^2 \rangle - \langle n_r \rangle^2$, where in the grand canonical ensemble the mean values are given by

$$\langle n_r \rangle = \frac{1}{\mathbb{Z}} \sum_r n_r e^{\beta n_r (\mu - \varepsilon_r)} = -\frac{1}{\beta} \frac{\partial \ln \mathbb{Z}}{\partial \varepsilon_r} = \frac{1}{e^{\beta(\varepsilon_r - \mu)} \pm 1}, \tag{12.13}$$

and

$$\langle n_r^2 \rangle = \left(\frac{1}{\beta^2 \mathbb{Z}} \right) \left(\frac{\partial^2 \mathbb{Z}}{\partial \varepsilon_r^2} \right) = \left(\frac{1}{\beta^2} \right) \left[\frac{\partial^2 \ln \mathbb{Z}}{\partial \varepsilon_r^2} + \left(\frac{\partial \ln \mathbb{Z}}{\partial \varepsilon_r} \right)^2 \right]. \tag{12.14}$$

(Use has been made of the identity $\partial^2 \ln \mathbb{Z} / \partial \varepsilon_r^2 = (1/\mathbb{Z})(\partial^2 \mathbb{Z} / \partial \varepsilon_r^2) - (1/\mathbb{Z}^2)(\partial \mathbb{Z} / \partial \varepsilon_r)^2$.)

The above expressions give $\sigma_{n_r}^2 = (1/\beta^2)(\partial^2 \ln \mathbb{Z} / \partial \varepsilon_r^2)$ and, with Equation 12.10, we obtain

$$\sigma_{n_r}^2 = -\left(\frac{1}{\beta} \right) \frac{\partial \langle n_r \rangle}{\partial \varepsilon_r} = \left(\frac{1}{\beta} \right) \left[\beta \left(1 - \frac{\partial \mu}{\partial \varepsilon_r} \right) e^{\beta(\mu - \varepsilon_r)} \right] \left[e^{\beta(\varepsilon_r - \mu)} \pm 1 \right]^{-2}. \tag{12.15}$$

If it is assumed that $\partial \mu / \partial \varepsilon_r \simeq 0$, generally a good approximation, it follows with the use of Equations 12.13 and 12.15 that

$$\sigma_{n_r}^2 = \frac{\left[e^{\beta(\mu - \varepsilon_r)} \pm 1 \right]}{\left[e^{\beta(\mu - \varepsilon_r)} \pm 1 \right]^2} \mp \frac{1}{\left[e^{\beta(\mu - \varepsilon_r)} \pm 1 \right]^2} = \langle n_r \rangle \mp \langle n_r^2 \rangle \quad \text{or} \quad \frac{\sigma_{n_r}^2}{n_r^2} \approx \left[\frac{1}{\langle n_r \rangle} \mp 1 \right], \tag{12.16}$$

where the minus sign applies to fermions and the plus sign to bosons. For fermion systems, we have seen that, for comparatively low T, $\langle n_r \rangle \approx 1$, except for states with energies close to μ. Equation 12.16 shows that, for fermions, the fluctuations are vanishingly small for most states, as might be expected. It is only for states near the Fermi energy that fluctuations are significant.

For bosons at low temperatures, $\langle n_r \rangle > 1$ for the lowest energy states and fluctuations become rather large, with the ratio in Equation 12.16 approaching unity. The large fluctuations in boson occupation of states are associated with the tendency of bosons to clump in the same state as mentioned in Section 12.1. At very high temperatures, when $\langle n_r \rangle \ll 1$ for both fermion and boson systems, the classical limit is obtained with $\sigma_{n_r}^2 / \langle n_r^2 \rangle \simeq 1/\langle n_r \rangle$. Although fluctuations in the particle number, for a *particular state*, can be very large for bosons, it is important to note that fluctuations in the total number of particles in the system remain very small. For a complete system of particles, we have $\sigma_N^2 / \langle N \rangle^2 - \langle N \rangle^2$, with $\langle N \rangle = \sum_r \langle n_r \rangle$, and $\langle N \rangle^2 \simeq \sum_r \langle n_r^2 \rangle$, because the probabilities of occupation of different states are statistically independent. In the classical limit, it follows that $\sigma_N^2/\langle N \rangle^2 = 1/\langle N \rangle$ and hence

$$\frac{\sigma_N}{\langle N \rangle} = \frac{1}{\sqrt{\langle N \rangle}}. \tag{12.17}$$

Equation 12.17 shows that, in the classical limit, fluctuations in the total particle number are very small for large systems. For an FD system at low temperatures, fluctuations in N will be extremely small because fluctuations are vanishingly small for each state of the system as shown by Equation 12.16, with $\langle n_r \rangle \approx 1$. Fluctuations in N for the BE systems at low temperatures are also small when averaging is carried out over a large number of states. A case of interest arises for bosons when $T \rightarrow 0$ K and the ground-state population becomes comparable to N. It is necessary to allow for $\partial \mu / \partial \varepsilon_r \neq 0$ in a description of this behavior in the low T limit. The BE condensation phenomenon, which occurs at very low temperatures, is discussed in Chapter 14.

12.6   THE CLASSICAL LIMIT

As shown above, the FD and the BE distributions are given by $\langle n_r \rangle = 1/e^{\beta(\varepsilon_r - \mu)} \pm 1$ (Equation 12.11), with the plus sign for FD statistics and the minus sign for BE statistics. The chemical potential μ has very different values for the Bose and Fermi systems. For a fixed number of particles N, the condition $\sum_r \langle n_r \rangle = N$ is used to determine μ. This is discussed in detail in Chapters 13 and 14. Figure 12.4 gives a schematic representation of the behavior of μ with temperature for the BE and FD systems.

For the FD statistics, μ is positive at low temperatures, decreases with an increase in temperature, and becomes negative at sufficiently high temperatures. For the BE statistics, μ is always negative but approaches zero as $T \rightarrow 0$ K. This behavior ensures that $\langle n_r \rangle$ is positive for all values of ε_r, including $\varepsilon_r = 0$. The values of μ for BE and FD statistics converge at high temperatures. This is indicative of an approach to the classical limit value given in Equation 5.17.

It is convenient to introduce an exponential function of μ defined as $\lambda = e^{\beta \mu}$ and termed the fugacity. The quantum distributions may then conveniently be written as

$$\langle n_r \rangle = \frac{1}{\lambda^{-1} e^{\beta \varepsilon_r} \pm 1}. \tag{12.18}$$

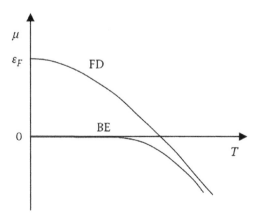

FIGURE 12.4   Schematic representation of the temperature dependence of the chemical potential μ for FD and BE statistics. At high T in the classical limit, the μ values converge.

In the classical limit, when μ is large and negative, $\lambda^{-1} \gg 1$ and $\langle n_r \rangle \ll 1$. It is reasonable to conclude that the ± 1 in the denominator is negligible compared with $\lambda^{-1}e^{\beta \epsilon r}$ in this limit, and Equation 12.18 becomes

$$\langle n_r \rangle = \lambda e^{-\beta \epsilon_r}. \tag{12.19}$$

The condition $\sum_r \langle n_r \rangle = N$ gives

$$\lambda = \frac{N}{\sum_r e^{-\beta \epsilon_r}}, \tag{12.20}$$

and on insertion of λ from Equation 12.20 into Equation 12.19, we recover the familiar canonical distribution form discussed in Chapter 10:

$$\langle n_r \rangle = \frac{Ne^{-\beta \epsilon_r}}{\sum_r e^{-\beta \epsilon_r}}. \tag{12.21}$$

The probability of occupation of state r is

$$P_r = \frac{\langle n_r \rangle}{N} = \frac{e^{-\beta \epsilon_r}}{\sum_r e^{-\beta \epsilon_r}}, \tag{12.22}$$

which is the canonical ensemble or Boltzmann probability given in Equation 10.29.

It is a simple matter to obtain an expression for $\ln Z$ in the classical limit from Equation 11.21, $\ln Z = \ln \mathbb{Z} - \beta\mu N$. On substitution for $\ln \mathbb{Z}$, from Equation 12.9, we obtain

$$\ln Z = \pm \sum_r \ln \left[1 \pm \lambda e^{-\beta \epsilon_r} \right] - \beta\mu N. \tag{12.23}$$

In the classical limit, $\lambda \ll 1$, and the logarithm on the right-hand side of Equation 12.23 may be expanded to give $\ln Z = \sum_r \lambda e^{-\beta \epsilon_r} - N \ln \lambda$.

Use of Equation 12.20 for λ leads to $\ln Z = N - N \ln N + N \ln \sum_r e^{-\beta \epsilon_r}$. Taking antilogarithms and with Stirling's approximation in inverse form ($N \ln N - N = \ln N!$), we obtain the partition function for N particles in an ideal gas:

$$Z = \frac{1}{N!} \left[\sum_r e^{\beta \epsilon_r} \right]^N = \frac{z^N}{N!}, \tag{12.24}$$

with z as the single-particle partition function. In Chapter 10, it was shown that for localized, distinguishable particles, $Z = z^N$. The additional factor $1/N!$ in Equation 12.24 takes account of the indistinguishability of particles in a gas and avoids overcounting of states. By taking the classical limit of the quantum distributions, the factor $1/N!$ is introduced quite naturally and does not have to be inserted in an *ad hoc* fashion, as was done, for example, in Section 4.5(b) in the (microcanonical ensemble) treatment of the accessible states for an ideal gas.

12.7   THE EQUATION OF STATE

The relationship (Equation 11.29) that connects the logarithm of the grand sum to the product PV, together with Equation 12.9, gives $PV = k_B T \ln \mathbb{Z} = \pm k_B T \sum_r \ln \left[1 \pm \lambda e^{-\beta \varepsilon_r} \right]$. If the sum over states is converted to an integral, by introducing the density of single-particle states, we obtain $PV = \pm 1/\beta \int_0^\infty \rho(\varepsilon) \ln \left[1 \pm \lambda e^{-\beta \varepsilon} \right] d\varepsilon$, and with $\rho(\varepsilon)$ from Equation 4.14, this leads to

$$PV = \pm \left(\frac{V}{4\pi^2 \beta} \right) \left(\frac{2m}{\hbar^2} \right)^{3/2} \int_0^\infty \varepsilon^{\frac{1}{2}} \ln \left[1 \pm \lambda e^{-\beta \varepsilon} \right] d\varepsilon. \tag{12.25}$$

Integration by parts with $\int_0^\infty [u \, dv = uv]_0^\infty - \int_0^\infty v \, du$, where $u = \ln [1 \pm \lambda e^{-\beta \varepsilon}]$ and $v = \frac{2}{3} \varepsilon^{2/3}$, leads to

$$PV = \frac{2}{3} \frac{V}{4\pi^2} \left(\frac{2m}{\hbar^2} \right)^{3/2} \int_0^\infty \frac{\varepsilon^{3/2} \, d\varepsilon}{\lambda^{-1} e^{\beta \varepsilon} \pm 1}. \tag{12.26}$$

It follows that

$$PV = \frac{2}{3} \langle E \rangle \tag{12.27}$$

because the mean energy of a quantum gas is given by

$$E = \int_0^\infty \varepsilon \rho(\varepsilon) \left(\frac{1}{\lambda^{-1} e^{\beta \varepsilon} \pm 1} \right) d\varepsilon = \frac{V}{4\pi^2} \left(\frac{2m}{\hbar^2} \right)^{3/2} \int_0^\infty \frac{\varepsilon^{3/2} \, d\varepsilon}{\lambda^{-1} e^{\beta \varepsilon} \pm 1}. \tag{12.28}$$

Equation 12.27 is a very simple result and shows that the pressure P is given by $\frac{2}{3} \langle E \rangle / V$, where $\langle E \rangle / V$ is the mean energy density in the quantum gas. $\langle E \rangle$ is a function of temperature and may be determined using Equation 12.28. Although the integral in Equation 12.28 is usually not simple to evaluate, it is possible to obtain the form of the equation of state in a fairly straightforward way for $\lambda < 1$ at high T.

Exercise 12.2

Show that for the Fermi and Bose gases at high temperatures and low densities, the equation of state takes the form $PV = Nk_B T \left[1 \mp (N/16V)(3/\pi)^{3/2} \left(h^2/3mk_B T \right)^{3/2} \right]$, where the minus sign is for bosons and the plus sign for fermions.

We change the variable in Equation 12.28 to $x = \beta \varepsilon$ and rearrange, with the use of Equation 12.27, to get $P = (1/6\pi)^2 \left(2m/\hbar^2 \right)^{3/2} \left(\lambda/\beta^{5/2} \right) \int_0^\infty \left[x^{3/2} e^{-x} dx / \left(1 \pm \lambda e^{-x} \right) \right]$. Expansion of the denominator with the aid of the binomial theorem for $\lambda < 1$ leads to

$$P \approx \frac{1}{6\pi^2} \left(\frac{2m}{\hbar^2} \right)^{3/2} \frac{\lambda}{\beta^{5/2}} \int_0^\infty x^{3/2} e^{-x} \left(1 \mp \lambda e^{-x} \right). \tag{12.29}$$

The resultant integrals have the form

$$\int_0^\infty x^n e^{-ax}\,\mathrm{d}x \frac{\Gamma(n+1)}{a^{n+1}}, \tag{12.30}$$

where the gamma function is defined as

$$\Gamma(n) = \int_0^\infty x^{n-1} e^{-x}\,\mathrm{d}x\,(n>0). \tag{12.31}$$

From a table of gamma functions that is given in Chapter 14, $\Gamma\left(\dfrac{3}{2}\right)=\left(\dfrac{1}{2}\right)\sqrt{\pi}$ and $\Gamma\left(\dfrac{5}{2}\right)=\dfrac{3}{4}\sqrt{\pi}$. The integral in Equation 12.29 may be evaluated in terms of $\Gamma\left(\dfrac{5}{2}\right)$, and we obtain

$$P \approx \frac{1}{6\pi^2}\left(\frac{2m}{\hbar^2}\right)^{3/2}\frac{\lambda}{\beta^{5/2}}\left[\frac{3\sqrt{\pi}}{4}\left(1\mp\frac{\lambda}{2^{5/2}}\right)\right]. \tag{12.32}$$

An expression for λ in the classical limit may be obtained from Equation 12.20 by conversion of the sum to an integral, that is, $\lambda\displaystyle\int_0^\infty \rho(\varepsilon)e^{-\beta\varepsilon}\,\mathrm{d}\varepsilon = N$. With the density of states from Equation 4.14 and change of variable to $x=\beta\varepsilon$, this becomes $\lambda\left[V/4\pi^2\left(2m/\hbar^2\right)^{3/2}1/\beta^{3/2}\displaystyle\int_0^\infty x^{1/2}e^{-x}\,\mathrm{d}x = N\right]$. Evaluation of the integral as $\Gamma\left(\dfrac{3}{2}\right)$ results in

$$\lambda = \left(\frac{2\pi\hbar^2}{m}\right)^{3/2}\left(\frac{N}{V}\right)\beta^{3/2}. \tag{12.33}$$

Substitution for λ in Equation 12.32 gives finally

$$PV = Nk_\mathrm{B}T\left[1\mp\frac{N}{16V}\left(\frac{3}{\pi}\right)^{3/2}\left(\frac{h^2}{3mk_\mathrm{B}T}\right)^{3/2}\right], \tag{12.34}$$

where the minus sign is for the BE statistics and the plus sign for the FD statistics. This is the required equation of state and has the form of the ideal gas equation, with correction terms due to quantum statistics. For fermions, the pressure is higher than for a classical gas. This is linked to the exclusion principle for fermions. For bosons, the pressure is reduced compared with that of classical particles.

It is useful to write Equation 12.34 in terms of the quantum volume on the basis of the thermal de Broglie wavelength $\lambda_T = 3mk_\mathrm{B}T$, introduced in Chapter 4, with $\lambda_\tau = h/\langle p\rangle$ and $P \approx 3mk_\mathrm{B}T$, from the equipartition theorem. In terms of the quantum volume $V_\mathrm{Q} = \lambda_T^3$ and the volume per particle $V_A = V/N$, Equation 12.34 becomes

$$PV = Nk_\mathrm{B}T\left[1\pm\alpha\left(\frac{V_\mathrm{Q}}{V_\mathrm{A}}\right)\right] \tag{12.35}$$

where α is a coefficient less than unity. The correction term becomes unimportant when $V_Q \ll V_A$ as discussed in Chapter 4 in our estimate of the condition for the classical approximation to hold. It is gratifying to obtain this result using a detailed quantum statistics approach.

From Equation 12.33, we can obtain an expression for μ in the classical limit. We have $\lambda = e^{\beta\mu}$,

and therefore, $\mu = k_B T \ln \lambda = k_B T \left[\ln (N/V) \lambda_T^3 - \frac{3}{2} \ln (2\pi/3) \right] = k_B T \left[\ln V_Q / V_A - \frac{3}{2} \ln(2\pi/3) \right]$,

where, as given above, $V_A = V/N$ and the quantum volume is $V_Q = (h / \langle p \rangle)^3$. This approximate expression for μ is almost the same as that obtained for the ideal gas in Chapter 5, differing only slightly in the numerical constant. For $V_Q < V_A$, the chemical potential is negative as expected.

The quantum distribution functions derived in this chapter are extremely useful in many areas of physics whenever what are called degenerate Fermi or Bose fluids are considered. Degenerate in this context means that the low-lying quantum states are heavily populated. Chapters 13 and 14 make use of the quantum distributions to predict the thermodynamic properties of ideal Fermi and Bose gases, respectively. Chapter 15 is concerned with photons in an electromagnetic cavity and phonons in solids. Photons and phonons obey the Planck distribution, which is similar to the BE distribution but with the chemical potential $\mu = 0$. Many of the results obtained in the following three chapters are compared with experimental observations for a number of systems that obey quantum statistics.

PROBLEMS CHAPTER 12

12.1   Consider three noninteracting particles, each of which has three accessible quantum states. Determine the possible states for this system if the particles are (a) fermions, (b) bosons, or (c) obey MB statistics.

12.2   A small system in contact with a heat bath at temperature T consists of two noninteracting particles, each of which has three quantum states with energies 0, ε, and 3ε. List the possible states of the small system, and give expressions for the partition function for the following situations: (a) the particles obey classical MB statistics, (b) the particles obey BE statistics, and (c) the particles obey FD statistics.

12.3   Obtain the grand potential for both FD gas and BE gas with the aid of the grand sum expression given in Equation 12.9. Use the grand potential to derive expressions for the entropy for the two quantum gases. Express your result in terms of the mean occupancy of the quantum states.

12.4   Consider a system of N noninteracting particles at temperature T in a container of volume V. If each particle has energy states $\varepsilon_n = n\varepsilon_0$ where $n = 0, 1, 2, 3, \ldots$, obtain expressions for the entropy per particle for distinguishable particles that obey MB statistics.

12.5   Calculate the thermal de Broglie wavelengths for helium-4 (boson) and helium-3 (fermion) gases at $T = 5$ K. Find the particle densities in these two cases for which quantum statistics must be used in considering properties of interest. Compare the pressures of the two gases at these calculated particle densities for equal-volume containers at 2 K.

12.6   Consider a Fermi gas at a finite temperature, and show that the high-energy tail of the distribution may be approximated by the Boltzmann distribution.

12.7   A two-dimensional film of particles has an area A. Assuming that the film can be treated as a two-dimensional quantum gas, obtain an expression for the area A times the force per unit length F exerted on a boundary. Make use of the two-dimensional density of states expression $\rho(\varepsilon)\, d\varepsilon = (Am/2\pi\hbar^2)\, d\varepsilon$. Compare and contrast the two-dimensional result with the three-dimensional expression as given by Equation 12.27.

13 Ideal Fermi Gas

13.1 INTRODUCTION

Armed with the Fermi distribution function and the density of states expression for particles in a box from Chapter 4, we are now for a position to give a detailed account of the properties of a non-interacting ideal Fermi gas. In particular, we shall focus on the degenerate Fermi gas for which it is predominantly the lowest energy levels that are occupied by fermions. The Fermi energy is obtained in terms of fundamental constants and the two-thirds power of the fermion density. Expressions for the specific heat, the magnetic susceptibility, and the pressure of a Fermi gas are obtained and discussed. The results are used to explain the observed properties of a number of systems that range from metals to stars.

13.2 THE FERMI ENERGY

Although an ideal Fermi gas made up of noninteracting fermions may appear to be an extreme idealization, there are a number of systems to which the theoretical results apply as a good approximation under certain conditions. Examples are electrons in metals, white dwarf stars, and liquid ^{3}He. The main reason why the theory provides useful predictions in these cases is that the fermion energies at the top of the Fermi–Dirac (FD) distribution are very large compared with other energies, specifically compared with the thermal energy $k_B T$. A numerical estimate of the Fermi energy for a representative fermion density is given below.

To obtain an expression for the Fermi energy ε_F, defined in Section 12.3, consider a system of N fermions of spin $\frac{1}{2}$ and mass m in a container of volume V at $T = 0$ K. In zero-applied magnetic field, the $m_s = \pm\frac{1}{2}$ spin states are degenerate, and it is convenient to introduce a degeneracy factor in expressions that involve the density of single-particle states. The fermions fill the lowest energy states up to the energy $\varepsilon_F = \mu$. By equating the number of particles to the number of states in the range from 0 to ε_F, an expression for ε_F is obtained. In general, for fermions, $N = \sum_r n_r = \sum_r 1/\left(e^{\beta(\varepsilon-\mu)} + 1\right)$. Conversion of the sum to an integral gives

$$N = 2\int_0^\infty \rho(\varepsilon)f(\varepsilon)\,d\varepsilon, \tag{13.1}$$

where $\rho(\varepsilon)$ is the density of states and the factor 2 is the spin degeneracy factor. $f(\varepsilon) = 1/(e^{\beta(\varepsilon-\mu)} + 1)$ is the Fermi function, which at $T = 0$ K has the form $f(\varepsilon) = 1$ for $\varepsilon \le \varepsilon_F$ and $f(\varepsilon) = 0$ for $\varepsilon > \varepsilon_F$. This may be seen in Figure 12.1. At $T = 0$ K, Equation 13.1 may therefore be written as

$$N = 2\int_0^{\varepsilon_F} \rho(\varepsilon)\,d\varepsilon. \tag{13.2}$$

From Equation 4.14, the density of single-particle states for particles in a box is $\rho(\varepsilon) = (V/4\pi^2)(2m/\hbar^2)^{3/2}\,\varepsilon^{1/2}$, and substitution in Equation 13.2 gives

$$N = \frac{V}{2\pi^2}\left(\frac{2m}{\hbar^2}\right)^{3/2}\int_0^{\varepsilon_F}\varepsilon^{1/2}\,d\varepsilon = \left(\frac{V}{3\pi^2}\right)\left(\frac{2m}{\hbar^2}\right)^{3/2}\varepsilon_F^{3/2}$$

and hence

$$\varepsilon_{\mathrm{F}} = \left(\frac{\hbar^2}{2m}\right)\left(\frac{3\pi^2 N}{V}\right)^{2/3}. \tag{13.3}$$

The Fermi energy is seen to depend on the two-thirds power of the particle density per unit volume.

Exercise 13.1

Estimate the Fermi energy for a metal in which the electron number density is $10^{23}\,\mathrm{cm}^{-3}$ ($10^{29}\,\mathrm{m}^{-3}$). Comment on the value obtained.

With the use of Equation 13.3, we obtain the Fermi energy as $\varepsilon_\mathrm{F} = 1.26 \times 10^{-18}\mathrm{J} = 7.9\mathrm{eV}$. This energy is comparable with the binding energy of electrons in atoms and much higher than typical thermal energies.

At a temperature $T = 100$ K, the *thermal* energy is $k_\mathrm{B}T = 1.38 \times 10^{-21}$ J $= 8.6\mathrm{meV}$. This is a thousand times smaller than the Fermi energy of the electron gas considered above. If we define the Fermi temperature as $T_\mathrm{f} = \varepsilon_\mathrm{F}/k_\mathrm{B}$, then for $\varepsilon_\mathrm{F} = 7.9\mathrm{eV}$, the Fermi temperature is $T_\mathrm{F} = 9.2 \times 10^4$ K.

Although it may seem extremely difficult to achieve electron densities of the order of $10^{23}\,\mathrm{cm}^{-3}$, nature has provided systems where such densities are found. Examples are metals where the ion cores give rise to screening effects and electron–electron interactions are comparatively small on the scale of the Fermi energy. It will be seen later that the ideal Fermi gas model provides a reasonably good prediction of the electronic contributions to the thermal properties of metals. Because the Fermi energy is so high, the Fermi distribution in metals is not greatly changed at ambient temperatures of 300 K compared with the distribution at $T = 0$ K. A few electrons at the top of the distribution are excited to higher states, as shown in Figure 12.2.

13.3   FERMI SPHERE IN MOMENTUM SPACE

It is instructive to use a momentum space, or k-space, representation in discussing a Fermi gas. In the derivation of the expression for the density of states $\rho(\varepsilon)$ given in Equation 4.14, use was made of quantum number space for the particle in a box situation. There is a correspondence between quantum number space and k-space, although it must be borne in mind that quantum numbers take positive values only, whereas k can take positive or negative values. The momentum p of a particle may be written in terms of its wave vector k as $p = \hbar k$. For a particle in a three-dimensional box, the wave function has the form

$$\chi(x,y,z) = A\, e^{i(k_x x + k_y y + k_z z)}. \tag{13.4}$$

Imposing boundary conditions such that the wave function has nodes, or, more generally, the same amplitude at the walls of the box, it is easily seen that

$$k_i = \left(\frac{2\pi}{L_i}\right)n_i \quad \text{with } i = x,y,z; \quad n_i = 1,2,3,\ldots \tag{13.5}$$

Consider a small volume in quantum number space $dn_x\, dn_y\, dn_z$, as shown in Figure 13.1.

From Equation 13.5, $dn_i = (L_i/2\pi)\, dk_i$ ($i = x, y, z$) and the volume $dn_x\, dn_y\, dn_z$, which is equal to the number of quantum states in this volume element, is given by

$$dn_x\, dn_y\, dn_x = \left(\frac{V}{(2\pi)^3}\right)dk_x\, dk_y\, dk_z \tag{13.6}$$

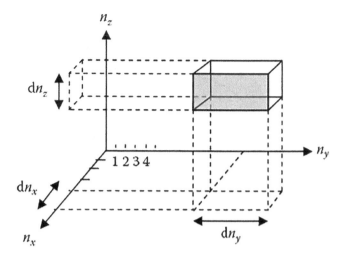

FIGURE 13.1 Volume elements dn_x, dn_y, and dn_z in quantum number space. The quantum numbers n_x, n_y, and n_z take integer values as shown.

where $V = L_xL_yL_z$. It follows that the density of states in k-space is simply

$$\rho_k = \frac{V}{(2\pi)^3}. \tag{13.7}$$

The energy density of states $\rho(\varepsilon) = (V/4\pi^2)\,(2m/\hbar^2)^{3/2}\,\varepsilon^{1/2}$ given by Equation 4.14 follows from Equation 13.7 with the use of the identity $\rho_k d^3k = \rho(\varepsilon)\,d\varepsilon$, or $\rho(\varepsilon) = \rho_k 4\pi k^2\,dk/d\varepsilon$, where $dk/d\varepsilon = (d\varepsilon/dk)^{-1} = m/(\hbar^2 k) = m/(\hbar\sqrt{2m\varepsilon})$. It is often convenient to use a spherical shell in k-space as a volume element.

Exercise 13.2

Obtain an expression for the radius of the Fermi sphere for an ideal Fermi gas at 0 K. Discuss the physical significance of the Fermi sphere representation.

   We define the Fermi momentum as $p_F = \hbar k_F$, with $p_F = \sqrt{2m\varepsilon_F}$ provided we assume the fermions to be nonrelativistic. It follows from Equation 13.3 that the radius of the Fermi sphere is

$$k_F = \left(\frac{3\pi^2 N}{V}\right)^{1/3}. \tag{13.8}$$

In momentum space, the Fermi sphere, or the Fermi surface, of radius k_F, as shown in Figure 13.2, is the boundary between occupied and empty states.

The number of discrete states in the Fermi sphere is equal to the number of states in the corresponding octant in quantum number space. Because the number of states is extremely large, they form a quasi-continuum, and this is why the sum over states may be replaced by an integral. At $T = 0$ K, the Fermi sphere is sharply defined, with all states inside the surface filled and all states outside the surface empty. As the temperature is raised, the surface of the sphere becomes less and less well defined as fermions are thermally excited from states just below the surface to states just above it. For electrons in solids, the Fermi surface is, in general, no longer spherical because of the effects the periodic lattice has on the dynamics of the charge carriers. Considerable effort has been devoted to determining the shape of the Fermi surface for metallic systems.

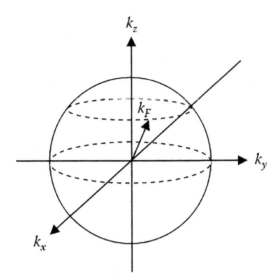

FIGURE 13.2   Fermi sphere for a system of fermions as represented in k-space. At $T = 0$ K, all states inside the Fermi sphere of radius k_F are occupied, and states outside the sphere are empty. As the temperature of the system is raised, some fermions near the Fermi surface are excited to higher energy states.

13.4   MEAN ENERGY OF IDEAL FERMI GAS AT $T = 0$ K

The mean energy of a Fermi gas is given by the integral

$$\langle E \rangle = 2 \int_0^\infty \varepsilon\, \rho(\varepsilon) f(\varepsilon)\, d\varepsilon, \tag{13.9}$$

where $\rho(\varepsilon)$ is the energy density of states and $f(\varepsilon)$ is the Fermi distribution function. The factor 2 is the spin degeneracy factor for spin $\frac{1}{2}$ fermions. At $T = 0$ K, the Fermi function has the rectangular form, as shown in Figure 12.1. In the calculation of the average energy $\langle E_0 \rangle$ at $T = 0$ K, the integral in Equation 13.9 may again be simplified in the same way as was done in Equation 13.2:

$$\langle E_0 \rangle = 2 \int_0^{\varepsilon_F} \varepsilon \rho(\varepsilon) d\varepsilon = \left(\frac{2V}{4\pi^2 \hbar^3} \right) (2m)^{3/2} \int_0^{\varepsilon_F} \varepsilon^{3/2}\, d\varepsilon,$$

and carrying out the integral gives

$$\langle E_0 \rangle = \left(\frac{V}{5\pi^2} \right) \left(\frac{2m}{\hbar^2} \right)^{3/2} \varepsilon_F^{5/2}. \tag{13.10}$$

With the use of Equation 13.3, the volume and other constants can be eliminated from Equation 13.10, and this simplification leads to the expression

$$\langle E_0 \rangle = \frac{3}{5} N\, \varepsilon_F. \tag{13.11}$$

It is instructive to consider the density of occupied states shown in Figure 13.3.

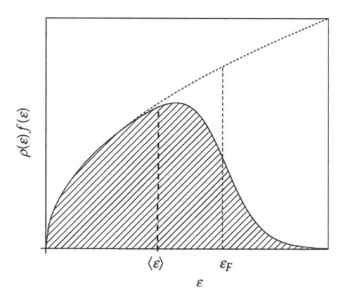

FIGURE 13.3 The density of occupied states for an ideal Fermi gas, plotted as a function of energy, at some finite temperature. The shaded region shows occupied states for $T > 0$ K. ε_F is the Fermi energy, and $\langle \varepsilon \rangle$ is the mean energy.

The product $\rho(\varepsilon)f(\varepsilon)$ may be written as

$$\rho(\varepsilon)f(\varepsilon) = \frac{V}{4\pi^2}\left(\frac{2m}{\hbar^2}\right)^{3/2}\left[\frac{\varepsilon^{1/2}}{e^{\beta(\varepsilon-\varepsilon_F)}+1}\right], \tag{13.12}$$

where it is convenient, for $T \ll T_F$, to take $\mu = \varepsilon_F$. For $T = 0$ K, the function increases as $\varepsilon^{1/2}$ and drops abruptly to zero for $\varepsilon = \varepsilon_F$. At finite temperatures, the density of occupied states develops a tail above ε_F and a decrease in the density just below ε_F. From Equation 13.11, the mean energy per particle is $\langle \varepsilon \rangle = \frac{3}{5}\varepsilon_F$. There are more high-energy occupied states than low-energy occupied states, and this imbalance results in $\langle \varepsilon \rangle$ having the comparatively high value of 0.6 ε_F.

13.5 APPROXIMATE EXPRESSIONS FOR THE HEAT CAPACITY AND MAGNETIC SUSCEPTIBILITY OF AN IDEAL FERMI GAS

To obtain an expression for the heat capacity of a Fermi gas, it is necessary to obtain an expression for the mean energy $\langle E(T) \rangle$ of the system and then to use $C_V = (\partial \langle E \rangle/\partial T)_V$. Equation 13.9 gives a general expression for $\langle E \rangle$ in terms of T, but the integral with the Fermi function is not simple to evaluate and is considered later in this chapter. It is possible to obtain an expression for C_V by making a fairly crude approximation based on the equipartition theorem. For a classical ideal gas, the equipartition theorem gives $\langle E \rangle = \frac{3}{2}Nk_BT$ and hence $C_V = \frac{3}{2}Nk_B$. For a Fermi gas, only particles in states near the Fermi energy ε_F are thermally excited to higher energy states. This suggests that only a small fraction of particles contribute to the specific heat. If we denote the small number of high-energy fermions by N_e, where

$$N_e \simeq N\left(\frac{k_BT}{\varepsilon_F}\right) = N\left(\frac{T}{T_F}\right), \tag{13.13}$$

it follows that the heat capacity is approximately given by

$$C_V \simeq \frac{3}{2} N_e k_{\mathrm{B}} = \frac{3}{2} N k_{\mathrm{B}} \left(\frac{T}{T_{\mathrm{F}}} \right). \tag{13.14}$$

Equation 13.14 predicts that, for a Fermi gas, the heat capacity should be proportional to T and that the value should be comparatively small because T_f is large, as discussed in Section 13.2. Experiments made at temperatures $T \ll T_f$ on systems that approximate an ideal Fermi gas confirm the linear T dependence predicted by Equation 13.14 and give values in reasonable agreement with this result.

Exercise 13.3

Obtain an approximate expression for the magnetic susceptibility χ of a Fermi gas in terms of the effective number of spins N_e that make a contribution to χ.

For an ideal paramagnet, the isothermal magnetic susceptibility $\chi = (\partial M/\partial H)_t$ has the form $\chi = C/T$, with Curie's constant given by Equation 10.41 as $C = (N\mu^2\mu_0)/(Vk_B)$. If we replace N by N_e from Equation 13.13, this gives an approximate expression for the susceptibility of a Fermi gas

$$\chi \approx \left(\frac{N\mu^2\mu_0}{Vk_{\mathrm{B}}T_{\mathrm{F}}} \right). \tag{13.15}$$

Equation 13.15 predicts that the magnetic susceptibility for a Fermi gas should be temperature independent and rather small because of the factor T_F in the denomination. These predictions have again been verified by experiment.

The temperature-independent susceptibility as given by Equation 13.15 contrasts with the Curie law susceptibility for localized spins. The susceptibility of a Fermi gas is called the Pauli susceptibility after Wolfgang Pauli, who first applied FD statistics to calculate this quantity. A more detailed discussion of the Pauli susceptibility is given later in this chapter. As the temperature of the gas is raised, the chemical potential μ decreases, and the approximate expressions given in Equations 13.14 and 13.15 will no longer apply. Eventually, at sufficiently high temperatures, classical expressions should be used. It is impractical to attain such temperatures in normal metals but in some systems, such as heavily doped semiconductor materials with delocalized carriers, ε_F can be quite low (~100 K). The classical limit can be easily reached in these systems.

13.6   SPECIFIC HEAT OF A FERMI GAS

It is possible to improve on the rather crude approximation to the specific heat given in Section 13.5 by the evaluation of the integral in Equation 13.9 to a good approximation at low T. To simplify expressions, it is convenient to write the density of states in terms of the Fermi energy ε_F given in Equation 13.3. With allowance for spin degeneracy, this gives for the mean energy

$$\langle E \rangle = 2 \int_0^\infty \varepsilon \rho(\varepsilon) \, f(\varepsilon) \mathrm{d}\varepsilon = \left(\frac{3N}{2\varepsilon_F^{3/2}} \right) \int_0^\infty \varepsilon^{3/2} f(\varepsilon) \mathrm{d}\varepsilon = \left(\frac{3N}{2\varepsilon_F^{3/2}} \right) I. \tag{13.16}$$

Evaluation of the integral, designated I, by parts leads to

$$I = \int_0^\infty \varepsilon^{3/2} f(\varepsilon) \mathrm{d}\varepsilon = f(\varepsilon) \frac{2}{5} \varepsilon^{5/2} \Big]_0^\infty - \int_0^\infty \frac{2}{5} \varepsilon^{5/2} \left(\frac{\mathrm{d}f(\varepsilon)}{\mathrm{d}\varepsilon} \right) \mathrm{d}\varepsilon. \tag{13.17}$$

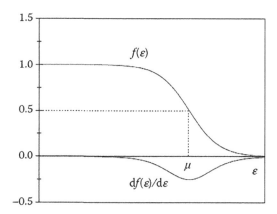

FIGURE 13.4 The Fermi function $f(\varepsilon)$ and its first derivative $\mathrm{d}f(\varepsilon)/\mathrm{d}(\varepsilon)$. At low temperatures, the first derivative is sharply peaked and negative at $\varepsilon = \mu$.

The first term vanishes at both limits. The derivative $\mathrm{d}f(\varepsilon)/\mathrm{d}\varepsilon$ is given by

$$\frac{\mathrm{d}f(\varepsilon)}{\mathrm{d}\varepsilon} = -\frac{\beta e^{\beta(\varepsilon-\mu)}}{\left(e^{\beta(\varepsilon-\mu)}+1\right)^2} = -\frac{1}{4}\beta \; \mathrm{sech}^2\left[\frac{1}{2}\beta(\varepsilon-\mu)\right]. \tag{13.18}$$

Figure 13.4 shows a plot of the Fermi function $f(\varepsilon)$ and its first derivative $\mathrm{d}f(\varepsilon)/\mathrm{d}\varepsilon$ for $T \ll T_{\mathrm{F}}$.

Because the Fermi function $f(\varepsilon)$ is approximately constant except in the vicinity of $\varepsilon = \mu$, the first derivative is zero everywhere except near μ, where it is sharply peaked, as shown in Figure 13.4, with the form given by the hyperbolic function in Equation 13.18. For $T \ll T_{\mathrm{F}}$, the function $\partial f/\partial \varepsilon$ is very sharply peaked. The factor $\varepsilon^{5/2}$ in the integral I expressed in Equation 13.17 is slowly varying over the small range where $\partial f/\partial \varepsilon \neq 0$. Expansion of $\varepsilon^{5/2}$ in a Taylor series about μ gives

$$\varepsilon^{5/2} = \mu^{5/2} + \frac{5}{2}\mu^{5/2}(\varepsilon-\mu) + \frac{15}{8}\mu^{5/2}(\varepsilon-\mu)^2 + \cdots \tag{13.19}$$

Inserting this expansion of $\varepsilon^{5/2}$ into Equation 13.17 and changing the variable to $x = \beta(\varepsilon - \mu)$ lead to

$$I = \frac{2}{5}\mu^{5/2}\int_{-\beta\mu}^{\infty}\frac{e^x}{\left(e^x+1\right)^2}\mathrm{d}x + \frac{\mu^{3/2}}{\beta}\int_{-\beta\mu}^{\infty}\frac{xe^x}{\left(e^x+1\right)^2}\mathrm{d}x + \frac{3}{4}\frac{\mu^{1/2}}{\beta^2}\frac{x^2e^x}{\left(e^x+1\right)^2}\mathrm{d}x + \cdots,$$

which may be written as

$$I = \frac{2}{5}\mu^{5/2}I_0 + \left(\frac{\mu^{3/2}}{\beta}\right)\left(\frac{\mu^{3/2}}{\beta}\right)I_1 + \frac{3}{4}\left(\frac{\mu^{1/2}}{\beta^2}\right)I_2 + \cdots. \tag{13.20}$$

The integrand in each integral has a sharp maximum at $x = 0$, and the lower limit in the integrals may be extended to $-\infty$ with negligible error because $\beta\mu \gg 1$. Now the function $e^x/\left(e^x+1\right)^2 = 1/\left(e^{x/2}+e^{-x/2}\right)^2 = \frac{1}{4}\mathrm{sech}^2(x/2)$ is an even function of x and integrals in an odd power of x, such as I_1, vanish. With standard integrals from Appendix A, we obtain

$$I_0 = \int_{-\infty}^{\infty}\left[\frac{e^x}{\left(e^x+1\right)^2}\right]\mathrm{d}x = 1 \quad \text{and} \quad I_2 = \int_{-\infty}^{\infty}\left[\frac{x^2e^x}{\left(e^x+1\right)^2}\right]\mathrm{d}x = \frac{\pi^2}{3}.$$

Substitution for I_0 and I_2 in Equation 13.20 gives

$$I = \frac{2}{5}\mu^{5/2} + \frac{1}{4}\pi^2\left(\frac{\mu^{1/2}}{\beta^2}\right) + \cdots. \tag{13.21}$$

The higher-order terms are negligible for sufficiently large β. Insertion of Equation 13.21 into Equation 13.16 results in

$$\langle E \rangle = \frac{3}{5}\left(\frac{N\mu^{5/2}}{\varepsilon_F^{3/2}}\right) + \frac{3}{16}\left(\frac{N\pi^2\mu^{1/2}}{\beta^2\varepsilon_F^{3/2}}\right) + \cdots.$$

At the temperatures of interest where the approximations made are valid, $\mu \simeq \varepsilon_F$, and it follows that

$$\langle E \rangle \simeq \frac{3}{5}N\varepsilon_F\left[1 + \frac{5}{12}\frac{\pi^2}{(\beta\varepsilon_F)^2} + \cdots\right]. \tag{13.22}$$

The heat capacity is obtained immediately:

$$C_V = \left(\frac{\partial\langle E(T)\rangle}{\partial T}\right)_V = \frac{3}{2}Nk_B\left(\frac{\pi^2}{3}\right)\frac{T}{T_F} + \cdots. \tag{13.23}$$

This expression for C_V differs only by a factor $\pi^2/3$ from the crude approximation given in Equation 13.14. It is customary to write

$$C_V = \gamma T \tag{13.24}$$

with

$$\gamma = \left(\frac{\pi^2}{2}\right)\left(\frac{Nk_B}{T_F}\right) = \left(\frac{\pi^2}{2}\right)\left(\frac{Nk_B^2}{\varepsilon_F}\right). \tag{13.25}$$

Using Equation 13.3 to eliminate ε_F gives the following alternative expression for γ

$$\gamma = \frac{1}{2}Nk_B^2\left(\frac{2m}{\hbar^2}\right)\left(\frac{\pi V}{3N}\right)^{2/3}. \tag{13.26}$$

γ involves the number of fermions, the fermion mass, and the $-\frac{2}{3}$ power of the fermion number density.

For electrons in metals, T_F is typically in the range of 10^4–10^5 K. This implies that the electron contribution to the specific heat of metals is extremely small and can be measured only at low temperatures when other contributions become unimportant. The specific heat of liquid helium-3, which is not an ideal Fermi gas but a Fermi *liquid* in which interactions between fermions cannot be assumed to be small, is shown as a function of temperature in Figure 13.5. Below 0.06 K (60 mK), the specific heat exhibits a linear dependence on T.

The weak interactions between helium-3 atoms are taken into account in what is called Fermi liquid theory. The heat capacity is predicted to vary linearly with temperature, as given by Equation 13.24 but with a modified value for γ. In Equation 13.26, which gives γ in terms of N, V, and m, it is necessary to introduce an effective mass m^* to allow for interactions between particles. With this modification, the Fermi liquid theory can account not only for the specific heat, but also

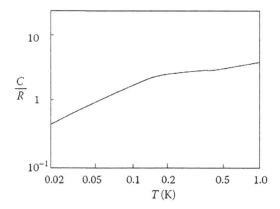

FIGURE 13.5 Specific heat, in units of the gas constant, as a function of temperature for liquid helium-3. For $T < 60$ mK, the specific heat shows a linear dependence on T.

for other thermodynamic properties of liquid helium-3. The experimentally deduced Fermi temperature T_F for liquid helium-3 is around 50 mK, which is lower than would be obtained from the Fermi gas expression as given by Equation 13.3.

13.7 PAULI PARAMAGNETISM

Consider an ideal Fermi gas located in a magnetic field B. The fermions have spin $S = \frac{1}{2}$ and an associated magnetic dipole moment $\mu_S = g\mu BS$. (For negatively charged electrons, μ_S is antiparallel to S, and a minus sign is included in the expression for μ_S.) The magnetic energy $g\mu_B B$ is of the order of 1 meV in a field of 10 T, whereas the Fermi energy is of the order of an electron volt. The magnetic interaction may therefore be viewed as a small perturbation. We choose $T = 0$ K to simplify integrals that have to be evaluated. Spin degeneracy is lifted by the applied magnetic field, and each single-particle state splits into a doublet with energy separation $2\mu_S B$. Figure 13.6 shows the modified Fermi distribution for the two classes of electron spins, moments up (i.e., μ_S parallel to B with low energy) and moments down.

The in-field Fermi distributions are energy-shifted by small amounts compared with the zero field distribution as shown. Figure 13.7 depicts the corresponding density of states curves.

The Fermi energy $\mu = \varepsilon_F$ remains fixed because fermions in higher energy states, with their moments antiparallel to the applied field, spin flip to lower energy unoccupied up states. This process results in a slight excess of up (+) moments compared with down (−) moments.

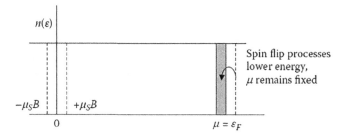

FIGURE 13.6 Fermi function for magnetic moment "up" and "down" fermions at $T = 0$ K in a magnetic field B. The up and down energy levels are slightly shifted with respect to each other as shown. The higher energy down moments flip orientation so that they can occupy the lower energy states. The Fermi level remains fixed.

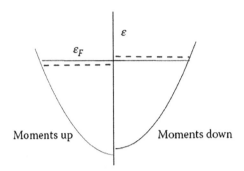

FIGURE 13.7   The density of states curves for moment up (spin down) and moment down electrons in a magnetic field B. A small field-induced relative displacement of the two curves is shown. Spin flip processes transfer electrons from the higher energy states above the Fermi energy ε_F to lower energy states just below ε_F.

The magnetization of the system is given by

$$M = \frac{\mathcal{M}}{V} = \frac{\mu_s}{V}(N_+ - N_-).$$
(13.27)

In the applied field B, the Fermi function may be written as

$$f(\varepsilon \pm \mu_s B) = \frac{1}{e^{\beta[(\varepsilon \pm \mu_s B) - \mu]} + 1},$$
(13.28)

where in the exponent in the denominator, the $+$ sign applies to up moments and the $-$ sign to down moments. With Equation 13.1 and the density of states, as given by Equation 4.14, the numbers of up and down moments are readily obtained.

$$N_\pm = \int_0^\infty \rho(\varepsilon \pm \mu_s B) f(\varepsilon) d\varepsilon.$$
(13.29)

At $T = 0$ K, $f(\varepsilon) = 1$ for $\varepsilon \le \varepsilon_F$ and $f(\varepsilon) = 0$ for $\varepsilon > \varepsilon_F$. At 0 K, Equation 13.29 is therefore written as

$$N_\pm = \int_{\pm \mu_s B}^{\varepsilon_F} \rho(\varepsilon \pm \mu_s B) d(\varepsilon \pm \mu_s B) = \frac{2}{3} \frac{V}{\pi^2} \left(\frac{2m}{\hbar^2}\right)^{3/2} (\varepsilon_F \pm \mu_s B)^{3/2}$$
(13.30)

or

$$N_\pm \simeq \frac{2}{3} \frac{V}{\pi^2} \left(\frac{2m}{\hbar^2}\right)^{3/2} \varepsilon_F^{3/2} \left(1 \pm \frac{3}{2} \frac{\mu_s B}{\varepsilon_F}\right).$$

Finally, to a good approximation,

$$N_\pm = \left(\frac{N}{2}\right)\left(1 \pm \frac{3}{2} \frac{\mu_s B}{\varepsilon_F}\right)$$
(13.31)

because $N = N_+ + N_-$ and $\mu_s B \ll \varepsilon_F$. Use of the Equation 13.31 results into Equation 13.27 gives for the magnetization

$$M = \frac{3}{2}\left(\frac{N}{V}\right)\left(\frac{\mu_s^2 B}{\varepsilon_F}\right).$$
(13.32)

With $B = \mu_0 H$, the Pauli susceptibility follows immediately:

$$\chi_p = \left(\frac{\partial M}{\partial H}\right)_T = \frac{3}{2}\left(\frac{N}{V}\right)\left(\frac{\mu_0 \mu_s^2}{\varepsilon_F}\right) = \frac{3}{2}\left(\frac{N}{V}\right)\left(\frac{\mu_0 \mu_s^2}{k_B T_F}\right). \tag{13.33}$$

This agrees with the crude estimate given in Equation 13.15 to within a factor $\frac{3}{2}$. The expression for χ is similar to the Curie law expression except that T is replaced by T_F. For $T \ll T_F$, the Pauli susceptibility is temperature independent as discussed in Section 13.5. It is possible to obtain a better approximation for χ_p at finite temperatures by the evaluation of the integral in Equation 13.29 by parts, in an approach similar to that adopted in Section 13.5. The calculation is fairly lengthy, and we simply quote the result:

$$\chi_p(T) \simeq \chi_p(0)\left[1 - \frac{\pi^2}{12}\left(\frac{T}{T_F}\right)^2\right]. \tag{13.34}$$

The finite temperature correction term is indeed very small for $T \ll T_F$.

13.8   THE PRESSURE OF A FERMI GAS

Equation 12.27 gives a general expression for the equation of state of a quantum gas in terms of the mean energy $\langle E \rangle$, $PV = \frac{2}{3}\langle E \rangle$. For a Fermi gas at low temperatures, where $T \ll T_F$, Equation 13.11 can be used to give an approximate expression for $\langle E \rangle$.

$$\langle E \rangle \approx \frac{3}{5}N\ \varepsilon_F = \frac{3}{5}N\left(\frac{\hbar^2}{2m}\right)\left(\frac{3\pi^2 N}{V}\right)^{2/3}. \tag{13.35}$$

From Equation 12.27, it follows that

$$P = \frac{2}{5}\left(\frac{\hbar^2}{2m}\right)\left(\frac{3\pi^2 N}{V}\right)^{5/3} = \frac{2}{5}\left(\frac{N}{V}\right)\varepsilon_F. \tag{13.36}$$

The Fermi pressure depends on the fermion density and the Fermi energy. P can be extremely large in typical Fermi gas systems, and this can be understood as a manifestation of the Pauli exclusion principle.

Exercise 13.4

Calculate the Fermi pressure for the electron system considered in Exercise 13.1. Explain how a metal remains a solid in spite of the large Fermi pressure.

We have from Exercise 13.1 $N/V = 10^{29}\,\text{m}^{-3}$ and $\varepsilon_F = 1.26 \times 10^{-18}\,\text{J}$. Substitution in Equation 13.36 gives for the Fermi pressure $P = 5 \times 10^4\,\text{MPa} = 5 \times 10^5\,\text{atm}$. This is a very large pressure, and in metals, the lattice of positive ion cores effectively acts as a confining box for the conduction electrons.

13.9   STARS AND GRAVITATIONAL COLLAPSE

Stars gradually burn up their light element nuclear fuel, which is the source of energy generation, through fusion processes. At the end of a star's life, various events, such as supernova formation, may occur. Gravitational collapse is an important process and can lead to white dwarf stars, neutron stars, or black holes dependent on the mass of stellar material involved.

A simple calculation on the basis of Newton's universal gravitation law provides an estimate of the gravitational pressure involved in gravitational collapse. The gravitational potential energy of a sphere of material of mass M and radius R is

$$U_G = -\frac{3}{5}\frac{GM^2}{R},$$ (13.37)

where G is the gravitational constant. Equation 13.37 is obtained from the change in energy, which occurs when a spherical shell of material of mass $4\pi r^2 \rho \, dr$ is added to a core of mass $\frac{4}{3}\pi r^3 \rho$ with integration over r from 0 to R to obtain the total potential energy involved. The total mass of the star is $M = \frac{4}{3}\pi R^3 \rho$, where ρ is the average density of stellar material.

The change in gravitational energy with volume is

$$dU_G = -P_G \, dV,$$ (13.38)

where P_G is the gravitational pressure. Use of Equation 13.37 gives

$$P_G = -\frac{dU_G}{dV} = -\left(\frac{dU_G}{dR}\right)\frac{dR}{dV} = -\frac{3}{20}\left(\frac{GM^2}{\pi R^4}\right).$$ (13.39)

White dwarf stars have high densities of the order of $10^4 - 10^7 \text{kg m}^{-3}$ and radii approximately 10^{-2} times that of the Sun. The temperature in the interior of these stars is of the order of 10^7 K, and the constituent atoms are ionized into nuclei and free electrons. The electrons form a degenerate electron gas.

The Fermi energy is estimated using Equation 13.3, with N/V given by

$$\frac{N}{V} = \left(\frac{M}{2m_n}\right)\left(\frac{1}{\frac{4}{3}\pi R^3}\right),$$ (13.40)

where M is the mass of the star and m_n the mass of a nucleon. The number of electrons is roughly half the number of nucleons. Typical values for a star similar to the Sun are $R = 7 \times 10^6 \text{m}$, $M = 2 \times 10^{30}\text{kg}$, and with $m_n = 1.67 \times 10^{-27}\text{kg}$, these values give $N/V = 4 \times 10^{33}\text{m}^{-3}$ and $\varepsilon_F = 3 \times 10^5\text{eV}$. The Fermi temperature for the electron gas is $T_F = 3.5 \times 10^9$ K. T_F is much higher than the temperature in the interior of a star, as quoted above, and this confirms that the electrons may be regarded as a degenerate Fermi gas with a well-defined Fermi surface. For the present discussion, we ignore relativistic effects, although a simple calculation shows that the Fermi energy is of the order of $m_e c^2$, and relativistic effects should strictly be taken into account.

The Fermi pressure due to the degenerate electron gas balances the gravitational pressure. When the two pressures given in Equations 13.36 and 13.39 are equated, the following mass–radius relation is obtained for white dwarf stars

$$M^{1/3}R = 2(3\pi)^{2/3}\left(\frac{\hbar^2}{Gm}\right)\left(\frac{3}{8m_n}\right)^{5/3} = 1.4 \times 10^{17}\,\text{kg}^{1/3}\text{m}.$$ (13.41)

The relationship is confirmed for white dwarfs with masses up to 1.4 solar masses. For larger masses, the Fermi pressure due to the electron gas is not great enough to prevent gravitational collapse.

Neutron stars are much smaller in radius than white dwarfs and are composed of a degenerate Fermi gas of neutrons. Protons and electrons no longer exist as distinct particles at the very high pressures that exist in the interior of these massive bodies. A very similar argument to that used for white dwarf stars can be applied to neutron stars, with the electron mass replaced by the neutron mass $m_n \sim 1000\, m_e$. The mass radius relation for neutron stars is

$$M^{1/3}R \sim 10^{14}\, \mathrm{kg}^{1/3}\mathrm{m}. \tag{13.42}$$

Neutron stars have high rotational frequencies, and a number have been identified as pulsars.

When the gravitational pressure is sufficiently large so that the radius of the star becomes less than the Schwarzschild radius $2GM/c^2$, a black hole is formed. All massive stars eventually form black holes as a result of gravitational collapse once nuclear fusion processes have run their course. Black holes vary in size depending on their mass and are typically extremely cold as discussed in Section 6.10.

PROBLEMS CHAPTER 13

13.1   For a Fermi gas, we may define a temperature T_0 at which the chemical potential of the gas is zero. Express T_0 in terms of the Fermi temperature T_F of the gas. Allow for spin degeneracy of the spin $\frac{1}{2}$ particles.

13.2   Use the two-dimensional density of states expression $\rho(\varepsilon) = (Am/2\pi\hbar^2)$ to obtain the chemical potential μ of a noninteracting two-dimensional Fermi gas of N fermions occupying an area A at temperature $T = 0$ K. For $T > 0$ K, show that μ is given to a good approximation by $\mu = \varepsilon_F - k_B T\, \ln\left(1 + e^{-T_F/T}\right)$.

13.3   Obtain expressions for the mean energy and the specific heat of the two-dimensional Fermi gas described in Question 13.2. Show that the specific heat depends linearly on T.

13.4   Consider an ideal Fermi gas in a fixed container at $T = 0$ K, and find expressions for the mean velocity $\langle v_x \rangle$ and the mean square velocity v_x^2 of the fermions along the x-direction. Express your answers in terms of the chemical potential μ. Assume that relativistic effects may be ignored.

13.5   A doped semiconductor contains N donor levels with energy, ε, with respect to the bottom of the conduction band. The donor levels may be either unoccupied as a result of thermal excitation of electrons into the conduction band or occupied by a single electron with spin up or down. Give an expression for the grand partition function for the donor system. Show that the mean number of electrons in donor sites at a given temperature is given by $n = N/[1 + (1/2)\exp\{-\beta(\mu + \varepsilon)\}]$.

13.6   For relativistic electrons in a white dwarf star, the dispersion relation is given by $\varepsilon = cp$. Use the dispersion relation to obtain the density of states for this relativistic gas as $\rho(\varepsilon) = [V/2\pi^2][\varepsilon^2/(c\hbar)^3]$, where V is the volume of the star. Derive expressions for the Fermi energy and the mean energy of the gas as a function of temperature. Make use of the integral formulae in Section 13.5.

13.7   An extreme relativistic electron gas similar to that described in Question 13.6 is contained in a volume V. Obtain an expression for the grand potential Ω_G (defined in Chapter 11) for the electron gas at temperature T. Use your result to obtain the entropy as a function of temperature for the relativistic gas.

13.8   By analogy with the spin $\frac{1}{2}$ case, derive expressions for the mean energy and the specific heat C_V of a hypothetical spin $\frac{3}{2}$ Fermi gas for $T \ll T_F$. Give a qualitative discussion of the magnetic susceptibility for this system.

13.9 Consider electrons emitted by a hot metal filament at temperature T into an evacuated space above the metal. Assuming that the electron density in the space close to the filament is low, show that the electron current density is given by $\left(4\pi emk_B^2 T^2 / h^3\right)e^{-(\phi / k_B T)}$, where ϕ is the work function of the metal. The work function is the minimum energy in excess of the chemical potential that electrons in the high-energy tail of the FD distribution must have to escape from the metal.

14 Ideal Bose Gas

14.1 INTRODUCTION

The properties of the ideal Bose gas are quite different to those of the ideal Fermi gas discussed in Chapter 13. At sufficiently low temperatures, bosons predominantly occupy the ground state with zero energy and momentum. We shall see that below a particular temperature, known as the Bose–Einstein (BE) condensation temperature, the ground-state occupancy becomes comparable with the number of bosons in the system. In this range, the de Broglie wavelength becomes macroscopically large of the order of the size of the container. Our discussion is based on the BE distribution together with the density of states expression for an ideal gas. The expressions that we obtain are useful in gaining insight into the superfluid transition in liquid helium-4. In addition, both superconductivity in metals and superfluidity in liquid helium-3, at sufficiently low temperatures, are associated with pairing interactions of fermions that lead to bosonic behavior.

14.2 LOW-TEMPERATURE BEHAVIOR OF THE CHEMICAL POTENTIAL

The chemical potential plays a key role in the behavior with temperature of the BE distribution given in Equation 12.11 $\langle n_r \rangle = 1/(e^{\beta(\varepsilon_r - \mu)} - 1)$. For a macroscopic system, the states r are very closely spaced in energy, and it is advantageous to rewrite Equation 12.11 as

$$n(\varepsilon) = \frac{1}{e^{\beta(\varepsilon - \mu)} - 1} = \frac{1}{\lambda^{-1} e^{\beta \varepsilon} - 1}, \tag{14.1}$$

where the energies ε form a quasi-continuum and $\lambda = e^{\beta \mu}$, as defined in Chapter 12. The chemical potential μ is determined by the use of the constraint,

$$N = \sum_{\varepsilon=0}^{\infty} n(\varepsilon) = \int_0^{\infty} \rho(\varepsilon) n(\varepsilon) \, d\varepsilon. \tag{14.2}$$

At low temperatures, the number of particles in the ground state, with $\varepsilon = 0$, becomes large, and care must be exercised in the evaluation of the integral. The density of states given by Equation 4.14 has the form $\rho(\varepsilon) \propto \varepsilon^{1/2}$, which gives $\rho(\varepsilon) = 0$ for $\varepsilon = 0$. The ground state is therefore excluded in the integral. For fermions, this introduces a negligible error because the exclusion principle prevents multiple occupancy of a particular state, but for bosons, this is not the case. At very low temperatures, the number of bosons in the ground state may be expected to become macroscopically large.

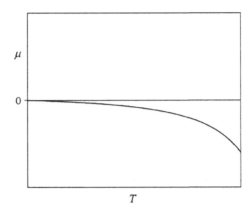

FIGURE 14.1 The curved line in the plot shows the behavior of the chemical potential μ for an ideal Bose gas as a function of temperature T near absolute zero. μ is always negative and approaches zero as T tends to zero.

Exercise 14.1

Use Equation 14.1 to obtain an approximate expression for the chemical potential of an ideal Bose gas close to $T = 0$ K. Assume that the majority of bosons are in the ground state at the temperatures of interest.

   If we consider just the ground state with $\varepsilon = 0$ in Equation 14.1 and take $n(0) \approx N_0$ to allow for the large ground-state occupancy at low temperatures, we obtain $N_0 \approx 1/(\lambda^{-1} - 1)$, or to a fair approximation

$$e^{-\beta\mu} = \lambda^{-1} = 1 + \frac{1}{N_0}. \tag{14.3}$$

For large N_0, it is clear that $\beta\mu$ must be very small (close to zero) and the exponential function may be expanded to give the simple and useful approximation

$$\mu \approx -\frac{k_B T}{N_0}. \tag{14.4}$$

If we assume that $N_0 \to N$ as $T \to 0$ K, Equation 14.4 predicts that, at sufficiently low temperatures, μ will tend to zero linearly with T from the negative side. This behavior is represented schematically in Figure 14.1.

   The above discussion of the chemical potential and the ground-state occupancy raises the question of how low the temperature must be for macroscopic occupation of the ground state to occur. This question is addressed below.

14.3   THE BOSE–EINSTEIN CONDENSATION TEMPERATURE

The integral in Equation 14.2 is used to obtain the number of bosons in states other than the ground state for which $\varepsilon = 0$ (i.e., excluding $\varepsilon = 0$). We denote the number of bosons in excited states by $N_e(T)$ and obtain

$$N_e(T) = \int_0^\infty \frac{\rho(\varepsilon)\, d\varepsilon}{\lambda^{-1} e^{\beta\varepsilon} - 1}. \tag{14.5}$$

With the density of single-particle states for particles in a box given by Equation 4.14 and $x = \beta\varepsilon$, Equation 14.5 becomes

$$N_e(T) = \frac{V}{4\pi^2}\left(\frac{2m}{\hbar^2}\right)^{3/2} \frac{1}{\beta^{3/2}} \int_0^\infty \frac{x^{1/2}\, dx}{\lambda^{-1} e^x - 1}. \tag{14.6}$$

At low temperatures, for $\beta\mu \ll 1$, as discussed in Section 14.2, it is permissible to take $\lambda^{-1} \simeq 1$ to simplify the integral, which may be evaluated in terms of Riemann zeta functions and the gamma functions used in Chapter 12. We have the general form for the integrals of interest:

$$\int_0^\infty \frac{x^{n-1}\,dx}{e^x - 1} = \Gamma(n)\zeta(n), \tag{14.7}$$

where

$$\zeta(n) = \sum_{m=1,\ 2,\cdots} \frac{1}{m^n}. \tag{14.8}$$

Values of the ζ and Γ functions are available in tables. For $n = \dfrac{3}{2}$ and $\dfrac{5}{2}$, the values of the two functions are given in Table 14.1.

With the aid of Equation 14.7 and Table 14.1, the integral in Equation 14.6 is evaluated to give

$$N_e(T) = 2.612 V \left(\frac{mk_B T}{2\pi\hbar^2} \right)^{3/2} \tag{14.9}$$

or

$$N_e(T) \propto \frac{V}{\lambda_T^{3'}}, \tag{14.10}$$

where $\lambda_T = h/\sqrt{3mk_B T}$ is the thermal de Broglie wavelength introduced in Chapter 4.

At low T, λ_T becomes macroscopically large, comparable with the size of the container, and the number of bosons in excited states decreases in a dramatic fashion. Equation 14.9 may be written as

$$N_e(T) = N \left(\frac{T}{T_0} \right)^{1/2}, \tag{14.11}$$

where

$$T_0 = \left(\frac{2\pi\hbar^2}{mk_B} \right) \left(\frac{1}{2.612} \frac{N}{V} \right)^{2/3} \tag{14.12}$$

is called the BE condensation temperature. N is the total number of bosons in the system. This expression is very similar in form to that obtained for the Fermi temperature T_F on the basis of Equation 13.3 in Chapter 13, $T_F = \varepsilon_F/k_B = (\hbar^2/2mk_B)(3\pi^2\,N/V)^{2/3}$.

TABLE 14.1

Values of Zeta Function $\zeta(n)$ and Gamma Function $\Gamma(n)$ for $n = \dfrac{3}{2}$ and $\dfrac{5}{2}$

N	$\zeta(n)$	$\Gamma(n)$
$\dfrac{3}{2}$	2.612	$\dfrac{1}{2}\sqrt{\pi}$
$\dfrac{5}{2}$	1.341	$\dfrac{3}{4}\sqrt{\pi}$

Exercise 14.2

Compare the Fermi temperature for an ideal Fermi gas of electrons with the BE condensation temperature for a high-density ideal Bose gas of helium-4 atoms. Approximate the Bose gas by liquid helium-4 with density $\rho_{He} = 0.125\,g\,cm^{-3}$. The mass of a helium-4 atom is $m_{He} = 6.65 \times 10^{-24}\,g$.

From Equation 14.12 and the expression for T_F given above, we find the ratio $T_0/T_F = 31.7\,(m_e/m_{He})(n_{He}/n_e)^{2/3}$, where n_{He} and n_e are the number densities for liquid helium and for electrons in a metal, respectively. The number density ratio is estimated as $n_{He}/n_e \sim 0.2$. In the evaluation of T_F for a Fermi gas of electrons, we use the electron mass and number density $n_e \sim 10^{23}\,cm^{-3}$. As shown in Section 13.2, these values give $T_F = 7.9\,eV = 9.16 \times 10^4\,K$. With the use of the density of liquid helium, we obtain $T_0 = 2.8\,K$. T_0 is much smaller than T_F because the atomic mass of helium is so much larger than the electron mass.

At temperatures well below T_0, bosons in an ideal Bose gas heavily populate the ground state. In contrast, fermions in a Fermi gas at low T populate the lowest states up to the Fermi level as a result of the Pauli exclusion principle. It follows from Equation 14.11 that the ground-state population for a Bose gas is given by

$$N_0(T) = N - N_e(T) = N\left[1 - \left(\frac{T}{T_0}\right)^{3/2}\right]. \tag{14.13}$$

Breakdown of Equation 14.13 occurs for $T \sim T_0$ because the approximation $\lambda^- = 1$ is no longer valid. We have for the ground-state population $N_0(T) = 1/(\lambda^{-1} - 1) = 1/(e^{-\beta\mu} - 1)$, and with Equation 14.13, this gives

$$\mu = -k_B T \ln\left\{1 + \frac{1}{N}\left[1 - \left(\frac{T}{T_0}\right)^{3/2}\right]^{-1}\right\}. \tag{14.14}$$

For $T \to 0$ K, the log function may be expanded, and this results in

$$\mu \simeq -\frac{k_B T}{N}\left[1 - \left(\frac{T}{T_0}\right)^{3/2}\right]^{-1}. \tag{14.15}$$

Equation 14.15 has a form similar to Equation 14.4 and again shows that $\mu \to 0$ as $T \to 0$ K. As mentioned above for $T \to T_0$, the approximation $\lambda \approx 0$ is no longer valid, and Equation 14.15 is not reliable. Figure 14.2 shows a schematic plot of the ratios $N_e(T)/N$ and $N_0(T)/N$ as a function of T/T_0 in the range $T \leq T_0$. The ground-state population grows dramatically as T/T_0 decreases.

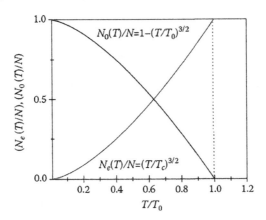

FIGURE 14.2 Population ratios for bosons in the excited states $N_e(T)/N$ and in the ground state $N_0(T)/N$ for an ideal Bose gas as a function of T/T_0. The ground-state population increases dramatically with decreasing T below T_0.

14.4 HEAT CAPACITY OF AN IDEAL BOSE GAS

The mean energy of a Bose gas is $\langle E(T) \rangle = \int_0^\infty \varepsilon \rho(\varepsilon) n \, d\varepsilon$. Let $\beta \varepsilon = x$, as before, and, with the density of states expression as given by Equation 4.14, this leads to

$$\langle E(T) \rangle = \frac{V}{4\pi^2} \left(\frac{2m}{\beta \hbar^2} \right)^{3/2} \frac{1}{\beta} \int_0^\infty \frac{x^{3/2}}{\lambda^{-1} e^x - 1} dx. \tag{14.16}$$

In the low-temperature limit, the integral is evaluated with the aid of Equation 14.7 and values of the gamma function and the Riemann zeta function from Table 14.1 to give

$$\langle E(T) \rangle = \left(1.341 \left(\frac{3}{4} \sqrt{\pi} \right) \right) \frac{V}{4\pi^2} \left(\frac{2m}{\beta \hbar^2} \right)^{3/2} \frac{1}{\beta}. \tag{14.17}$$

With T_0, from the definition of Equation 14.12, we obtain

$$\langle E(T) \rangle = 0.77 B k_B T \left(\frac{T}{T_0} \right)^{3/2} \tag{14.18}$$

and the heat capacity follows immediately

$$C_V = \left(\frac{\partial \langle E(T) \rangle}{\partial T} \right)_V = 1.93 N k_B \left(\frac{T}{T_0} \right)^{3/2}. \tag{14.19}$$

Figure 14.3 shows $C_V / N k_B$ as a function of T/T_0.

It must be emphasized that the approximation $\lambda \approx 1$ fails for $T/T_0 \rightarrow 1.0$ and, in this temperature range, the integral in Equation 14.16 must be evaluated using a better approximation for λ. Nevertheless, the main features as shown in Figure 14.3 are qualitatively correct.

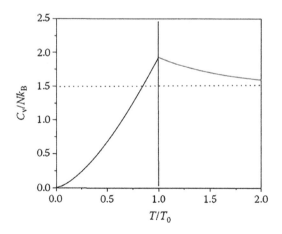

FIGURE 14.3 $C_V / N k_B$ plotted as a function of T/T_0 for an ideal Bose gas. The approximations made in the derivation of Equation 14.19 break down as T/T_0 approaches 1.0. The classical monatomic ideal gas value $C_V / N k_B = 3/2$ is shown for comparison.

14.5 THE PRESSURE AND ENTROPY OF A BOSE GAS AT LOW TEMPERATURES

We can easily obtain an expression for the pressure of an ideal Bose gas at low temperatures where B-E condensation occurs. For $T < T_0$, it is convenient to use the general form of the equation of state for quantum gases given in Equation 12.27 $PV = \frac{2}{3}\langle E(T) \rangle$, and with Equation 14.18, we get

$$PV = 0.51 N k_B T \left(\frac{T}{T_0} \right)^{3/2}. \tag{14.20}$$

Equation 14.20 shows that, for $T < T_0$, the pressure tends to zero as $T^{5/2}$. Particles in the ground state have zero momentum and therefore exert no pressure on the walls of the container. This behavior contrasts with that of a Fermi gas at low temperatures where, as we have seen in Chapter 13, very large pressures are found.

Exercise 14.3

Obtain an expression for the entropy of an ideal Bose gas at low temperatures where the condition $T < T_0$ is satisfied. Discuss the temperature dependence of the entropy in this limit.

At sufficiently low temperatures, the entropy of an ideal Bose gas is obtained with the use of the heat capacity expression (Equation 14.19). This gives the straightforward integral

$$S = \int_0^T \frac{dQ}{T} = \int_0^T \frac{C_V\, dT}{T} = 1.931 \frac{N k_B}{T_0^{3/2}} \int_0^T T^{1/2}\, dT = 1.3 N k_B \left(\frac{T}{T_0} \right)^{3/2}. \tag{14.21}$$

Equation 14.21 shows that below T_0, the entropy of the gas tends to zero as $T^{3/2}$. This is because the entropy is associated with particles in excited states. Near $T = 0$ K, almost all of the particles are in the ground state, and the system is highly ordered. The entropy expression (Equation 14.21) is valid only for $T < T_0$ because of the approximation $\lambda \approx 1$ made in the derivation of Equation 14.19.

14.6 THE BOSE–EINSTEIN CONDENSATION PHENOMENA IN VARIOUS SYSTEMS

a. *Liquid Helium-4.* Atoms of ⁴He are bosons, and because of the weak interactions between helium atoms, the properties of liquid helium-4 may be expected to have some resemblance to those of an ideal Bose gas. Using Equation 14.12 with the atomic mass of ⁴He and the density of liquid helium gives $T_0 = 3.13$ K. Liquid helium exhibits a phase transition at 2.17 K, with a lambda-shaped peak in the specific heat curve. Figure 14.4 shows the specific heat at constant volume and under saturated vapor pressure for liquid helium-4 as a function of temperature.

The phase transition temperature in liquid helium-4 is denoted by T_λ because of the shape of the specific heat curve. Below $T \simeq 0.6$ K, the specific heat follows a T^3 law, which suggests that phonon excitations are important. Phonon contributions to the specific heat of substances are discussed in Chapter 16. At temperatures below T_λ, liquid helium-4 exhibits superfluid properties, such as nonviscous flow through very small channels that prevent the flow of a normal fluid. The properties of the superfluid phase are accounted for phenomenologically by means of a two-fluid model in which normal and superfluid components coexist. Although it is necessary to allow for weak interatomic interactions, the model is consistent with the BE condensation phenomenon, with a significant number of atoms in the ground state and the remainder in excited states, as shown in Figure 14.2. A detailed theory of the superfluid transition in liquid helium-4 requires allowance for the interparticle interactions.

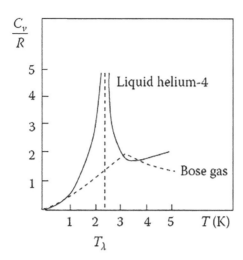

FIGURE 14.4 The specific heat of liquid helium-4 (measured under the vapor pressure) shown as a function of temperature with the lambda transition at 2.17 K. The theoretically predicted curve for an ideal Bose gas on the basis of Figure 14.3 is included for comparison.

b. *Superconductors, Liquid Helium-3, and Alkali Atom Systems in Traps.* The phenomenon of superconductivity in metals is related to BE condensation, although the electron gas consists of fermions. At sufficiently low temperatures, interaction of the electrons with the lattice that involve a phonon mechanism or some other electron coupling mechanism leads to the formation of what are called Cooper pairs of electrons. The pairs have net spin zero, or in some cases integral spin, and behave as bosons. Although superconductivity in many conventional metals is well understood, high-temperature superconductors, specifically the cuprates and recently discovered iron-based pnictides, are less well understood and are the subject of ongoing research.

Liquid helium-3, which as pointed out in Chapter 13 is a good example of a Fermi liquid below 60 mK, is found to exhibit superfluid properties at sufficiently low temperatures with a magnetic field-dependent transition below 2 mK. Pairing interactions of the helium-3 atoms are important, and the Cooper pairs obey BE statistics. The detailed theory of the superfluid phases in helium-3 is not simple, but we can qualitatively understand the transition as a BE condensation phenomenon.

Beautiful experiments have been carried out on vapors of alkali atoms such as rubidium-87 that are cooled to very low temperatures by means of sophisticated experimental techniques that involve laser cooling followed by evaporative cooling. Several thousand atoms can be cooled and magnetically confined at temperatures below 1 μK. Coupling of the electron and nuclear spins through the hyperfine interaction can lead to bosonic properties of the alkali atoms, and BE condensation phenomena have been observed in these weakly interacting Bose gas systems.

PROBLEMS CHAPTER 14

14.1 Calculate the predicted BE condensation temperature for liquid helium-4 in the ideal Bose gas limit. The molar volume of liquid helium is $27.6 \, \text{cm}^3 \text{mol}^{-1}$.

14.2 Use the expression for the mean energy of a Bose gas near the BE condensation temperature T_0 to obtain a value for the average de Broglie wavelength λ_T at T_0 for helium-4 (~3 K). Describe the behavior of λ_T below the condensation temperature emphasizing the role of the ground state and the excited states in determining λ_T. Compare λ_T with the interparticle spacing obtained from the molar volume given in Question 14.1.

14.3 By what factor should the temperature of an ideal Bose gas be lowered below the conden-sation temperature for the ground state to be populated by one-tenth of the bosons?

14.4 The BE condensation was first observed in an atomic system in 1995 *87Rb atoms were confined in a magneto-optical trap and using special techniques were cooled to below 1 μK. Evidence for BE condensation was obtained at a temperature of 100 nK with the atoms confined to a volume of $10^{-15}\,m^3$. Estimate the required number of ^{87}Rb atoms in the trap for condensation to be observed. The confining potential used in the experiments was a three-dimensional harmonic potential but simplify the problem by assuming a box potential.

14.5 The BE condensation in atomic hydrogen gas was found in 1998* at a temperature of roughly 50 μK. The atomic hydrogen was produced in a cryogenic discharge process and confined in a magnetic trap. Estimate the hydrogen atom density required to observe BE condensation.

14.6 A Bose gas in a monolayer two-dimensional film has area A. Obtain an expression for the mean energy of the gas at low temperatures. Assume that the chemical potential $\mu \sim 0$. Show that the specific heats for two-dimensional Fermi and Bose gases are identical and depend linearly on T. You may make use of the integral $\int_0^\infty x/(e^x-1)\,\pi^2/6.$

14.7 An ideal two-dimensional Bose gas as described in Question 14.6 should not undergo BE condensation. Justify this statement with the use of the two-dimensional density of states.

14.8 Obtain an approximate expression for the Helmholtz potential of an ideal Bose gas with the use of the relationship between the grand partition function and the partition function given in Chapter 11. Explain why this expression is of limited use below the condensation temperature.

14.9 A Bose gas consists of particles with internal degrees of freedom. Obtain an expression for the BE condensation temperature of the gas.

14.10 Obtain an expression for the grand potential of an ideal Bose gas. Use the grand potential to obtain forms for the entropy and the pressure of the gas. Show that the expression for the pressure reduces to the form $PV = \frac{2}{3}E$ given in Equation 12.27.

15 Photons and Phonons: The "Planck Gas"

15.1 INTRODUCTION

In 1900, Max Planck introduced the concept of the quantum of electromagnetic radiation, later called the photon, into physics. He did this to explain the spectral properties of electromagnetic radiation emitted through a small aperture by a constant-temperature "black body" enclosure. This marked the start of quantum physics, which led to the development of quantum mechanics in the 1920s. The energy ε of a photon of frequency v is given by Planck's famous expression, $\varepsilon = hv$, or alternatively $\varepsilon = \hbar\omega$, where ω is the angular frequency.

Inside a constant-temperature enclosure, photons are continually absorbed and emitted by the walls. The number of photons in the enclosure is not fixed but fluctuates around some average number for any chosen frequency. This fluctuation in the number of photons represents an important difference from the situation in the fermion and boson systems considered in Chapters 13 and 14, where the number of particles N is fixed. The chemical potential for photons is not defined because there is no constraint on N. It follows that μ should be omitted in the photon distribution, and this is shown in Section 15.3. Photons have spin 1 and are bosons. Because they travel at the speed of light, there are two and not three allowed spin orientations. Classically, electromagnetic radiation is considered to be a transverse wave with two polarization directions. Putting $\mu = 0$ in the Bose–Einstein distribution given in Equation 14.1 leads to the Planck distribution for photons,

$$\langle n_r \rangle = \frac{1}{e^{\beta\varepsilon_r} - 1} = \frac{1}{e^{\beta h\varepsilon_r - 1'}}, \tag{15.1}$$

where ε_r is the energy of photons in state r. For $\beta\varepsilon_r \ll 1$, the average number of photons found in state r becomes very large because the denominator in Equation 15.1 becomes very small.

Because the chemical potential is zero, it follows that the grand partition function $\mathbb{Z}$ and the partition function Z are identical for a photon gas. This can be seen from the approximate relationship as given in Equation 11.21, $\ln Z \simeq \ln \mathbb{Z} - \beta\mu N$. Putting $\mu = 0$ shows that $\mathbb{Z} \simeq Z$. The exact identity can, of course, be shown using the definitions for $\mathbb{Z}$ and Z. The Planck distribution also applies to quanta linked with other types of fields. For example, the number of elementary excitations, called phonons, associated with lattice vibrations of a particular frequency in a solid, is given by Equation 15.1. This result permits an expression for the specific heat of an insulating solid to be derived in a straightforward way. Other excitations, such as magnons, which are associated with spin waves in magnetic materials, also obey the Planck distribution.

15.2 ELECTROMAGNETIC RADIATION IN A CAVITY

Consider an evacuated enclosure of volume V with walls that conduct electrically and with no free charges or currents present. Classically, electromagnetic radiation inside the enclosure is governed by the wave equation that, for the associated electric field E, has the form

$$\nabla^2 E - \varepsilon_0\mu_0 \frac{\partial^2 E}{\partial t^2}. \tag{15.2}$$

ε_0 and μ_0 are the permittivity and permeability of free space, respectively, with $\varepsilon_0\mu_0 = 1/c^2$, where c is the speed of light in free space. The solution of Equation 15.2 may be written in terms of spatial and time-dependent functions as

$$E(r,t) = E(r)e^{-i\omega t}. \tag{15.3}$$

Substitution of E from Equation 15.3 into Equation 15.2 gives

$$\nabla^2 E(r) + \varepsilon_0\mu_0\omega^2 E(r) = 0. \tag{15.4}$$

Because the walls of the container are conducting, allowed solutions are standing waves with nodes at the walls. To simplify the situation, let the container be cubical with edges of length L. The solution of Equation 15.4 is analogous to the solution of the Schrödinger equation for particles in a box discussed in Section 4.2. A possible solution to Equation 15.4 is

$$E = E_0\sin\left(q_x x + q_y y + q_z z\right), \tag{15.5}$$

with wave numbers $q_x = 2\pi/\lambda_x$, $q_y = 2\pi/\lambda_y$, $q_z = 2\pi/\lambda_y$, and $q^2 = q_x^2 + q_y^2 + q_z^2 = \omega^2/c^2$. To satisfy the boundary conditions, the following relationships must apply: $q_x = (2\pi/L)n_x$, $q_y = (2\pi/L)n_y$, and $q_z = (2\pi/L)n_z$. The n_x, n_y, and n_z take integral values and are closely related to particle in a box quantum number. A three-dimensional standing wave of angular frequency ω and with wave vector q is called a cavity mode. The number of modes in the range from q to $q+dq$ is

$$\rho_q \delta^3 q = \delta n_x\ \delta n_y \delta n_z = \frac{L^3}{(2\pi)^3}\delta q_x\ \delta q_y\ \delta q_z,$$

or, infinitesimally,

$$\rho_q d^3 q = \left(\frac{V}{(2\pi)^3}\right)d^3 q. \tag{15.6}$$

The density of modes ρ_q is identical to the density of quantum states for a particle in a box.

It is useful to obtain an expression for the number of modes $\rho_\omega\, d\omega$ in the range from ω to $\omega + d\omega$. For $d^3 q = 4\pi q^2\, dq$, which corresponds to a shell of radius q and thickness dq in q-space, and with allowance for two polarization directions, we obtain

$$\rho_\omega d\omega = 2\rho_q 4\pi q^2\left(\frac{dq}{d\omega}\right)d\omega. \tag{15.7}$$

With $\omega = cq$, it follows that

$$\rho_\omega d\omega = \left(\frac{V}{\pi^2 c^3}\right)\omega^2 d\omega. \tag{15.8}$$

Equation 15.8 together with the Planck distribution permits the laws for black body radiation to be derived directly.

15.3   THE PLANCK DISTRIBUTION

In Section 15.1, the Planck distribution was obtained simply by letting $\mu = 0$ in the Bose–Einstein distribution. It is straightforward to derive the Planck distribution from the partition function. For a system of photons in a cavity at temperature T, the partition function Z is given by

$$Z = \sum_{n_1, n_2, n_3 \cdots} e^{-\beta n_1 \hbar \omega_1} e^{-\beta n_2 \hbar \omega_2} e^{-\beta n_3 \hbar \omega_3} = \Pi_r \left[\sum_{n_r} e^{-\beta n_r \hbar \omega_r} \right], \qquad (15.9)$$

where n_r is the number of photons of energy $\varepsilon_r = \hbar \omega_r$ associated with a mode of frequency ω_r. In contrast to the situation of a fixed number of particles in a container, there is no constraint on the n_r values. This means that, in evaluating the sum $\displaystyle\sum_{n_r = 1, 2, \cdots} e^{-\beta n_r \hbar \omega_r}$, the upper limit on n_r can be allowed to be very large, tending to infinity. The summation is easily evaluated as an infinite geometric series (see Appendix A):

$$\sum_{n_r} e^{-\beta n_r \hbar \omega_r} = 1 + e^{-\beta \hbar \omega_r} + e^{-2\beta \hbar \omega_r} + \cdots = \frac{1}{1 - e^{-\beta \hbar \omega_r}}. \qquad (15.10)$$

Equation 15.10 together with the expression for Z in Equation 15.9 gives

$$\ln Z = -\sum_r \ln \left(1 - e^{-\beta \hbar \omega_r} \right). \qquad (15.11)$$

The mean number of photons in mode r is given by

$$\langle n_r \rangle = -\left(\frac{1}{\beta} \right) \frac{\partial \ln Z}{\partial \varepsilon_r} = \left(\frac{1}{\beta \hbar} \right) \frac{\partial \ln Z}{\partial \omega_r} = \frac{1}{\left(e^{\beta \hbar \omega_r} - 1 \right)}, \qquad (15.12)$$

which is the Planck distribution for $\langle n_r \rangle$ as noted previously in Equation 15.1. The mean energy associated with photons in mode r is $\langle \varepsilon_r \rangle = \langle n_r \rangle \hbar \omega_r = \hbar \omega_r / \left(e^{\beta \hbar \omega_r} - 1 \right)$. Fluctuations in the number of photons in mode r may be obtained with the use of the expression for bosons given in Equation 12.16

$$\frac{\sigma_{n_r}^2}{\langle n_r \rangle^2} = \left(1 + \frac{1}{\langle n_r \rangle} \right) = e^{\beta \hbar \omega_r} \qquad (15.13)$$

Note that fluctuations in n_r become very large when $\langle n_r \rangle$ becomes small. This result is important, for example, when dealing with low-intensity radiation.

15.4   THE RADIATION LAWS

With the results obtained in Sections 15.2 and 15.3, specifically Equations 15.8 and 15.12, we can write down an expression for the total energy $U_\omega \, d\omega$ associated with cavity radiation in the frequency range ω to $\omega + d\omega$, in a cavity at temperature T

$$U_\omega \, d\omega = \langle n_\omega \rangle \hbar \omega \rho_\omega \, d\omega = \left(\frac{Vh}{\pi^2 c^3} \right) \frac{\omega^3 \, d\omega}{e^{\beta \hbar \omega} - 1}. \qquad (15.14)$$

Planck's equation for black body radiation is obtained immediately as $u_\omega \, d\omega$, the energy per unit volume in the range from ω to $\omega + d\omega$:

$$u_\omega \, d\omega = \left(\frac{\hbar}{\pi^2 c^3} \right) \frac{\omega^3}{\left(e^{\beta \hbar \omega} - 1 \right)} \, d\omega. \qquad (15.15)$$

In terms of frequency, this becomes

$$u_\nu d\nu = \left(\frac{8\pi h}{c^3}\right)\frac{\nu^3}{\left(e^{\beta h\nu}-1\right)}d\nu \tag{15.16}$$

and, in terms of wavelength, using $|d\nu| = (c/\lambda^2)d\lambda$, we get

$$u_\lambda d\lambda = \left(\frac{8\pi\ hc}{\lambda^5}\right)\frac{1}{\left(e^{\beta hc/\lambda}-1\right)}d\lambda. \tag{15.17}$$

By allowing radiation to escape through a small hole in the side of the cavity, it is possible to measure the spectral properties of black body radiation. The curves as shown in Figure 15.1 are based on Equation 15.16 and agree well with the experiment.

It is convenient to rewrite Equation 15.15 in terms of the variable $x = \beta\hbar\omega = \hbar\omega/k_B T$,

$$u_\omega d\omega = u_x dx = \left(\frac{k_B^4}{\pi^2 c^3\hbar^3}\right)T^4\frac{x^3 dx}{e^x-1} \tag{15.18}$$

or $u_x\ dx = CT^4[x^3/(e^x-1)]dx$, where $C = k_B^4/\pi^2 c^3\hbar^3$.

Figure 15.2 shows a plot of the function $x^3/(e^x-1)$ versus x.

A maximum in the function as shown in Figure 15.2 is observed at $x_{max}\approx 2.8$. It follows that ω_{max}/T is a constant for black body radiation, or, for two different cavity temperatures,

$$\frac{\omega_{1\ max}}{T_1} = \frac{\omega_{2\ max}}{T_2}. \tag{15.19}$$

This is Wien's displacement law for the frequency at which the maximum energy density occurs as a function of temperature. The total radiation energy per unit volume in the cavity is given by

$$\overline{u}_{total} = \int_0^\infty u_\omega\ d\omega = \int_0^\infty u_x\ dx = CT^4\int_0^\infty\frac{x^3 dx}{e^x-1}.$$

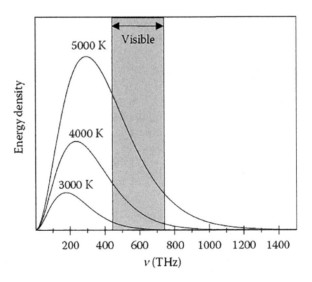

FIGURE 15.1 Spectral energy density as a function of frequency for black body radiation obtained using Planck's equation at three temperatures. The visible region of the spectrum is shown as a band.

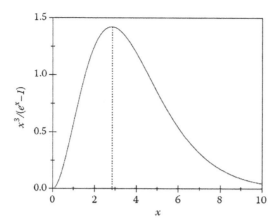

FIGURE 15.2   The plot shows the universal function $x^3/(e^x - 1)$ versus x. The maximum occurs near $x = 2.8$. Wien's displacement law for black body radiation follows from this plot.

From Appendix A, the integral is given by $\int_0^\infty x^3\, dx/(e^x - 1) = \pi^4/15$, and we obtain for the energy density

$$\overline{u}_{\text{total}} = \frac{\pi^4 C}{15} T^4. \tag{15.20}$$

Equation 15.20 is an expression of the Stefan–Boltzmann law. The related Stefan–Boltzmann law for the intensity of radiation emitted through a small hole in the wall of a cavity may be derived from Equation 15.20 and is set as a problem at the end of this chapter.

   An important and beautiful illustration of black body radiation is the cosmic background microwave radiation that resulted from the Big Bang formation of the universe billions of years ago. The radiation has cooled to a temperature of 2.725 K with a peak in the spectral distribution at 160.2 GHz. Figure 15.3 shows a log–log plot of the spectral density as a function of frequency for the cosmic background radiation. The observations made with far infrared instruments on the COBE satellites starting in the 1990s, together with various other observations, are fitted to high precision over several orders of magnitude in frequency using the Planck distribution. The plotted experimental points are found to lie perfectly on the curve shown.

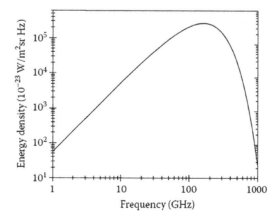

FIGURE 15.3   Spectral density as a function of frequency for microwave cosmic background radiation at 2.725 K calculated with Planck's equation. Experimental values from precise satellite measurements lie exactly on the curve. The peak in the spectrum corresponds to a wavelength of 1.9 mm.

Roughly 370,000 years after the Big Bang, the hot plasma had expanded and cooled to 3000 K so that neutral atoms could form. This allowed radiation to propagate freely in space, and over the subsequent 14.4 billion years, the radiation has cooled adiabatically with the expansion of the universe to reach the present temperature below 3 K.

15.5   RADIATION PRESSURE AND THE EQUATION OF STATE FOR RADIATION IN AN ENCLOSURE

Electromagnetic radiation in equilibrium in a cavity exerts a pressure $\langle P \rangle$ on the walls. From the expression for ln Z as given in Equation 15.11, $\langle P \rangle$ is obtained with the aid of the relation, $\langle P \rangle = (1/\beta) \partial \ln Z/\partial V$. To simplify the notation, we put $\langle P \rangle = P$ and find

$$P = -\left(\frac{1}{\beta}\right)\frac{\partial}{\partial V}\left[\sum_r \ln\left(1 - e^{\beta\hbar\omega_r}\right)\right] = -\sum_r \frac{\hbar e^{-\beta\hbar\omega_r}}{\left(1 - e^{\beta\hbar\omega_r}\right)}\frac{\partial\omega_r}{\partial V} = -\sum_r \langle n_r\rangle\frac{\partial\varepsilon_r}{\partial V}. \quad (15.21)$$

$\varepsilon_r = \hbar\omega_r$ is the photon energy in mode r. Because $\lambda_r \propto V^{1/3}$, from the boundary conditions, it follows that $\partial\varepsilon_r/\partial V = -\frac{1}{3}\varepsilon_r/V$, and the pressure is therefore

$$P = \frac{1}{3V}\sum_r \langle n_r\rangle\varepsilon_r = \frac{1}{3}\bar{u}_{total}, \quad (15.22)$$

where $\bar{u}_{total}$ is the energy density of the radiation given by Equation 15.20.

Exercise 15.1

The expression $P = \frac{1}{3}\bar{u}_{total}$ as given in Equation 15.22 for photons in a cavity is similar to the result as given in Equation $P = \frac{2}{3}\langle E\rangle/V$, obtained for boson and fermion quantum gases in Chapter 12. Account for the difference in the numerical coefficient in these two expressions.
For the particles in a quantum gas, $\varepsilon \propto V^{-2/3}$ rather than $V^{-1/3}$ as used above for photons. The change in the exponent leads to a different numerical coefficient in the expression for the pressure of a gas of particles compared with that for photons.

Exercise 15.2

Obtain the pressure exerted by black body radiation on the walls of an enclosure at a temperature of 1000 K.
Substitution from Equation 15.20 into Equation 15.22 leads to

$$P = \left(\frac{\pi^2 k_B^4}{15C^3\hbar^3}\right)T^4 = \frac{1}{3}\alpha T^4. \quad (15.23)$$

The coefficient α involves fundamental constants and in SI units has the value $\alpha = 2.5\times10^{-16}$ Nm^{-2} K^{-4}. At a temperature of 1000 K, the mean pressure on the walls of an enclosure because of electromagnetic radiation is 2.5×10^{-4} Pa or 2.5×10^{-9} atm. This is a very small pressure compared with the pressure exerted by gas molecules on the walls of an enclosure at ambient temperature. The equation of state for a photon *gas* given in alternative forms in Equations 15.22 and 15.23 is rather simple and has some interesting consequences. The first T dS equation given by Equation 7.66 is $T\,dS = C_V\,dT + T\,(\partial P/\partial T)_V\,dV$. For photons, we have $(\partial P/\partial T)_V = dP/dT = \frac{4}{3}\alpha T^3$

and $C_V (\partial E/\partial T)_V = 4\alpha V T^3$, where we have taken $E = U_{total} = V\bar{u}_{total}$. Substituting for $(\partial P/\partial T)_V$ and C_V in Equation 7.66, we obtain

$$T\,dS = 4\alpha V T^3\,dT + \frac{4}{3}\alpha T^4\,dV. \tag{15.24}$$

Exercise 15.3

Consider a reversible adiabatic change of volume of a cavity in which no entropy change occurs. This may be accomplished using an idealized piston–cylinder-type cavity that has perfectly reflecting walls. Obtain an expression for the final temperature of radiation in the cavity in terms of the initial temperature and the initial and final volumes of the cavity.

For this process, Equation 15.24 becomes $0 = 4\alpha V T^3\,dT + \frac{4}{3}\alpha T^4\,dV$ or $dT/T = -\frac{1}{3}\,dV/V$.

Integration gives $\ln\left(T_i/T_f\right) = -\frac{1}{3}\ln\left(V_f/V_i\right)$, and hence,

$$T_f = \left(\frac{V_i}{V_f}\right)^{1/3} T_i. \tag{15.25}$$

The ratio of the final to the initial temperature is given by the cube root of the inverse volume expansion factor. The temperature inside the cavity may be measured by having a small perfectly absorbing thermometer inside the cavity that reaches equilibrium with the radiation. Equation 15.25 helps us understand how the cosmic background radiation has cooled to below 3 K as a result of the enormous expansion of the universe.

15.6 PHONONS IN CRYSTALLINE SOLIDS

In Section 8.2, the law of Dulong and Petit for the specific heat of solids in the classical high-temperature limit was obtained with the use of the equipartition of energy theorem. A simple model for solids was introduced in which the atoms are considered to be joined together by springs which obey Hooke's law. We now outline the theoretical treatment of this model and show that it can account for the specific heats of solids over a wide range of temperature.

Formally, the total potential energy U of a solid containing N ions or atoms that can oscillate with temperature-dependent amplitude about their low-temperature equilibrium positions may be written as a Taylor expansion about the low-temperature value U_0:

$$U = U_0 + \sum_{i\alpha}\left(\frac{\partial U}{\partial u_{i\alpha}}\right)_0 u_{i\alpha} + \frac{1}{2}\sum_{\substack{i\alpha\\j\beta}} u_{i\alpha}\left(\frac{\partial^2 U}{\partial u_{i\alpha}\,\partial u_{j\beta}}\right)_0 u_{j\beta} + \cdots, \tag{15.26}$$

where $u_{i\alpha}$ is the displacement of atom i from equilibrium along direction α. The total potential energy U_0 is the sum of pair potentials when all atoms are in their equilibrium positions. The linear term in Equation 15.26 vanishes because the potential energy is a minimum in equilibrium and, to a good approximation, only the zero- and second-order terms in Equation 15.26 need be retained. This is known as the *harmonic* approximation

$$U = U_0 + \frac{1}{2}\sum_{\substack{i\alpha\\j\beta}} u_{i\alpha} D_{\substack{i\alpha\\j\beta}} u_{j\beta} \text{ with } D_{\substack{i\alpha\\j\beta}} = \left(\frac{\partial^2 U}{\partial u_{i\alpha}\,\partial u_{j\beta}}\right)_0.$$

Classical mechanics shows that for a system of particles, it is possible to choose a set of independent, generalized coordinates q_i to specify the positions of all particles in the system, with the number of generalized coordinates giving the number of degrees of freedom of the system. We transform from the set of $3N$ displacements $u_{i\alpha}$ to a set of generalized coordinates q_r using a linear transformation

$$u_{i\alpha} = \sum_{i-1}^{3\alpha} A_{i\alpha} q_r.$$

(15.27)

With a proper choice of the coefficient $A_{i\alpha}$, the Hamiltonian for the system of particles may be written as

$$\mathcal{H} = U_0 + \frac{1}{2} \sum_{r=1}^{3N} m\left(\dot{q}_r^2 + \omega_r^2 q_r^2\right).$$

(15.28)

Equation 15.28 has no cross terms between different coordinates, and the transformed Hamiltonian corresponds to $3N$ independent harmonic oscillators. The frequencies ω_r are called the normal mode frequencies. From quantum mechanics (see Appendix C), the energy eigenvalues for the harmonic oscillators are $\varepsilon_r = \left(n_r + \frac{1}{2}\right)\hbar\omega_r$, where $n_r = 0, 1, 2, \ldots$, is the set of quantum numbers for oscillator r. The total lattice energy in the harmonic approximation may be written in the simple form

$$E = U_0 + \sum_{r=1}^{3N} \left(n_r + \frac{1}{2}\right)\hbar\omega_r.$$

(15.29)

In calculating the normal mode frequencies for a set of $3N$ oscillators, various approximations may be used in obtaining expressions for E, and hence the specific heat. Two famous models, the Einstein and the Debye models, which were briefly introduced in Chapter 8, are discussed in detail in Sections 15.8 and 15.9. In Equation 15.29, the quantum number n_r is identified with the number of elementary excitations called phonons, of energy $\hbar\omega_r$, associated with lattice vibrational mode r. Phonons in condensed matter are analogous to photons in electromagnetic radiation in a number of ways. Like photons, phonons are bosons and obey the Planck distribution. Phonons travel with the velocity of sound and can have three polarization directions, two transverse and one longitudinal. We distinguish below between optic and acoustic phonons. Expressions for the energy and specific heat of a solid are obtained from the partition function for the set of lattice oscillators.

15.7   THE SPECIFIC HEAT OF A SOLID

From Equation 15.29, the partition function associated with lattice vibrations in a solid is

$$Z = e^{-\beta U_0} \prod_{r=1}^{3N} \sum_{n_r} e^{-\beta\left(n_r + \frac{1}{2}\right)\hbar\omega_r}.$$

This gives

$$Z = e^{-\beta U_0} \prod_r \left[\frac{e^{-(1/2)\beta\hbar\omega_r}}{1 - e^{-\beta\hbar\omega_r}}\right],$$

(15.30)

where the geometric series sum result from Appendix A has been used to carry out the summation. The mean energy of the solid is obtained directly with the use of the general canonical distribution expression as given by Equation 10.34:

$$\langle E \rangle = -\frac{\partial \ln Z}{\partial \beta} = U_0 + \sum_{r=1}^{3N} \hbar\omega_r \left[\frac{1}{e^{\beta\hbar\omega_r} - 1} + \frac{1}{2} \right], \tag{15.31}$$

and the heat capacity follows immediately

$$C_v = \left(\frac{\partial \langle E \rangle}{\partial T} \right)_V = -k_B\beta^2 \frac{\partial}{\partial \beta} \sum_{r=1}^{3N} \left[\frac{\hbar\omega_r}{e^{\beta\hbar\omega_r} - 1} \right]. \tag{15.32}$$

To obtain an explicit expression for C_V, it is necessary to evaluate the summation in Equation 15.32. Various approximations and models may be used to do this. In the high-temperature limit, where $e^{\beta\hbar\omega_r} \ll 1$ for all r, the denominator may, to a good approximation, be written as $(e^{\beta\hbar\omega r} - 1) \approx \beta\hbar\omega_r$. In this limit, Equation 15.32 becomes

$$C_V = 3Nk_B, \tag{15.33}$$

and for $N = N_A$, we obtain for the molar-specific heat $c_V = 3R$, which is the familiar Dulong–Petit law discussed in Chapter 8. More generally, the sum in Equation 15.32 may be replaced by an integral over all wave vectors q:

$$C_v = -k_B\beta^2 \frac{\partial}{\partial \beta} \sum_s \int_0^{qm} \frac{\hbar\omega_{q,s}}{e^{\beta\omega_{q,s}} - 1} \rho_q d^3q, \tag{15.34}$$

where $\omega_{q,s}$ is the frequency of a lattice mode with wave vector q in branch s. The upper limit q_m is discussed below for the Debye model. For monatomic solids, we consider two transverse acoustic modes and one longitudinal mode, and the sum over s will introduce a factor 3. For ionic and other solids consisting of two different types of ions, or different atoms, it is necessary to consider both optic and acoustic modes. For acoustic modes, neighbor atoms tend to vibrate in phase, whereas in optic modes, neighbor atoms are out of phase. Further details may be found in books on solid-state physics. The integral in Equation 15.34 is in principle carried out over the first Brillouin zone, with the zone boundary along a given direction in q-space determined by the condition that the wavelength of a lattice mode becomes equal to the spacing between lattice planes along this direction in the solid. The relatively simple Einstein and Debye models lead to specific heat expressions that can be compared with the experiment. Although the assumptions that are made simplify the situation, the predictions are in fair agreement with the experiment. Details of the Einstein and Debye models are presented in Sections 15.8 and 15.9, respectively.

15.8 THE EINSTEIN MODEL FOR THE SPECIFIC HEAT OF SOLIDS

In the integral over $\omega_{q,s}$, Einstein made the very simple assumption that all $3N$ lattice modes have the same frequency ω_E regardless of wave vector. Figure 15.4 shows the ω–q dispersion curve for an Einstein solid.

The form shown in Figure 15.4 *is* actually better suited to high-frequency optic modes compared to acoustic modes. From Equation 15.32 with $\omega_r = \omega_E$ for all r, the heat capacity is readily shown to be

$$C_v = 3Nk_B \left(\frac{\hbar\omega_E}{k_BT} \right)^2 \frac{e^{\beta\hbar\omega_E}}{\left(e^{\beta\hbar\omega_E} - 1 \right)^2}. \tag{15.35}$$

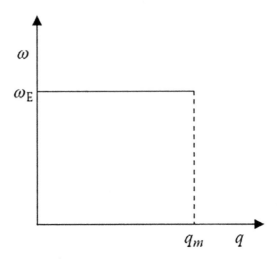

FIGURE 15.4 Dispersion curve ω versus q for the Einstein solid. All $3N$ oscillators have the same Einstein frequency for wave vectors in the accessible range 0 to q_m.

If we define the Einstein temperature as $\theta_E = \hbar\omega_E/k_B$, then for $N = N_A$, Equation 15.35 gives the specific heat as

$$C_V = 3R\left(\frac{\theta_E}{T}\right)^2 \frac{e^{\theta_E/T}}{\left(e^{\theta_E/T} - 1\right)^2}. \tag{15.36}$$

Figure 15.5 shows the predicted form for the specific heat as a function of the reduced temperature θ_E/T. With θ_E as an adjustable parameter, it is possible to obtain fairly good agreement with measured specific heats for many solids. In the high-temperature limit, where $\theta_E/T \ll 1$, the law of Dulong and Petit $c_V = 3R$ is obtained, as expected.

At low temperatures, where $\theta_E/T \gg 1$, the specific heat becomes

$$c_V \approx 3R\left(\frac{\theta_E}{T}\right)^2 e^{-\frac{\theta_E}{T}}. \tag{15.37}$$

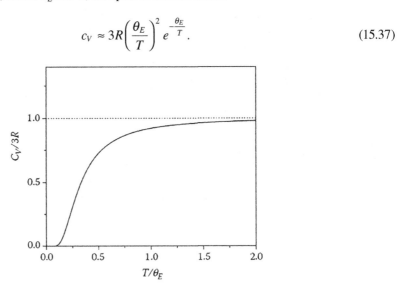

FIGURE 15.5 The Einstein molar-specific heat, plotted in the scaled form $c_V/3R$, as a function of the reduced temperature T/θ_E. The Einstein temperature θ_E is used to obtain the best fit to the measured specific heat curve for a particular solid.

It follows that c_V is predicted to go to zero exponentially as $T \rightarrow 0$ K. This prediction is not in good agreement with experimental results for solids. The reason for the failure of the Einstein model in the low T limit is that the model does not allow for low-frequency modes, which are of dominant importance at low temperatures. The Debye model is much more successful in this respect and correctly predicts the form of the specific heat at low temperatures. For many solids, θ_E in the range of 200–500 K is found to provide a reasonable agreement between theory and experiment at temperatures that are not too low. For hard solids such as diamond, which have high effective "spring constants," the Einstein temperature is much higher than for more ductile solids. The Einstein frequency is defined as $\omega_E = k_B \theta_E / \hbar$, and for $\theta_E \simeq 300$ K, this expression gives $\omega_E = 6 \times 10^{12} s^{-1}$, which is in the infrared. At high temperature, we recover the Dulong–Petit law, as noted previously.

15.9 THE DEBYE MODEL FOR THE SPECIFIC HEAT OF SOLIDS

Debye suggested the use of a continuum model in calculating the specific heat of a solid. The wave equation for the propagation of elastic waves in a continuous medium is similar in form to Equation 15.2

$$\nabla^2 u - \left(\frac{1}{v^2}\right)\frac{\partial^2 u}{\partial t^2} = 0 \qquad (15.38)$$

u is the displacement of an element in the solid, and v is the phase velocity of the wave. We again allow for the elastic waves to have three directions of polarization, two transverse and one longitudinal. The velocities of the transverse and longitudinal waves are, in general, not the same. When the dimensions of the solid are much larger than the wavelength, the expression for the density of modes for elastic waves will be the same as that for electromagnetic waves given by Equation 15.7. Both expressions are equivalent to the particle in a box density of states.

Debye introduced the assumption that the dispersion relation for acoustic waves is of the form $\omega = vq$. This is a good approximation for long-wavelength acoustic modes but does not apply very well to modes where the wavelength is comparable with the lattice spacing. Figure 15.6 illustrates the form of the Debye linear dispersion relation.

The form of the dispersion relation as shown in Figure 15.6 is to be contrasted with the form as shown in Figure 15.4 for the Einstein model. Debye introduced an upper cutoff wave vector q_D, and equivalently a cutoff frequency ω_D, for each polarization on the basis of the condition $\frac{4}{3}\pi q_D^3 \rho_q = N$,

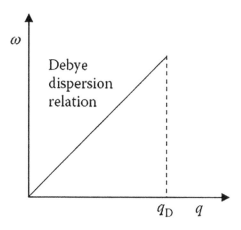

FIGURE 15.6 Linear dispersion relation $\omega = vq$ used in the Debye model. The high-frequency cutoff has wave vector q_D and frequency ω_D.

with the density of states $\rho_q = V/(2\pi)^3$ from Equation 15.6 and V as the volume of the solid. In the Debye model, the Brillouin zone is approximated by a sphere of radius q_D. This procedure gives $q_D = (6\pi^2 N/V)^{1/3}$. The heat capacity expression follows from Equation 15.34:

$$C_v = k_B \beta^2 \frac{\partial}{\partial\beta} \frac{3V}{(2\pi)^3} \int_0^{q_D} \frac{\hbar v q}{e^{\beta\hbar v q} - 1} \left(4\pi q^2\right) dq,$$

where the volume element in q-space is taken to be a spherical shell of radius q and thickness dq. A factor 3 is introduced to allow for the three branches s corresponding to the three polarization directions. Performing the differentiation with respect to β and with introduction of the variable $x = \beta\hbar v q$, we obtain

$$C_v = \frac{3k_B V}{2\pi^2 \beta^3 \hbar^3 v^3} \int_0^{x_D} \frac{x^4 e^x}{\left(e^x - 1\right)^2} dx. \tag{15.39}$$

The Debye temperature is defined as $\theta_D = \hbar\omega_D/k_B = \hbar v q_D / / k_B$, and Equation 15.39 may be written as

$$C_v = 9Nk_B \left(\frac{T}{\theta_D}\right)^3 \int_0^{\frac{\theta_D}{T}} \frac{x^4 e^x}{\left(e^x - 1\right)^2} dx. \tag{15.40}$$

At high temperatures when $T > \theta_D$, the exponential factors in the integral in Equation 15.40 may be expanded and to a good approximation

$$C_v \approx 9Nk_B \left(\frac{T}{\theta_D}\right)^3 \int_0^{\frac{\theta_D}{T}} x^2 \, dx = 3Nk_B. \tag{15.41}$$

This is once again the Dulong–Petit law. At low temperatures where $T \ll \theta_D$, the upper limit in the integral in Equation 15.40 may, to a good approximation, be extended to infinity. Integration by parts results in the same integral that was encountered in the derivation of the Stefan–Boltzmann law as given by Equation 15.20:

$$\int_0^\infty \frac{x^4 e^x}{\left(e^x - 1\right)^2} dx = 4 \int_0^\infty \frac{x^3}{\left(e^x - 1\right)} = \frac{4\pi^4}{15}.$$

In the low-temperature limit for $N = N_A$, the specific heat therefore becomes

$$c_v \approx 9R \left(\frac{4\pi^4}{15}\right) \left(\frac{T}{\theta_D}\right)^3 = \frac{12}{5} \pi^4 R \left(\frac{T}{\theta_D}\right)^3. \tag{15.42}$$

This predicts that $c_v \propto T^3$ at low temperatures, which is in very good agreement with the experimental values obtained for insulating solids. For metals, allowance must be made for the conduction electron contribution $c_V = \gamma T$ that is of dominant importance at sufficiently low temperatures ($T < 10$ K).

The importance of the Planck distribution has been shown in this chapter in treating two important applications: black body radiation and the specific heats of solids. Photons and phonons along with other excitations such as magnons in magnetic materials are governed by the Planck distribution. Together with the Fermi–Dirac and Bose–Einstein quantum distributions, considered in Chapters 13 and 14, respectively, the Planck distribution plays a vital role in explaining phenomena in many body physics.

PROBLEMS CHAPTER 15

15.1   Black body radiation exists in an enclosure of volume V at temperature T. Use the partition function to obtain expressions for the Helmholtz potential F and the entropy S of the radiation.

15.2   Find the pressure exerted on the walls of the enclosure for the black body radiation of Question 1. Give a numerical estimate of the radiation pressure for $T = 1000$ K.

15.3   Fluctuations in the number of photons in cavity mode r for black body radiation in an enclosure are given by $\sigma_{nr}/\langle n_r \rangle = [1/\langle n_r \rangle + 1]^{1/2}$. Describe the behavior of these fluctuations for mode r, of frequency ω_r, as a function of temperature.

15.4   The energy density in the range from ω to $\omega + d\omega$ for black body radiation is given by the Planck expression $CT^4 \, (x^3 dx/(e^x - 1))$, where $C = k_B^4/\pi^2 c^3 \hbar^3$ and $x = \hbar\omega/k_B T$. Show that this distribution has a maximum for $x = 2.82$. (Use numerical or graphical methods to solve the equation that you obtain.) If the temperature of the cavity is increased from 1000 to 1200 K, find the shift in frequency of the maximum in the energy density.

15.5   Derive the Stefan–Boltzmann law $I(T) = \sigma T^4$ for the total intensity of radiation emitted through a small hole in the wall of a black body enclosure at temperature T. Show that the constant σ is given by $\sigma = \mu^2 k_B^4/60 c^2 \hbar^3$. Start your calculation with the expression for the energy density given in Equation 15.18, and consider radiation striking an area A in time t.

15.6   The cosmic background microwave radiation spectrum is well fit by the Planck equation for black body radiation with a temperature close to 3 K. Models for the evolution of the universe predict that, following the Big Bang, condensation of particles occurred as the universe expanded and cooled. After 300 million years, the temperature had dropped sufficiently that electrons became bound in atoms and the universe became transparent to electromagnetic radiation. Estimate the radiation temperature at this time, assuming the average ionization energy is close to that of hydrogen eV. Use this temperature and the present radiation temperature to estimate by how much the universe has expanded in the past 14 billion years.

15.7   Obtain expressions for the phonon contribution to the specific heat of a two-dimensional lattice of atoms of mass m in the low T and high T limits by making use of both the Einstein and Debye models.

15.8   Evaluate the partition function for a solid in the Einstein approximation, and use your result to obtain an expression for the entropy of the solid as a function of temperature.

15.9   The spins in a ferromagnetic solid are coupled by an exchange interaction. At low temperatures, elementary excitations called spin waves are important, and for a cubic lattice, the dispersion relation may to a good approximation be written as $\omega = Kq^2$, where ω is the frequency, q is the wave vector, and K is a constant. Use this form to obtain an expression giving the temperature dependence of the spin wave contribution to the low-temperature specific heat of the solid. Spin waves have a single polarization for each q value. In integration over q-space, you may assume that the temperature is sufficiently low that the upper limit in the integral may be put equal to infinity.

Section IIC

The Classical Ideal Gas, Maxwell–
Boltzmann Statistics, Nonideal Systems

16 The Classical Ideal Gas

16.1 INTRODUCTION

Following our discussion of the quantum gas cases in Chapters 13 and 14, we are now in a position to consider the classical ideal gas in some detail. For simplicity, we shall initially consider systems of particles without internal degrees of freedom, such as monatomic gases, but this restriction will be listed later in this chapter. In Chapter 12, we considered the classical limit of the Bose–Einstein and Fermi–Dirac distributions. In the classical limit, the partition function for a system of N particles

in a gas is given by Equation 12.24, $Z = (1/N!)\left[\sum_r e^{-\beta\varepsilon_r}\right]^N = (1/N!)\, z^N$, with z as the single-particle

partition function. The factor $1/N!$ allows for the indistinguishability of particles, as discussed in Chapters 4 and 12. Because internal energy contributions are not considered, the states r, with energy eigenvalues ε_r, correspond to the particle in a box state considered in Chapter 4. Equation 4.5 gives the energy eigenvalues as $\varepsilon_r = \left(\pi^2\hbar^2/2m\right)\left(1/V^{2/3}\right)\left(n_x^2 + n_y^2 + n_z^2\right)$, where the quantum numbers n_x, n_y, and n_z take integral values.

As noted in Chapter 4, in the classical limit, a large number of eigenstates are populated, albeit in a sparse way, with the mean number of particles in a given state $\langle n_r\rangle \ll 1$. It is instructive to make use of the familiar classical limit inequality $V_Q \ll V_A$, where $V_Q = h^3/(3mk_BT)^{3/2}$ is the quantum volume and $V_A = (V/N)$ is the volume per particle. The inequality may be rewritten in the following form:

$$\left(\frac{N}{V}\right)^{2/3}\left(\frac{\hbar^2}{3m}\right) \ll k_B T \tag{16.1}$$

which is convenient for the comparison that is made below. From Equation 4.5, the spacing between adjacent energy levels with quantum numbers (n_x, n_y, n_e) and (n_x+1, n_y, n_z) is to a good approximation

$$\Delta\varepsilon_r \approx \left(\frac{n_x}{4}\right)\left(\frac{\hbar^2}{mV^{2/3}}\right). \tag{16.2}$$

Comparison of Equations 16.1 and 16.2 suggests that $k_B T \gg \Delta\varepsilon_r$, so that the thermal energy per degree of freedom is very much larger than the spacing of the energy levels, as depicted in Figure 16.1.

The energy states are so closely spaced that they form a quasi-continuum. Because the factor $e^{-\beta\varepsilon_r}$ in Equation 12.24 varies slowly with ε_r, it is permissible to replace the sum by an integral in the evaluation of the partition function Z. This approach is made use of in Section 16.2. Once the partition function has been determined, we use the bridge relationship $F = -k_B T \ln Z$ to obtain the Helmholtz potential and from F the entropy of the ideal monatomic gas. For polyatomic molecules, allowance must be made for internal degrees of freedom, and this topic is discussed in Section 16.5. This chapter concludes with a proof of the equipartition theorem and a brief discussion of the Maxwell velocity distribution for a classical gas.

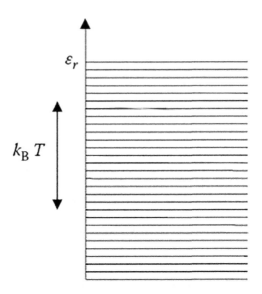

FIGURE 16.1 Schematic representation of a range of energy levels for a particle in a box. For gases in the classical limit, the thermal energy per degree of freedom $\sim k_B T$ is much larger than the spacing between single-particle levels.

16.2 THE PARTITION FUNCTION FOR AN IDEAL CLASSICAL GAS

If the gas particles do not possess internal energies, the single-particle partition function for an ideal gas may be written as

$$z = \sum_r e^{-\beta \varepsilon_r} = \sum_{n_x, n_y, n_z} e^{-\beta \left(\pi^2 \hbar^2 / 2mV^{2/3}\right)\left[n_x^2 + n_y^2 + n_z^2\right]}. \tag{16.3}$$

To simplify the expression, introduce $K = (\pi^2 \hbar^2 / 2mV^{2/3})$ so that the partition function may be written as

$$z = \sum_{n_x=1}^{\infty} e^{-\beta K n_x^2} \sum_{n_y=1}^{\infty} e^{-\beta K n_y^2} \sum_{n_z=1}^{\infty} e^{-\beta K n_x^2}.$$

Each of the identical summations may, to a good approximation, be replaced by an integral over $\sum_{n=1}^{\infty} e^{-\beta K n^2} \simeq \int_0^{\infty} e^{-\beta K n^2} \, dn$, with the lower limit taken as zero, with negligible error, because so many states contribute to the integral. The integral is given in Appendix A, $\int_0^{\infty} e^{-x^2} \, dx = \sqrt{\pi}/2$, and, with $x = (\beta K)^{1/2} n$, it follows that $\left(\int_0^{\infty} e^{-\beta K n^2} \, dx = \frac{1}{2} \sqrt{\pi/\beta K} \right)$. The single-particle partition function is therefore given by

$$z = \left(\frac{\pi}{4\beta K}\right)^{3/2} = V \left(\frac{m}{2\pi \hbar^2 \beta}\right)^{3/2}. \tag{16.4}$$

The partition function for a gas of N molecules follows from Equation 12.24:

$$Z = \left(\frac{V^N}{N!}\right)\left(\frac{m}{2\pi\hbar^2\beta}\right)^{3N/2} \tag{16.5}$$

and hence,

$$\ln Z = N\left[\ln V - \ln \beta + \frac{3}{2}\ln\left(\frac{m}{2\pi\hbar^2}\right)\right] - N \ln N + N, \tag{16.6}$$

where the use has been made of Stirling's formula for $\ln N!$

Exercise 16.1

Show that the single-particle partition function z may, to a good approximation, be written in terms of the quantum volume $V_Q = (h^2/3mk_BT)^{3/2}$ as $z \approx V/V_Q$. Obtain an expression for the partition function for N particles in terms of the quantum volume and the atomic volume.

From Equation 16.4, the single-particle partition function is given by

$$z = V\left(\frac{m}{2\pi\hbar^2\beta}\right)^{3/2} = V\left(\frac{3mk_BT}{\hbar^2}\right)^{3/2}\left(\frac{2\pi}{3}\right)^{3/2} \approx \left(\frac{V}{V_Q}\right).$$

For N particles, we write the N particle partition function as

$$\ln Z \approx N \ln\left(\frac{V}{V_Q}\right) - N \ln N + N = N\left[\ln\left(\frac{V_A}{V_Q}\right) + 1\right].$$

An advantage of the simple form for $\ln Z$ in terms of $V_A = V/N$ and V_Q is that it is memorable and easy to write down. Furthermore, the form emphasizes the importance of the classical limit condition $V_A \gg V_Q$.

16.3 THERMODYNAMICS OF AN IDEAL GAS

From Equations 10.49 and 16.6, an expression for the Helmholtz potential for an ideal gas is immediately obtained

$$F = -k_BT \ln Z = -Nk_BT\left[\ln V + \frac{3}{2}\ln k_BT + \frac{3}{2}\ln\left(\frac{m}{2\pi\hbar^2}\right)\right] + Nk_BT(\ln N - 1). \tag{16.7}$$

The entropy is as follows:

$$S = -\left(\frac{\partial F}{\partial T}\right)_V = Nk_B\left[\ln\left(\frac{V}{N}\right) + \frac{3}{2}\ln\left(\frac{mk_BT}{2\pi\hbar^2}\right) + \frac{5}{2}\right]. \tag{16.8}$$

This is the Sackur–Tetrode equation, given previously in the equivalent form in Equation 5.15. The entropy is obtained correctly as an extensive quality because the factor $1/N!$, which takes account of the indistinguishability of particles, is introduced in a natural way in taking the classical limit of the quantum distributions. For constant particle density (i.e., V/N constant) and T constant, Equation 16.8 shows that S is proportional to N. The entropy expression does not allow for possible spin states of the ideal gas particles. If the particles have angular momentum specified by quantum number J, then an additional term $Nk_B \ln(2J+1)$ must be added to the entropy, corresponding to the $(2J+1)^N$ possible spin states.

Exercise 16.2

Use the expression for the Helmholtz potential F to obtain expressions for thermodynamic quantities of interest, including pressure, specific heat, and chemical potential. (Several of these results have been discussed previously in Section 5.4.)

The pressure is given by $P = -(\partial F/\partial V)_T = Nk_BT/V$, which is the ideal gas equation of state.

The mean energy is $\langle E \rangle = -\partial \ln Z/\partial \beta = \frac{3}{2}Nk_BT$ or, per particle, $\langle E \rangle = \langle E \rangle/N = \frac{3}{2}k_BT$, in agreement with the equipartition of energy theorem, with three translational degrees of freedom for each particle.

The heat capacity of the gas follows directly: $C_V = (\partial \langle E \rangle/\partial T)_V = \frac{3}{2}Nk_B$, or for 1 mol $C_V = \frac{3}{2}R$, which is the familiar result for the specific heat of a monatomic ideal gas.

Finally, the chemical potential is given by

$$\mu = -T\left(\frac{\partial S}{\partial N}\right)_{T,V} = k_BT\left[\ln\left(\frac{N}{V}\right) + \frac{3}{2}\ln\left(\frac{2\pi\hbar^2}{mk_BT}\right) - \frac{3}{2}\right].$$

As noted in Chapter 11, μ is proportional to T and depends on the particle concentration N/V. We have seen that $\mu < 0$ in the classical limit of the quantum distributions. μ may again be written in a simple form by making use of the quantum volume V_q and the atomic volume V_A. To a good approximation, we obtain $\mu \simeq k_BT\left[\ln\left(V_Q/V_A\right) - \frac{3}{2}\left[\ln(2\pi/3)+1\right]\right]$, with $V_Q \ll V_A$ in the classical limit, as emphasized previously. Representative values for a classical gas at standard temperature and pressure lead to $V_Q/V_A \sim 10^{-4}$ so that $\mu < 0$. In general, for given T, large values for V_A, which correspond to a low particle density, lead to relatively large negative values for μ. For systems not in equilibrium, where there is a gradient in μ, particles will tend to move from regions of high concentration to regions of low concentration. When V_Q and V_A become comparable, the classical limit no longer applies and the appropriate quantum statistics expressions must be used.

Exercise 16.3

Use the expression for μ derived above to obtain a simple classical limit form for the fugacity $\lambda = e^{\beta\mu}$, which was defined in Chapter 11 in our discussion of the quantum distributions.

From the expression for μ given above, we obtain $\lambda \approx 0.2(V_Q/V_A)$. It is clear that $\lambda \ll 1$ for a classical gas, as expected from the discussion given in Chapter 12.

16.4   CLASSICAL MECHANICS DESCRIPTION OF THE IDEAL GAS

In dealing with an ideal gas in the classical limit, it is appropriate to attempt a description of the system using classical mechanics rather than quantum mechanics. The concept of classical phase space was introduced in Chapter 4. There the microcanonical ensemble approach was used with the energy of the system held constant. In the canonical ensemble case, energy is exchanged with a heat bath at temperature T.

The energy of a gas particle of mass m and momentum p may be classically written as $\varepsilon = \mathcal{H}$ $K + U = p^2/2m$ because the potential energy $U = 0$ for particles in an ideal gas. Making use of phase space ideas and converting the sum over states into an integral, the partition function for a particle may be written as

$$z = \frac{1}{h_0^3}\int_V \int_{-\infty}^{\infty} e^{-\beta\left(p^2/2m\right)}d^3q\,d^3\boldsymbol{p}, \tag{16.9}$$

where $(1/h_0{}^3)\, \mathrm{d}^3\boldsymbol{q}\, \mathrm{d}^3\boldsymbol{p}$ is the number of cells in phase space, with $\boldsymbol{q}$ in the range from $\boldsymbol{q}$ to $\boldsymbol{q}+\mathrm{d}\boldsymbol{q}$ and $\boldsymbol{p}$ in the range from $\boldsymbol{p}$ to $\boldsymbol{p}+\mathrm{d}\boldsymbol{p}$. $h_0{}^3$ is the volume of an elementary cell in 6D phase space and is written as $h_0{}^3 = \delta p_x \delta p_y \delta p_z \delta x\, \delta y\, \delta z$ in Cartesian coordinates. Figure 16.2 shows the phase space representation for a single particle in a one-dimensional box of length L.

For N particles in a three-dimensional container, it is necessary to use the 6N-dimensional phase space representation introduced in Chapter 4. As noted there, it is impossible for us to visualize such a space, but this does not prevent us from using the representation. From our single-particle expression, we can see that the integral over the spatial coordinates in Equation 16.9 is straightforward because the integral is independent of these coordinates. It follows that

$$z = \frac{V}{h_0{}^3} \int_{-\infty}^{\infty} \int_{-\infty}^{\infty} \int_{-\infty}^{\infty} e^{-\beta\left[\left(p_x^2+p_y^2+p_z^2\right)/2m\right]}\, \mathrm{d}p_x\, \mathrm{d}p_y\, \mathrm{d}p_z = \frac{V}{h_0^3}\left[\int_{-\infty}^{\infty} e^{-\beta\left(p_x^2/2m\right)}\, \mathrm{d}p_x\right]^3, \quad (16.10)$$

where the use has been made of the equivalence of x, y, and z. With $u = (\beta/2m)^{1/2}\, p_x$ and a standard integral from Appendix A, $\int_{-\infty}^{\infty} e^{-\mu^2}\, \mathrm{d}u = \sqrt{\pi}$, we get $z = (V/h_0{}^3)(2\pi m/\beta)^{3/2}$. For a system of N particles, using Equation 12.24, the partition function is

$$Z = 1/N!\left[\frac{V}{h_0^3}\left(\frac{2\pi}{\beta}\right)^{3/2}\right]^N. \quad (16.11)$$

This agrees with the expression given in Equation 16.5 for Z, obtained using the quantum mechanical description, provided we put $h_0 = h$. The arbitrariness of the size of a cell in classical phase space is replaced by a fixed value involving Planck's constant in the quantum mechanical description. The thermodynamic results derived previously may be obtained from the partition function given in Equation 16.11, and the classical and quantum mechanical descriptions are equivalent in the classical limit. The classical phase space approach was used by pioneers in the field before the development of quantum mechanics in the 1920s.

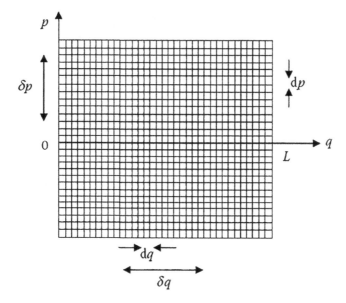

FIGURE 16.2 Phase space representation of the accessible states for a single particle in a one-dimensional box in the canonical ensemble. Not all microstates are equally probable.

16.5   IDEAL GAS OF PARTICLES WITH INTERNAL ENERGIES

The discussion given above for a classical gas of particles may readily be extended to include particles, such as polyatomic molecules, with internal degrees of freedom. The total energy of such a particle with independent energy contributions may, to a good approximation, be written in the additive form,

$$\varepsilon_r = \varepsilon^{\text{tr}} + \varepsilon^{\text{rot}} + \varepsilon^{\text{vib}} + \varepsilon^{\text{el}}, \tag{16.12}$$

where ε^{tr}, ε^{rot} ε^{vib}, and ε^{el} represent the translational, rotational, vibrational, and electronic energy contributions, respectively. We shall assume that at temperatures of interest, molecules will be found in their electronic ground states, with probabilities close to unity. This is because the thermal energy per degree of freedom $\sim k_{\text{B}}T \sim 25$ meV for $T = 300$ K, whereas electronic excitation energies are typically much higher than this on the order of electron volts. The single-particle partition function is given by

$$z = \sum_r e^{-\beta \varepsilon_r} = \sum_{\text{tr}} \sum_{\text{rot}} \sum_{\text{vib}} \sum_{\text{el}} e^{-\beta\left(\varepsilon^{\text{tr}}+\varepsilon^{\text{rot}}+\varepsilon^{\text{vib}}+\varepsilon^{\text{el}}\right)}$$

$$= \sum_r e^{-\beta \varepsilon^{\text{tr}}} \sum_{\text{rot}} e^{-\beta \varepsilon^{\text{rot}}} \sum_{\text{vib}} e^{-\beta \varepsilon^{\text{vib}}} \sum_{\text{el}} e^{-\beta \varepsilon^{\text{el}}}, \tag{16.13}$$

where the summations are over all translational, rotational, vibrational, and electronic energy states. Because of the properties of the exponential function, factorization of the partition function occurs. It follows that

$$\ln z = \ln z^{\text{tr}} + \ln z^{\text{rot}} + \ln z^{\text{vib}} + \ln z^{\text{el}}. \tag{16.14}$$

For a system of N such particles, the partition function is given, as before, by $\ln Z = N \ln z - N \ln N + N$.

To proceed, consider a system of diatomic molecules, each with moment of inertia $I = \dfrac{1}{2}\mu_R r_0^{\,2}$ about an axis passing through the center of mass and perpendicular to the bond joining the atoms. The reduced mass μ_R is given by $\mu_R = m_1 m_2/(m_1 + m_2)$, with m_1 and m_2 as the atomic masses and r_0 the equilibrium internuclear separation. The interatomic potential has the familiar form as shown in Figure 16.3.

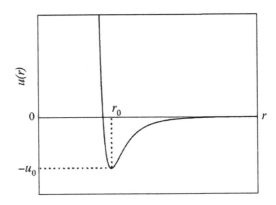

FIGURE 16.3   The plot shows the form of the interatomic potential as a function of separation of the atoms in a diatomic molecule. The distance r_0 corresponds to the equilibrium separation of the two atoms.

We now consider in some detail the vibrational, rotational, and translational energy states that occur in the partition function as given in Equation 16.13.

a. The vibrational motion may be treated in a way similar to that used in Section 15.6 in dealing with lattice vibrations. Expanding the potential function in a Taylor series about the minimum at r_0 and omitting the first-order term which vanishes at the potential minimum give

$$u(r) = -u_0 + \frac{1}{2}\left(\frac{\partial^2 u}{\partial r^2}\right)_{r_0}(r - r_0)^2 + \cdots. \tag{16.15}$$

For small-amplitude vibrations, as an approximation, we retain terms up to second order in the expansion. This corresponds to the harmonic approximation introduced for vibrational modes in solids in Chapter 15. With $q_r = (r - r_0)$, the energy associated with vibrational motion is classically given by

$$\varepsilon^{\text{vib}} = \frac{1}{2}\mu_R \dot{q}_r^2 + \frac{1}{2}\mu_R \omega^2 q_r^2, \tag{16.16}$$

where $\omega = \sqrt{k/\mu_R}$, with $k = \left(\partial^2 u/\partial r^2\right)_{r_0}$ being the effective spring constant.

The electronic energy ε^{el} corresponds to $-u_0$ in Equation 16.15. From Equation 16.16, it is seen that the vibrational motion may be treated as simple harmonic motion. Quantum mechanically, the energy eigenvalues for a simple harmonic oscillator are

$$\varepsilon^{\text{vib}} = \left(n + \frac{1}{2}\right)\hbar\omega, \text{ with } n = 0, 1, 2, \cdots. \tag{16.17}$$

b. The translational motion of the molecule gives rise to particle in a box energy eigenvalue as before but now involving the total mass $(m_1 + m_2)$ of the molecule

$$\varepsilon^{\text{tr}} = \left(\frac{\pi^2 \hbar^2}{2V^{2/3}(m_1 + m_2)}\right)\left(n_x^2 + n_y^2 + n_z^2\right), \tag{16.18}$$

with n_x, n_y, $n_z = 1, 2, 3, \ldots$.

c. Finally, we consider the rotational motion with the rotational energy given by

$$\varepsilon^{\text{rot}} = \frac{1}{2}I\omega_r^2 = \frac{1}{2}\frac{L^2}{I}, \tag{16.19}$$

where I is the moment of inertia and $L = I\omega_r$ is the angular momentum about an axis of rotation. For a diatomic molecule, two axes perpendicular to the internuclear axis must be considered. The quantum mechanical expression for the energy eigenvalues is

$$\varepsilon^{\text{rot}} = \frac{1}{2}\left(\frac{\hbar^2}{I}\right)L(L+1), \tag{16.20}$$

where the operator L^2 has eigenvalues $\hbar^2 L(L+1)$ with $L = 0, 1, 2, \ldots$. Allowance must be made for a degeneracy factor $(2L+1)$ for each eigenstate corresponding to the $(2L+1)$ values of the quantum number M_L.

We are now in a position to evaluate all of the molecular partition function factors in Equation 16.14.

Translational motion. Equation 16.5 gives for the translational single-particle partition function,

$$z^{\text{tr}} = \left(\frac{V}{\beta^{3/2}} \right) \left[\frac{m_1 + m_2}{2\pi\hbar^2} \right]^{3/2} \tag{16.21}$$

and

$$\left\langle E^{\text{tr}} \right\rangle = -N \frac{\partial \ln z^{\text{tr}}}{\partial \beta} = \frac{3}{2} \frac{N}{\beta} = \frac{3}{2} N k_B T. \tag{16.22}$$

Rotational motion. From Equation 16.19 with allowance for degeneracy, the single-particle rotational partition function is given by

$$z^{\text{rot}} = \sum_{L=0,\,1,2} (2L+1) e^{\left(-\beta\hbar^2/2I \right)L(L+1)} = 1 + 3e^{\left(-\beta\hbar^2/I \right)} + 5e^{-3\beta\left(\hbar^2/2I \right)} + \dots. \tag{16.23}$$

At high temperatures for $(\beta\hbar^2/2I)L(L+1) \ll 1$, the sum over L may be replaced by an integral. With $x = L(L+1)$, we obtain

$$z^{\text{rot}} = \int_0^{\infty} e^{\left(-\beta\hbar^2/2I \right)x} \, dx = \frac{2I}{\beta\hbar^2} \tag{16.24}$$

and hence,

$$\left\langle E^{\text{rot}} \right\rangle = \frac{N}{\beta} = N k_B T. \tag{16.25}$$

This result is correct if the molecule is heteronuclear, such as HCl. For homonuclear molecules, such as O_2, where the two atoms are indistinguishable, it is necessary to introduce a factor 1/2 in the expression for z^{rot} to avoid overcounting of states.

Vibrational motion. From Equation 16.17, we obtain for the vibrational partition function

$$z^{\text{vib}} = \sum_{n=0}^{\infty} e^{-\beta[n+(1/2)]\hbar\omega} = \frac{e^{-1/2\beta\hbar\omega}}{1 - e^{-\beta\hbar\omega}} \tag{16.26}$$

where the use has been made of the geometric series summation formula in Appendix A. The mean vibrational energy for the system is

$$\left\langle E^{\text{vib}} \right\rangle = N \left\langle \varepsilon^{\text{vib}} \right\rangle = -N \frac{\partial \ln z^{\text{vib}}}{\partial \beta} = \frac{1}{2} N \hbar\omega \left[1 + \frac{2}{e^{\beta\hbar\omega} - 1} \right]. \tag{16.27}$$

Note that $1/2 \, (N\hbar\omega)$ is the sum of the ground-state vibrational energies of the N molecules.

Electronic contribution. For temperatures that are not too high, almost all the molecules in the gas are in their electronic ground states with energy $-u_0$, as given in Equation 16.15, and, to a good approximation, we have

$$z^{\text{el}} = e^{\beta u_0}. \tag{16.28}$$

It follows that

$$\left\langle E^{\text{el}} \right\rangle = -N u_o. \tag{16.29}$$

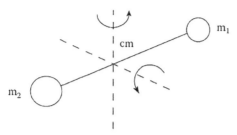

FIGURE 16.4 The rotational degrees of freedom are shown for a diatomic molecule. Rotations occur about two orthogonal axes through the center of mass. Rotational motion about the bond axis does not need to be considered because the moment of inertia about this axis is very small and the corresponding rotational levels are therefore very widely spaced.

In general, for diatomic molecules, the various energy contributions are ranked as follows: $\langle \varepsilon^{\text{el}} \rangle \gg \langle \varepsilon^{\text{vib}} \rangle \gg \langle \varepsilon^{\text{rot}} \rangle \gg \langle \varepsilon^{\text{tr}} \rangle$, with energies covering the range from electron volts to milli-electron volts. The fact that $\langle \varepsilon^{\text{vib}} \rangle \gg \langle \varepsilon^{\text{rot}} \rangle$ allows us to treat the vibrational and the rotational motions as uncoupled from each other. Consequently, at temperatures below 300 K, only the translational and rotational energy contributions are strongly temperature dependent for diatomic gases. It is convenient to characterize vibrational and rotational energies by characteristic temperatures $\theta^{\text{vib}} = \hbar \omega / k_{\text{B}}$ and $\theta^{\text{rot}} = \hbar^2 / 2 k_{\text{B}} I$, respectively. As an example, for nitrogen, these temperatures are $\theta^{\text{vib}} = 3 \times 10^3$ K and $\theta^{\text{rot}} = 3$ K. The total energy of a diatomic gas at normal temperatures may therefore be written, to a good approximation, as

$$\langle E \rangle = \frac{3}{2} N k_{\text{B}} T + N k_{\text{B}} T + \frac{1}{2} N \hbar \omega - N u_0. \tag{16.30}$$

The heat capacity is $C_v = 5/2 \, N k_{\text{B}}$, or per mole $c_v = 5/2 \, R$.

This is a result that can be written down immediately using the equipartition of energy theorem, assuming five degrees of freedom, three translations and two rotational. As noted in Chapter 1, the reason that classically there are only two rotational degrees of freedom is that the moment of inertia about the long axis of the molecule is extremely small and the characteristic temperature θ^{rot} for this motion is therefore extremely high. Figure 16.4 illustrates the situation.

16.6 PROOF OF THE EQUIPARTITION OF ENERGY THEOREM

We have stated the equipartition of energy theorem and have used it in a number of situations. The theorem states that in the classical limit, the mean energy associated with each independent quadratic energy term or degree of freedom as introduced in Section 16.5 is given by $1/2 \, (k_{\text{B}} T)$.

The proof of the theorem, using the canonical ensemble expressions in the classical limit, is straightforward. Consider a system of N particles that may be described using position and momentum coordinates $(q_1 \ldots q_n; p_1 \ldots p_n)$. Let the energy of the system take the form $E = \sum_{i=1}^{N} \varepsilon_i (q_i, p_i)$, where $\varepsilon_i (q_i, p_i)$ is the energy of particle i. We focus on the mean energy contribution $\langle \varepsilon_i (p_i) \rangle$ of particle i. The classical phase space approach of Section 16.4 gives

$$\langle \varepsilon_i (p_i) \rangle = \frac{\int \varepsilon_i e^{-\beta E} dq_1 \cdots dp_1 \cdots}{\int e^{-\beta E} dq_1 \cdots dp_1 \cdots} = \frac{\int \varepsilon_i e^{-\beta \varepsilon_i} dp_i \int e^{-\beta \sum \varepsilon_j} dq_1 \cdots dp_1 \cdots}{\int e^{-\beta \varepsilon_i} dp_i \int e^{-\beta \sum \varepsilon_j} dq_1 \cdots dp_1 \cdots}$$

and this reduces to

$$\langle \varepsilon_i \rangle = \frac{\int \varepsilon_i e^{-\beta \varepsilon_i} \, dp_i}{\int e^{-\beta \varepsilon_i} \, dp_i} = -\frac{\partial}{\partial \beta}\left[\ln\left(\int e^{-\beta \varepsilon_i} \, dp_i\right)\right]. \tag{16.31}$$

(To simplify the notation, we have put $\varepsilon_i \, (p_i) = \varepsilon_i$ in the integrals.) The result given in Equation 16.31 is perfectly general for the independent energy contribution ε_i. We now choose ε_i to have a quadratic dependence on p_i. This is important in evaluating the integral. Inserting $\varepsilon_i = \alpha p_i^2$ for ε_i in Equation 16.31 leads to $\langle \varepsilon_i \rangle = -\partial/\partial\beta \ln\left[\int_{-\infty}^{\infty} e^{-\beta(\alpha p_i^2)} \, dp_i\right]$. With a change of variable in the integral to $x = (\beta\alpha)^{1/2} p_i$ and the use of a standard integral from Appendix A, we obtain

$$\langle \varepsilon_i \rangle = \frac{\partial}{\partial \beta}\ln\left[\left(\frac{1}{\sqrt{\beta\alpha}}\right)\int_{-\infty}^{\infty} e^{-x^2} \, dx\right] = -\frac{\partial}{\partial \beta}\left(\ln\sqrt{\frac{\pi}{\beta\alpha}}\right) = \frac{1}{2\beta}.$$

This gives $\langle \varepsilon_i \rangle = (1/2) k_B T$ as required to prove the theorem. Application to a given system involves counting the number of independent quadratic degrees of freedom in the system. The total mean energy is obtained directly by multiplying this number by $(1/2) k_B T$.

16.7 THE MAXWELL VELOCITY DISTRIBUTION

In our discussion of ideal gas systems, attention has largely been focused on average values and, in particular, on the mean energy of particles. The molecules in a gas have a distribution of velocities, and we now determine the form of this distribution. The probability for a particle to have translational energy ε is given in the canonical ensemble by $p\,(\varepsilon) = e^{-\beta\varepsilon}/z$, where z is the translational single-particle partition function. The translational energy levels form a quasi-continuum, as shown in Figure 16.1, with density given by the particle in a box density of states $\rho(\varepsilon)$ in Equation 4.14. For a gas at temperature T, the probability $P(\varepsilon)$ for a particle to have energy in the range from ε to $\delta\varepsilon$ is

$$P(\varepsilon)\delta\varepsilon = p(\varepsilon)\rho(\varepsilon)\delta\varepsilon = \left(\frac{e^{-\beta\varepsilon}}{z}\right)\left(\frac{V}{4\pi^2\hbar^3}\right)(2m)^{3/2}\,\varepsilon^{1/2}\delta\varepsilon. \tag{16.32}$$

For $\varepsilon = \frac{1}{2}mv^2$, where $v^2 = \left(v_x^2 + v_y^2 + v_z^2\right)$, it follows that the probability of particles having speeds in the range from v to $v + dv$ is

$$P(v)\,dv = \left(\frac{1}{z}\right)\left(\frac{Vm^3}{2\pi^2\hbar^3}\right)v^2 e^{-(1/2)\beta mv^2}\,dv,$$

or in compact form,

$$P(v)\,dv = C_M v^2 e^{-(1/2)\beta mv^2}\,dv, \tag{16.33}$$

with $C_M = 1/z \, (Vm^3/2\pi^2\hbar^3)$. From Equation 16.4, $z = V(m/2\pi\beta\hbar^2)^{3/2}$, and this gives $C_M = \sqrt{c/\pi}\,(\beta \, m)^{3/2}$. The constant C_M may alternatively be obtained using the normalization condition, $\int_0^{\infty} P(v)\,dv = 1$. Equation 16.33 with the expression for C_M gives the Maxwell speed distribution in terms of the temperature T as

$$P(v)\,dv = 4\pi \left(\frac{m}{2\pi k_B T}\right)^{3/2} v^2 e^{\left(-mv^2/2k_B T\right)}\,dv. \tag{16.34}$$

Exercise 16.4

Obtain expressions for the most probable speed, the mean square speed, the root mean square speed, and the mean speed for particles that obey the Maxwellian speed distribution. Compare the values obtained and comment on the ratios. Obtain the root mean square speed for molecules in nitrogen gas at 300 K.

The condition $dP(v)/\,dv = 0$ gives the most probable speed $\tilde{v}$ at the maximum in the distribution, $\tilde{v} = \sqrt{2k_B T/m}$.

With the use of a standard integral from Appendix A, the mean square speed is evaluated as
$\langle v^2\rangle = \int_0^\infty v^2 P(v)\,dv = 3k_B T/m.$

The root mean square speed v_{RMS} is immediately given by $v_{RMS} = \sqrt{\langle v^2\rangle} = \sqrt{3k_B T/m}$.

Finally, the mean speed is obtained as $\langle v\rangle = \int_0^\infty vp(v)\,dv = \sqrt{3k_B T/\pi m}$.

It is readily seen that $V_{RMS} > \langle v\rangle > \tilde{v}$, with $V_{RMS} : \langle v\rangle : \tilde{v}$ given by $\sqrt{3} : \sqrt{8/\pi} : \sqrt{2}$. This trend in the various average speed values is a result of the long tail at high speeds in the Maxwell distribution, as shown in Figure 16.5.

For nitrogen at a temperature of 300 K, substitution in $v_{RMS} = \sqrt{3k_B T/m}$ gives $v_{RMS} \sim 5\times 10^2\ ms^{-1}$.

The Maxwell velocity distribution is readily obtained from the speed distribution. Figure 16.6 shows a representation of velocity space with Cartesian coordinates.

The speed distribution includes all velocities, independent of direction, which have magnitudes in the range from v to $v + dv$, and this corresponds to velocity vectors with their tips in a spherical shell of volume $4\pi v^2\,dv$, as shown in Figure 16.6. The velocity distribution corresponds to velocities in the range from $\boldsymbol{v}$ to $\boldsymbol{v}+d\boldsymbol{v}$, and a volume element $d^3v = dv_x dv_y dv_z$. It follows from Equation 16.34 that

$$P(\boldsymbol{v})\,d^3\boldsymbol{v} = \left(\frac{1}{4\pi v^2}\right) P(v)\,d^3\boldsymbol{v} = \left(\frac{m}{2\pi k_B T}\right)^{3/2} e^{\left(-mv^2/2k_B T\right)} d^3\boldsymbol{v}. \tag{16.35}$$

The magnitude of the root mean square velocity is $v_{RMS} = (k_B T/m)^{1/2}$. This follows immediately from the equipartition theorem or can be obtained by evaluating the mean square velocity with the use

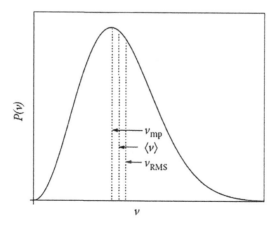

FIGURE 16.5 The Maxwell speed distribution showing the most probable speed $v_{mp} = \tilde{v}$, the mean speed $\langle v\rangle$, and the root mean square speed v_{RMS}. The long tail of the distribution at high speeds accounts for the trend $v_{RMS} > \langle v\rangle\ \tilde{v}$.

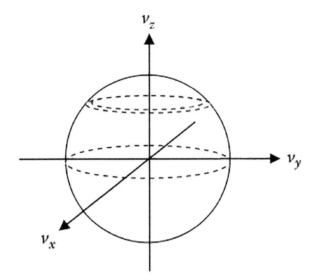

FIGURE 16.6 Velocity space representation that uses the three orthogonal velocity components, denoted by v_x, v_y, and v_z for particles in a gas. Particles whose velocity vectors end on the spherical surface shown all have the same speed.

of the distribution in Equation 16.35. Figure 16.7 shows the velocity distribution for velocities in the x-direction.

Elegant experiments using molecular beams have verified the Maxwell velocity distribution. Figure 16.8 illustrates schematically the time-of-flight methods used to determine molecular velocities. Particles from the source chamber, which is maintained at a fixed temperature, escape through a small nozzle and enter the vacuum chamber as a collimated beam. The high-vacuum region contains a rotating slotted wheel assembly, as depicted in Figure 16.8. For particles to be detected, they must pass through slots on the successive rotating wheels. The magnitude of the velocity of detected molecules is given by $v = L/t$, where $t = \theta/\omega$, with θ being the angular offset of the slots on the two rotating wheels and ω the angular velocity of the coupled wheels. Agreement between the experimental measurements and the theoretical predictions on the basis of the Maxwell distribution is found to be excellent.

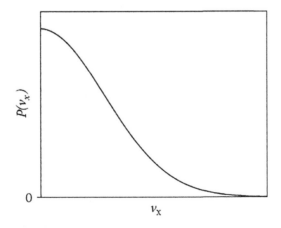

FIGURE 16.7 The plot shows the Maxwell velocity distribution function for a classical gas. The maximum in the distribution occurs for $\langle v_x \rangle = 0$, and the distribution is symmetrical about this value.

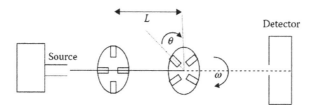

FIGURE 16.8 Time-of-flight method used to measure molecular velocities in a high-vacuum apparatus. The slotted rotating disks are used to select a particular small range of speeds that can reach the detector. The arrangement can be used to scan over a range of speeds. The distribution that is obtained is determined by the temperature of the source from which the molecules escape through a small aperture.

Up to this point in this book, we have used ideal systems in developing and applying statistical and thermal physics concepts and expressions. In Chapter 17, nonideal gases and spin systems are considered. For nonideal gases, allowance is made for both potential energy and kinetic energy contributions. As we shall see, this leads to considerable complication in evaluating the partition function, and approximations are introduced to simplify the discussion. For nonideal spin systems, we again use approximate methods or alternatively simplified model systems. As might be expected, the quantum fluids require more elaborate treatments, and we shall simply outline approaches used for weakly interacting Fermi and Bose systems.

PROBLEMS CHAPTER 16

16.1 Monatomic molecules adsorbed on a surface of area A move freely and may be treated as a classical two-dimensional ideal gas. Obtain an expression for the partition function and hence the specific heat per mole for this system.

16.2 Helium-4 vapor is in equilibrium with liquid helium at its normal boiling point of 4.2 K. Treating the vapor as an ideal gas, obtain an expression for the entropy and the chemical potential for this phase.

16.3 Obtain the rotational molar specific heat for carbon monoxide gas in the low- and high-temperature limits. Does the high-temperature limit apply at 300 K? The moment of inertia of the CO molecule for rotation about its center of mass is 1.3×10^{-39} g cm^2.

16.4 The linear CO_2 molecule has four vibrational modes with associated vibrational temperatures $\theta_{vib} = \hbar \omega_{vib}/k_B$ of 3498 K, 1908 K and two degenerate modes at 954 K. Obtain the molar specific heat for this gas at a temperature of 400 K. Assume that the rotational and translational degrees of freedom can be treated in the high-temperature limit.

16.5 One mole of nitrogen gas is heated to 800°C. Make use of the measured vibrational energy level spacing of 0.3 eV to determine the vibrational contribution to the specific heat for N_2 at this temperature.

16.6 Compare the rotational and vibrational contributions to the specific heat for chlorine gas at 300 and 1200 K. For Cl_2, $\hbar \omega_{vib}/k_B = 810$ K and $\hbar^2/2Ik_B = 0.35$ K.

16.7 Obtain expressions for the entropy and chemical potential of a classical diatomic gas at temperatures for which the molecules are in their vibrational ground state. Compare with expressions for a classical monatomic gas.

16.8 Give expressions for the most probable kinetic energy and the mean kinetic energy for a monatomic gas obeying the Maxwell speed distribution. Compare your expressions with the equipartition theorem kinetic energy value.

16.9 A gas is contained in a vessel with a small hole in the side through which molecules can escape into an evacuated space. This process is called effusion and is used in molecular beam experiments. Obtain an expression for the number of molecules with speeds in the range from v to $v+dv$, emerging from the hole of area A in a solid angle $d\Omega$.

17 Nonideal Systems

17.1 INTRODUCTION

Most of the systems considered in the development of the subject so far have been ideal systems, specifically ideal gases and ideal paramagnets. A notable exception is the harmonic solid, introduced in Chapter 15 in treating lattice dynamics, phonons, and specific heats of solids. In that discussion, the Einstein and Debye models are used in evaluating the partition function for the lattice dynamic modes of a solid. A number of nonideal systems may be described in the classical approximation. Examples considered in this chapter are nonideal gases and nonideal spin systems. The results obtained are useful in describing a variety of systems.

Nonideal Bose fluids, such as liquid helium-4, and Fermi fluids, such as liquid helium-3, require quantum statistics for a description of their properties. Powerful approaches such as Fermi liquid theory have been developed for treating quantum fluids, but a detailed discussion of these topics is beyond the scope of this book. The discussion of quantum fluids is therefore fairly brief, and classical systems are considered first.

17.2 NONIDEAL GASES

Consider a real gas of N molecules. The classical Hamiltonian for N particles, allowing for interparticle interactions, is given by Equation 4.2, $\mathcal{H} = K + U = \sum_{i=1}^{N} p_i^2/2m + \sum_{i>j=1}^{N} u(r_{ij})$, where $u(r_{ij})$ is the pair potential for particles i and j. As pointed out in Section 7.8, the intermolecular potential is often represented by the empirical 6–12, or Lennard–Jones, potential, with the form given by Equation 7.63, $u(r) = u_0[(r_0/r)^{12} - 2(r_0/r)^6]$. The attractive part of the potential is due to fluctuating electric dipole–dipole interactions between molecules. The short-range repulsive interaction is a result of the Pauli exclusion principle, which opposes interpenetration of the molecular electron distributions.

The partition function for a system for which the Hamiltonian has the form given in Equation 4.2 is

$$Z = \frac{1}{h^{3N}N!} \int \cdots \int e^{-\beta\left[\sum_{i=1}^{N}\frac{p_i^2}{2m} + \sum_{i>j=1}^{N} u(r_{ij})\right]} \left[d^3r_1 \ldots d^3r_N; d^3p_1 \ldots d^3p_N\right].$$

Adopting the approach used in Section 16.4, this may be written as a product of factors involving momentum and position coordinates separately:

$$Z = \frac{1}{h^{3N}N!} \left[\int_{-\infty}^{\infty} \cdots \int_{-\infty}^{\infty} e^{-\beta\sum_{i=1}^{N}\frac{p_i^2}{2m}} d^3p_1 \ldots d^3p_N\right] \left[\int_V \cdots \int_V e^{-\beta \sum u_{ij}} d^3r_1 \ldots d^3r_N\right]. \tag{17.1}$$

The limits in the momentum integrals are $\pm\infty$, and spatial integration is over the container volume. Equation 17.1 becomes

$$Z = \left(\frac{Z_{\text{ideal}}}{V^N}\right) Q_N. \tag{17.2}$$

Z_{ideal} is the familiar ideal gas partition function given by Equations 16.5 and 16.11, $Z_{ideal} = (V^N/N!)$ $(2\pi m/\beta h^2)^{3N/2}$, whereas Q_N depends on the particle coordinates and is called the configurational partition function. It is not a simple matter to evaluate Q_N because the potential function involves the interparticle separation $\boldsymbol{r}_{ij} = \boldsymbol{r}_j - \boldsymbol{r}_i$ and not the particle coordinates $\boldsymbol{r}_i, \boldsymbol{r}_j$ separately. It is therefore necessary to use approximations in obtaining an expression for Q_N. The properties of the exponential function allow Q_N to be written in the form

$$Q_N = \int_V \cdots \int_V \prod_{i>j=1}^{N} e^{-\beta u_{ij}} \, \mathrm{d}^3 r_1 \ldots \mathrm{d}^3 r_N, \tag{17.3}$$

where we put $u_{ij} = u(r_{ij})$ to simplify the notation. The condition $i > j$ ensures that each pair of interacting particles is considered only once. At low-particle densities, where the r_{ij} are on average large, $u_{ij} \to 0$ so that $e^{-\beta u_{ij}} \to 1$ and $Q_N \to V^N$. In the low-density limit, we recover the ideal gas partition function. It is convenient, for reasons that will become clear, to consider the exponential function $\left(e^{-\beta u_{ij}} - 1\right)$. As the density of the gas is increased, the function $\left(e^{-\beta u_{ij}} - 1\right)$ will increase from zero as a result of the decrease in r_{ij}/r_0 before tending to the value -1, as shown in Figure 17.1.

If we put $W_{ij} = \left(e^{-\beta u_{ij}} - 1\right)$ and insert this form in Equation 17.3, we obtain

$$Q_N = \int \cdots \int \prod_{i>j} \left(1 + W_{ij}\right) \mathrm{d}^3 r_1 \ldots \mathrm{d}^3 r_N. \tag{17.4}$$

It is advantageous to deal with integrals that involve W_{ij} because they converge for $r_{ij} \to \infty$, whereas integrals of $e^{-\beta u_{ij}}$ do not. The product in Equation 17.4 may be written as

$$\prod_{i>j} \left(1 + W_{ij}\right) = 1 + \sum_{i>j} W_{ij} + \sum_{i>j} \sum_{k>l} W_{ij} W_{kl} + \sum_{i>j} \sum_{k>l} \sum_{m>n} W_{ij} W_{kl} W_{mn} + \ldots. \tag{17.5}$$

The summations are over all molecules in the container, with the condition that identical clusters be counted once only. It should be remembered that the molecules are indistinguishable and the labels used are simply a convenience. The resultant overcount of states is taken care of by the factor $1/N!$ in Equation 17.1. Insertion of Equation 17.5 into Equation 17.4 gives

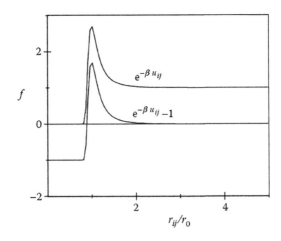

FIGURE 17.1 The functions $e^{-\beta u_{ij}}$ and $e^{-\beta u_{ij}} - 1$, which involve the interparticle potential u_{ij}, plotted versus r_{ij}/r_0. The function $e^{-\beta u_{ij}}$ tends to unity for large r_{ij}/r_0, whereas $e^{-\beta u_{ij}} - 1$ tends to zero.

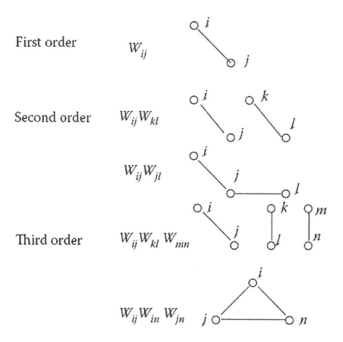

FIGURE 17.2 Cluster diagrams which represent the first-, second-, and third-order terms in the function W_{ij}, which occurs in integrals for the configurational partition function as given in Equation 17.6.

$$Q_N = V^N + \int \dots \int \left(\sum W_{ij} \right) d^3 r_1 \dots d^3 r_N + \int \dots \int \left(\sum \sum W_{ij} W_{kl} \right) d^3 r_1 \dots d^3 r_N + \dots. \quad (17.6)$$

Because the $W_{ij} \rightarrow 0$ in the dilute gas limit, it is appropriate to regard the various terms in W_{ij}, as correction terms of decreasing importance as the order increases. It is helpful to give a pictorial representation of terms that involve W_{ij}, W_{kl}, ... > 0. Examples of such cluster diagrams are shown in Figure 17.2.

It is readily shown that the number of ways in which n molecules in a cluster can be chosen from a total number of molecules N is $N (N - 1) \dots (N - (n - 1))/n!$. For example, for two molecule clusters, the first molecule may be chosen in N ways, the second in $(N - 1)$ ways, and a factor $1/2$ must be inserted to avoid counting the same cluster twice. Extension of this approach to n molecules leads to the general result given above. Consider the leading correction term in Equation 17.6 involving W_{ij}. This may be simplified by writing the integral as $\int \dots \int W_{ij} d^3 r_1 \dots d^3 r_N = V^{N-2} \int \int W_{ij} d^3 r_i d^3 r_j$.

Now W_{ij} depends on r_{ij} independent of orientation, and Figure 17.3 illustrates that, for a given separation r, molecule j must lie in a spherical shell of volume $4\pi r^2 dr$, provided wall effects can be ignored. This is a good assumption for a container whose dimensions are typically very large compared with molecular dimensions.

From the definition of W_{ij} in Equation 17.4, it follows that

$$\int \left(W_{ij} d^3 r_i d^3 r_j \right) = V \left[4\pi \int_0^\infty \left(e^{-\beta u(r)} - 1 \right) r^2 \ dr \right].$$

The configurational partition function is therefore written as

$$Q_N = V^N + \frac{N(N-1)}{2} V^{N-1} \int_0^\infty 4\pi \left(e^{-\beta u(r)} - 1 \right) r^2 \ dr + \cdots, \quad (17.7)$$

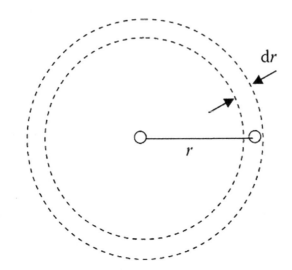

FIGURE 17.3 Spherical volume element that is available to molecule j for a given separation of a pair of molecules i and j in a gas, provided they are not near the walls of the container.

and if the form of $u(r)$ is chosen, the definite integral in Equation 17.7 can be evaluated. Specific cases are dealt with later. For the moment, we put

$$4\pi \int_0^\infty \left(e^{-\beta u(r)} - 1\right)r^2 dr = 2v_R, \tag{17.8}$$

with v_R as a volume of the order of the molecular volume. Note that for $\beta u(r) \ll 1$, which is a good approximation in many real gases at temperatures of interest, expansion of the exponential function in Equation 17.8 gives

$$2v_R = 4\pi \int_0^\infty \left(e^{-\beta u(r)} - 1\right)r^2 \ dr \approx -4\pi\beta \int_0^\infty u(r)r^2 \ dr. \tag{17.9}$$

This form for $2v_R$ leads to simple integrals provided $u(r)$ remains finite as $r \to \infty$. Examples are considered in Section 17.3. In terms of the volume v_R, Equation 17.7 takes the form

$$Q_N = V^N \left[1 + \frac{N^2 v_R}{V} + \cdots\right], \tag{17.10}$$

where the approximation $N(N-1) \approx N^2$ has been used because N is very large. The quantity Nv_R/V is of the order of the ratio of a molecular volume to that of the volume per molecule in the container and we expect $Nv_R/V < 1$, provided departures from ideal gas conditions are not too large. With the use of the binomial theorem, Equation 17.10 becomes

$$Q_N \approx \left[V\left(1 + \frac{Nv_R}{V}\right)\right]^N.$$

From Equations 17.2 and 16.5, the partition function for the gas is

$$Z = \frac{1}{N!}\left[\left(\frac{2\pi m}{\beta h^2}\right)^{3/2} V\left(1 + \frac{Nv_R}{V}\right)\right]^N. \tag{17.11}$$

This gives the Helmholtz potential $F = -k_B T \ln Z$ as

$$F = -Nk_B T \left[\ln V + \ln\left(1 + \frac{Nv_R}{V}\right) + \frac{3}{2}\ln\left(\frac{2\pi m}{h^2}\right) \right] + Nk_B T - Nk_B T \ln N. \tag{17.12}$$

As a good approximation, we take $\ln(1 + Nv_R/V) \approx Nv_R/V$, and the pressure is given by

$$P = -\left(\frac{\partial F}{\partial V}\right)_{T,N} = \frac{Nk_B T}{V}\left[1 - \frac{Nv_R}{V}\right]. \tag{17.13}$$

Equation 17.13 is similar in form to the leading terms in the virial expansion introduced in Chapter 1 and given in Equation 1.3:

$$P = \frac{Nk_B T}{V}\left[1 + \left(\frac{N}{V}\right)B(T) + \left(\frac{N}{V}\right)^2 C(T) + \cdots\right].$$

$B(T)$ is called the second virial coefficient, $C(T)$ the third virial coefficient, and so on. Careful experiments have yielded values for the coefficients for a large number of gases. Comparison of Equation 17.13 with Equation 1.3 shows that $v_R = -B(T)$ if higher-order terms in the expansions are ignored. Explicit comparisons of theory and experiment involve the intermolecular potential for a particular gas. From the expression for the Helmholtz potential given in Equation 17.12, various thermodynamic quantities, such as the entropy, can be obtained.

17.3 EQUATIONS OF STATE FOR NONIDEAL GASES

Equation 17.13 together with Equation 17.8 provides the basis for the determination of the equation of state for a given intermolecular potential function. It is necessary to choose a specific form for the intermolecular potential to evaluate the integral for v_R.

Exercise 17.1

Obtain the equation of state for a gas with a hard-core potential of the form as shown in Figure 17.4. Give a physical interpretation of the equation of state for this gas.

From Equation 17.8, $2v_R = -4\pi \int_0^{r_0} r^2 dr = -(4\pi/3)r_0^3 = -v_0$. Insertion of this value for v into Equation 17.13 gives

$$P = \frac{Nk_B T}{V}\left[1 - \frac{Nv_0}{2V}\right]. \tag{17.14}$$

FIGURE 17.4 Hard-core intermolecular pair potential with $u(r) = \infty$ for $r \le r_0$ and $u(r) = 0$ for $r > r_0$.

If we assume that $Nv_0/RV < 1$ and use the binomial theorem, Equation 17.14 may be written as

$$P \approx \frac{Nk_BT}{\left(V - \dfrac{1}{2}Nv_0\right)}. \tag{17.15}$$

Equation 17.15 is of the form of the ideal gas equation of state with the volume reduced by $\dfrac{1}{2}Nv_0$, which is just the sum of the hard-core volumes divided by two. The volume reduction effect leads to a pressure increase compared with the pressure of an ideal gas at the same volume and temperature. (Note that it is implicitly assumed that an ideal gas has a negligible hard-core radius.)

Exercise 17.2

Show that the van der Waals equation of state introduced in Section 1.4 and used in Section 7.8(b) can be derived using the intermolecular potential in Figure 7.5, which is represented by $u(r) = -u_0(r_0/r)^6$ $(r > r_0)$; $u(r) = \infty$ $(r \leq r_0)$.

The potential has a hard-core combined with a long-range attractive potential that is similar to the Lennard–Jones long-range form. Insertion of the potential into Equation 17.8 and the use of the high-temperature expansion for the long-range part give

$$2v_R = -4\pi \int_0^{r_0} r^2 \, dr + 4\pi\beta u_0 \int_{r_0}^{\infty} \left(\frac{r_0}{r}\right)^6 r^2 \, dr = -\left(\frac{4\pi}{3}\right)r_0^3 + \left(\frac{4\pi}{3}\right)r_0^3 \beta u_0$$

or

$$2v_R = -v_0\left(1 - \beta u_0\right),$$

where $v_0 = (4/3)\pi r_0^3$. With this value for $2v_R$, Equation 17.13 becomes

$$P = \frac{Nk_BT}{V}\left[1 + \left(\frac{Nv_0}{2V}\right)\left(1 - \beta u_0\right)\right].$$

If the temperature-dependent terms are grouped together and we use the binomial expansion in exactly the same way as for the simple hard-core potential considered above, then

$$\left(P + \frac{1}{2}\frac{N^2u_0v_0}{V^2}\right)\left(V - \frac{1}{2}Nv_0\right) = Nk_BT. \tag{17.16}$$

Equation 17.16 is of the form of the van der Waals equation $(P + a/V^2)(V - b) = Nk_BT$, with $a = (1/2)N^2u_0V_0$ and $b = (1/2)NV_0$. The long-range attractive part of the intermolecular potential increases the effective pressure, whereas the hard-core volume reduces the total volume available to molecules.

Other intermolecular pair potentials will lead to somewhat different equations of state. The examples given above show that the small cluster approach that involves the leading correction term in the expression for the configurational partition function, as given in Equation 17.6, provides useful results for real gases at moderate densities. For high densities, it is necessary to consider higher-order correction terms in Equation 17.6 and the calculations become more complicated.

17.4 NONIDEAL SPIN SYSTEMS: THE HEISENBERG–HAMILTONIAN

Ideal spin systems have been considered in Chapters 4 and 5. For a system of noninteracting electron spins j in a magnetic field $\boldsymbol{B}$ along the z-direction, the Hamiltonian may be written as

$$\mathcal{H} = -\sum_j \boldsymbol{\mu}_j \cdot \boldsymbol{B} = g\mu_B B \sum_j S_{jz}, \tag{17.17}$$

where $\boldsymbol{\mu} = -g\mu_B \boldsymbol{S}$ is the magnetic dipole operator with $\boldsymbol{S}$ as the spin operator. (As noted in Chapter 4 for negatively charged particles, μ and $\boldsymbol{S}$ are antiparallel.) The energy eigenvalues are depicted in Figure 4.6 for N identical spins $S = 1/2$, and the partition function obtained from Equation 10.37 is $Z = \left[2 \cosh \dfrac{1}{2}\beta\, g\, \mu_B B \right]^N$. Interactions between spins in a ferromagnet or antiferromagnet are dominated by what is termed the exchange interaction. The exchange Hamiltonian for spins is called the Heisenberg–Hamiltonian and has the form

$$\mathcal{H} = \sum_{ij} J_{ij} \boldsymbol{S}_i \cdot \boldsymbol{S}_j, \tag{17.18}$$

where J_{ij} is the exchange interaction between spins i and j. The exchange interaction is quantum mechanical in origin, with $2J_{ij}$ as the energy difference for spins i and j in parallel and antiparallel spin orientations. Ferromagnetism ($J_{ij} < 0$) corresponds to parallel spin orientations for interacting spins having a lower energy than antiparallel orientations, whereas for antiferromagnetism ($J_{ij} > 0$), the situation is reversed.

With the exchange interaction present, the Hamiltonian becomes

$$\mathcal{H} = \sum_{i,j} J_{ij} \boldsymbol{S}_i \cdot \boldsymbol{S}_j + g\mu_B B \sum_j S_{jz}. \tag{17.19}$$

Because the exchange coupling is nonlinear in the spin operators, it is not possible to obtain general solutions for the eigenvalues and approximation methods must be used. Mean field theory, which is based on the Heisenberg–Hamiltonian, provides a useful description of the properties of magnetic systems as described in Section 17.5.

17.5 MEAN FIELD THEORY FOR MAGNETIC SYSTEMS

17.5.1 FERROMAGNETISM

The Weiss mean field approach considers a single representative spin situated in a field made up of the external magnetic field and an internal effective field produced by all the other spins. The exchange interaction is strongest between the nearest-neighbor spins and, as a simplifying approximation, we neglect interactions with more distant spin neighbors. Let there be n nearest neighbors of the representative spin, as shown in Figure 17.5.

If we assume that the nearest-neighbor exchange interactions are equal and given by J, the Hamiltonian for a single representative spin in the mean field approximation is

$$\mathcal{H} = \left[g\mu_B B + nJ \langle S_z \rangle \right] S_z = g\mu_B \left[B + B_m \right] S_z, \tag{17.20}$$

where $B_m = (nJ/g\,\mu_B)\langle S_z \rangle$ is defined as the mean field and $\langle S_z \rangle$ is the average z component of a single spin in the system. To simplify the analysis, we consider the special case of spin $\dfrac{1}{2}$ particles

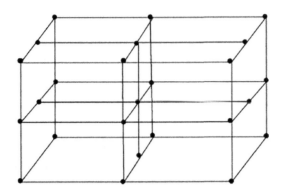

FIGURE 17.5 Nearest-neighbor spin sites for a simple cubic lattice for which the number of nearest neighbors $n = 6$. Interacting spins are located at the sites indicated by dark circles.

although for many ferromagnets $S > 1/2$. The mean field single-particle partition function for a spin $S = 1/2$, with $g = 2$, is obtained from Equation 10.36

$$z = 2 \cosh\left[\beta \mu_B (B + B_m)\right] \tag{17.21}$$

The magnetic moment of the complete system is given by

$$N\langle \mu \rangle = \langle \mathcal{M} \rangle = \left(\frac{N}{\beta}\right)\frac{\partial \ln z}{\partial B} = N\mu_B \tanh\left[\beta\mu_B (B + B_m)\right]. \tag{17.22}$$

Consider the special case $B = 0$ for which Equation 17.22, with $\langle \mu \rangle = - g \mu_B \langle S_z \rangle$, may be rewritten as

$$\langle S_z \rangle = -\frac{1}{2}\tanh\left[\frac{1}{2}\beta\, nJ\langle S_z \rangle\right]. \tag{17.23}$$

$\langle S_z \rangle$ is present on both sides of Equation 17.23, which is solved in a self-consistent way. For a ferromagnet with $J < 0$, we put $x = -(1/2)\,\beta n|J|\langle S_z \rangle$, and this gives

$$\left(\frac{2}{\beta n|J|}\right)x = \tanh x. \tag{17.24}$$

For a particular value of β, the solution to Equation 17.24 is obtained graphically, as shown in Figure 17.6, from the intersection of plots versus x of the two functions of x that occur in the equation.

From Figure 17.6, it is clear that, in the high-temperature limit, $x = 0$, which corresponds to $\langle S_z \rangle = 0$, always being a solution. The intersection for nonzero x is of greater interest because this solution implies $\langle S_z \rangle \neq 0$ for $B = 0$. The condition for nonzero solutions to occur in Equation 17.24 is that the initial slope of the function on the right-hand side must be greater than that of the function on the left-hand side. For small x, we expand the hyperbolic function in the form

$$\tanh x = x - \frac{1}{3}x^3 + \frac{2}{15}x^5 + \ldots, \tag{17.25}$$

and obtain for the initial slope of the right-hand side of Equation 17.24 $[d/dx(\tanh)]_{x\to 0} = 1$. The initial slope of the left-hand side of Equation 17.24 is $[d/dx(2x/\beta n \mid J \mid)]_{x\to 0} = 2/\beta n|J|$. The condition for finite solutions is therefore $(2/\beta n|J|) \leq 1$ or

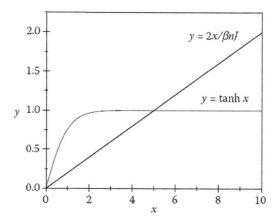

FIGURE 17.6 Graphical method used in the solution of Equation 17.24. The point of intersection of the two curves shown as a function of $x = (1/2)\beta n |J| \langle S_z \rangle$ gives the required solution.

$$T \le \frac{n|J|}{2k_B}. \tag{17.26}$$

Equation 17.26 implies that there is a critical transition temperature for the system given by

$$T_C = \frac{n|J|}{2k_B}. \tag{17.27}$$

Below T_C, there is a finite magnetic moment in the mean field approximation. For ferromagnetic systems, we identify T_C with the Curie temperature. Equation 17.27 together with Equation 17.24 gives the relationship $(T/T_C)x = \tanh x$, and solving for x graphically leads to $\langle S_z \rangle$ and hence the magnetization as a function of T. It is convenient to plot the reduced magnetization M/M_0 versus the reduced temperature T/T_C, as shown in Figure 17.7. M_0 is the magnetization at 0 K.

The mean field theory predicted the behavior of the magnetization at temperatures below the Curie point as shown in Figure 17.7, is broadly consistent with experimental results. However, small but important departures from theory are found, first, at very low temperatures and, second, as T_c is approached. The sources of these departures from the mean field theory predictions are briefly discussed below.

For small x, the use of Equation 17.25 in Equation 17.24 leads to $2/(\beta n|J|) = 1 - (1/3) x^2 + \ldots$, and with the definitions of x and T_C, this reduces to

$$\langle S_z \rangle \approx \sqrt{3} \frac{T}{T_C} \left(1 - \frac{T}{T_C} \right)^{\frac{1}{2}}, \tag{17.28}$$

Or, in terms of the magnetization,

$$\frac{\langle M \rangle}{\langle M_0 \rangle} = 2\sqrt{3} \frac{T}{T_C} \left(1 - \frac{T}{T_C} \right)^{1/2}, \tag{17.29}$$

where $\langle M_0 \rangle = N \mu_B/V$ for our spin $\frac{1}{2}$ particles. Equation 17.29 predicts that the magnetization decreases to zero as $T \to T_C$ with the critical exponent $\beta = 1/2$, in agreement with the Landau theory predictions discussed in Chapter 9. As noted in Chapter 9, this exponent prediction is not confirmed by careful magnetization measurements on 3D ferromagnets at temperatures just below T_C, which

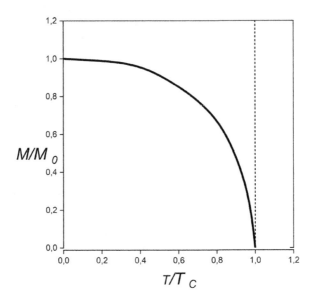

FIGURE 17.7 Mean field theory predicted behavior of the reduced magnetization M/M_0 as a function of reduced temperature T/T_c for a spin 1/2 ferromagnet. M_0 is the magnetization at 0 K and T_c the Curie temperature.

yield an exponent $\beta = 1/3$. It is not unexpected that the mean field theory should break down near T_C because in this temperature range, fluctuations in the order parameter become large and the mean field model assumptions are no longer valid. In spite of being a fairly crude model, the mean field approach provides a physical insight into ferromagnetic and antiferromagnetic transitions. Note that in antiferromagnetic systems, there are two ordered sublattices that interpenetrate. It is necessary to consider the sublattice magnetization rather than the total magnetization in antiferromagnets.

Exercise 17.3

Generalize the mean field discussion given above to allow for a nonzero external magnetic field. Obtain an expression for the magnetic susceptibility for $T > T_C$, and show that this has the Curie–Weiss form.

For $T \geq T_C$ and B small, expansion of the tanh function, with the retention of the lowest-order term, and use of Equation 17.27 allow Equation 17.23 to be written as

$$\langle \mathcal{M} \rangle = N\,\mu_B \left[\beta\mu_B \left(B + \frac{n|J|}{2\mu_B} \right)\langle S_z \rangle \right] = N\mu_B^2 \beta B + \left(\frac{T_C}{T} \right)\langle \mathcal{M} \rangle. \qquad (17.30)$$

With $\langle M \rangle = \langle \mathcal{M} \rangle / V$, this is rearranged to give

$$\langle M \rangle = \left(\frac{N\mu_B^2 B}{2Vk_B} \right)\frac{1}{(T - T_C)} \qquad (17.31)$$

For $T > T_C$ taking $B \approx \mu_0 H$, the magnetic susceptibility per unit volume is

$$\chi = \frac{N}{Vk_B}\frac{\mu_0\mu_B^2}{(T - T_C)}. \qquad (17.32)$$

This is the familiar Curie–Weiss law introduced in Chapter 1.

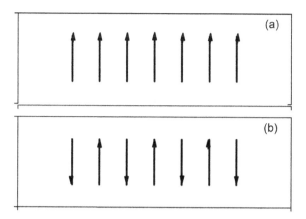

FIGURE 17.8 Spin configurations along a symmetry axis in a 3D cubic crystal for (a) a ferromagnet and (b) an antiferromagnet.

17.5.2 ANTIFERROMAGNETISM

In antiferromagnetic systems, the exchange interaction J in the Heisenberg–Hamiltonian is positive, and at low temperatures, this interaction leads to spin order with nearest neighbors alternating in spin-up and spin-down orientations. Figure 17.8b depicts the spin configuration along a symmetry axis in a 3D crystalline solid.

In applying the mean field approach to antiferromagnets, it is convenient to consider two interpenetrating sublattices: one with *up* spins, designated sublattice 1, and the other with *down* spins, designated sublattice 2. The two sublattices have opposed magnetizations.

The mean field Hamiltonian for a representative spin in a sublattice of an antiferromagnet is very similar to that for a ferromagnet given in Section 17.5.1. If we assume that an applied field B acts parallel to the chosen z-axis directed along the spin-up direction, and that B is lower than the mean field, then the Hamiltonian given in Equation 17.20 takes the following modified form, for a representative spin S_1 in sublattice 1

$$\mathcal{H} = g\mu_B \lceil B + B_{m2} \rceil S_{z1}. \tag{17.33}$$

A similar Hamiltonian applies to sublattice 2 with the subscripts 1 and 2 interchanged and $+B_{m2}$ replaced by $-B_{m1}$. The mean field $B_{m2} = (n\lfloor J \rfloor / g\mu_B)\langle S_{z2}\rangle$, which is produced by the n nearest-neighbor (antiparallel) spins, acts parallel to the applied field. For a system of spins $S = 1/2$ with $g = 2$, the magnetic moment of a sublattice in the antiferromagnetically ordered phase is very similar to that for a ferromagnet as given in Equation 17.22 but with N replaced by $N/2$ and with allowance for the sign change in the mean field between sublattice 1 and sublattice 2. For sublattice 1, the moment is given by

$$\langle \mathcal{M}_1 \rangle = \frac{1}{2} N\mu_B \tanh\lceil \beta\mu_B (B + B_{m2}) \rceil \tag{17.34}$$

Apart from the new subscripts, Equation 17.34 has the same form as Equation 17.22. An equation for $\mathcal{M}_2$ similar to Equation 17.34 applies to sublattice 2 with $+B_{m2}$ replaced by $-B_{m1}$. For $B = 0$, and introducing the variable $x_1 = (1/2)\beta n|J|\langle S_{z1}\rangle$, Equation 17.34 becomes

$$\frac{2}{\beta n|J|} x_1 = \tanh x_1 \tag{17.35}$$

Apart from the subscript change, Equation 17.35 is the same as Equation 17.24 for a ferromagnet. The antiferromagnetic ordering temperature, called the Néel temperature T_N, is obtained using the expansion of the tanh function given in Equation 17.25, in the small x-limit. This procedure leads to

$$T_N = \frac{n|J|}{2k_B} \tag{17.36}$$

Symmetry requires that this zero external field expression for T_N applies to both sublattices. Note that Equation 17.36 has the same form as Equation 17.27 for the Curie temperature T_C of a ferromagnet in zero applied field. Inserting the expression for T_N given in Equation 17.36 into Equation 17.35 gives

$$\frac{T}{T_N}x = \tanh x \tag{17.37}$$

Following the graphical procedure for obtaining solutions to Equation 17.24, as given in Section 17.5.1, and applying it to Equation 17.37 gives values for x, and hence $\langle S_z \rangle$, as a function of T/T_N. Figure 17.9 shows the plots of the reduced magnetization M/M_0 versus the reduced temperature for the two sublattices. As pointed out above, the two sublattice magnetizations are always equal in magnitude and opposed in direction.

Exercise 17.4

Obtain an expression for the magnetic susceptibility per unit volume of an antiferromagnet in its paramagnetic phase at temperatures above the Néel temperature.
The approach is similar to that used in Exercise 17.3 for a ferromagnet. As a starting point, consider the expression for the magnetic moment as given in Equation 17.30. In the paramagnetic phase, we do not have ordered spins and the need for subscripts falls away. In low applied fields, the tanh function can be expanded retaining the first-order term. This procedure gives

$$\langle M \rangle = N\mu_B^2 \beta (B - B_m) = N\mu_B^2 \beta B - \left(\frac{T_N}{T}\right)\langle M \rangle \tag{17.38}$$

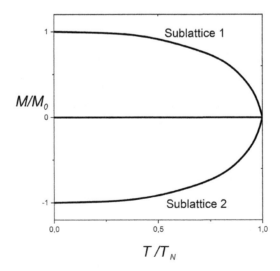

FIGURE 17.9 Sublattice magnetizations as a function of temperature in the Néel phase of a spin 1/2 antiferromagnet as predicted by the mean field theory. M_0 is the magnetization in the zero temperature limit and T_N the Néel temperature.

where the use has been made of Equation 17.36 for T_N. Equation 17.38 is very similar to Equation 17.30 for ferromagnetic systems and differs only in the change of sign of the term involving T_N/T in place of T_C/T. The change in sign comes about because the mean field produced by neighbor spins acts to oppose the applied field in the paramagnetic phase. Rearranging Equation 17382 leads to the following expression for the magnetic susceptibility per unit volume

$$\chi = \frac{N}{Vk_B}\frac{\mu_0\mu_B^2}{(T+T_N)} = \frac{C}{(+T_N)}. \tag{17.39}$$

This is the familiar Curie–Weiss form for the susceptibility of a paramagnet. Figure 17.10 shows the mean field predictions for the temperature dependence of the susceptibility χ in the paramagnetic phase of (a) a ferromagnet and (b) an antiferromagnet based on Equations 17.32 and 17.39, respectively.

The divergence of χ in the paramagnetic phase of a ferromagnet as $T \to T_C$ from above contrasts with the modest predicted variation of χ in an antiferromagnet as $T \to T_N$. Note that the dashed curve for the antiferromagnet in Figure 17.10b, which shows a divergent behavior for $T \to -\theta$, (with $\theta = T_N$), represents the extrapolated form predicted by Equation 17.39 at (unattainable) temperatures below 0 K.

As pointed out in Section 17.5.1, the mean field theory applied to the Heisenberg model provides a useful but incomplete description of the magnetic behavior of systems in which short-range exchange interactions are important. These comments apply to both ferromagnets and antiferromagnets. There are limitations to the validity of mean field theory which experimental and extensive theoretical works have revealed. First, the mean field theory does not allow for critical fluctuations as the magnetic transition temperature is approached. Second, the effects of spin waves are not considered in determining changes in the magnetization at low temperatures, where M/M_0 approaches unity. Third, no distinction is made in the mean field theory between 3D spin systems which exhibit a long-range order and lower-dimensional systems in which no long-range order occurs even at temperatures approaching 0 K.

While spin systems with dimensions lower than three do not exhibit a long-range order, they reveal interesting and challenging electronic properties at low temperatures. The cuprate superconductors such as YBCO ($YBa_2Cu_3O_7$) have stimulated an interest in materials in which antiferromagnetic short-range spin correlations are important. Spin configurations in these systems are qusai-2D. Other strongly correlated electron systems include 2D triangular lattice kagome systems,

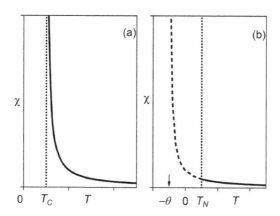

FIGURE 17.10 Mean field theory predictions of the temperature dependence of the magnetic susceptibility χ for (a) a ferromagnet at temperatures in the paramagnetic phase above the Curie temperature T_C and (b) an antiferromagnet in the paramagnetic phase above the Néel temperature T_N. The plots showing the divergence of χ at (a.) T_C and at (b.) $-\theta = -T_N$ are, respectively, based on the forms given in Equations 17.32 and 17.39.

in which magnetic frustration occurs, and 1D spin chains and spin ladders, which serve as model systems with interesting properties. Section 17.6 deals with the 1D Heisenberg chain with anti-ferromagnetic coupling between the nearest-neighbor spins, while Section 17.7 is concerned with the 1D Ising chain with ferromagnetic interactions between neighbors. The spin couplings in the Heisenberg and Ising Hamiltonians are quite different.

17.6 ANTIFERROMAGNETIC HEISENBERG CHAINS

The magnetic properties of 1D antiferromagnetic spin chains reveal novel features and, in particular, the existence of a magnetic field-induced quantum phase transition at 0 K. Systems of experimental interest include molecular chains along which Cu^{2+} ions with electron spin $S = \frac{1}{2}$ are incorporated at regular intervals. To avoid 3D ordering at very low temperatures involving weak interchain interactions, the chains should be well separated from each other with intrachain interactions of dominant importance. Superexchange interactions, which involve nonmagnetic atoms, which act as links in the copper chains, provide the J coupling between spins. Copper pyrazine dinitrate (CuPzN) is a good example of a close to ideal Heisenberg 1D chain system. Experimental measurements of the thermodynamic properties of CuPzN are in good agreement with theoretical predictions. The present discussion focuses on quantum criticality in Heisenberg spin ½ chain systems in which the nearest-neighbor exchange interactions are dominant, resulting in a Hamiltonian of the form given in Equation 17.19.

At low temperatures, the magnetization of a Heisenberg chain is found to increase with the applied field before saturating at a high-field plateau value with all spins aligned parallel to the field. Magnetic susceptibility versus temperature plots obtained over a range of fields exhibit Curie–Weiss-like increases in χ as the temperature T is lowered, similar to that given in Figure 17.10b, before reaching a field-dependent maximum at T_m and then decreasing smoothly at lower temperatures. Extrapolation of the corresponding T_m versus B plot to $0K$ gives a critical field B_c called the quantum critical point (QCP). Zero-temperature magnetic field-induced quantum phase transitions in 1D spin chains belong to a different class of continuous phase transitions than thermally induced transitions.

For a Heisenberg antiferromagnetic chain of spins at 0 K in zero applied field, the ground state is a macroscopic singlet. With a gradual increase in temperature and field, the system exhibits a quantum liquid phase in which spin ½ excitations, called spinons, travel along chains preventing the development of long-range spin order. Spinons are visualized as delocalized domain walls that are created in pairs and then propagate independently. Application of a magnetic field results in the creation of magnon quasi-particles associated with antiferromagnetic spin waves and an accompanying increase in magnetization. Adding a magnon effectively flips the orientation of a single spin in a chain from down to up in the applied field. The zero-temperature magnetization saturates at the QCP with all spins in the induced quasi-ferromagnetic state.

Exercise 17.5

Estimate the quantum critical field for an antiferromagnetic spin ½ Heisenberg chain with the nearest-neighbor exchange interaction $J - 1.38 \times 10^{-22}$ J or, expressed as a temperature, $J/k_B = 10K$. Take the Bohr magneton as $\mu_B = 9.27 \times 10^{-24}$ J/T.

At temperatures close to 0 K, the change in energy of a representative down spin that is produced by applying the critical field and changing the spin orientation from down to up is $\Delta E - 2J$.

It follows that the critical field is given by $B_c = 2J/g\mu_B = \left(2 \times 1.38 \times 10^{-22}\right)/\left(2 \times 9.27 \times 10^{-24}\right) = 14.95T$.

This fairly high magnetic field needed to reach the QCP of a Heisenberg chain is found in systems such as CuPzN.

It can be shown that the occupation of magnon states in a 1D Heisenberg antiferromagnetic chain in an applied field below the QCP field and at a temperature approaching 0 K is given by Fermi–Dirac statistics involving the chemical potential μ. Magnons occupy the set of energy states below the Fermi level in a fashion similar to the behavior of fermions in an ideal Fermi gas. The Fermi level increases in energy with an increase in applied field as the magnon number goes up. The chemical potential plays an important role in considering the properties of these chains.

In order to gain a deeper insight into quantum criticality, it is necessary to consider the behavior of μ for a spin chain as a function of applied field at low temperatures. Using the energy gap expression $\Delta E = 2J$ from Exercise 17.5, the chemical potential for a sin ½ chain for $B \leq B_c$ is given by

$$\mu = 2J - g\mu_B B = g\mu_B (B_C - B). \tag{17.40}$$

Note that μ varies from a value $g\mu_B B_c$ to zero as B is raised over the range from 0 to B_c. The magnetization therefore increases with field before saturating with all spins aligned along B.

The schematic phase diagram in Figure 17.11 is based on the exchange interaction $J/k_B = 10\,\mathrm{K}$ used in Exercise 17.5 plotted using the scaled and dimensionless variables $k_B T/J$ and B/B_c. The three low-temperature phases are identified as quantum critical (QC), Tomonaga–Luttinger liquid (TLL), and gapped (G). The straight lines I and II, which form a V-shaped region and meet at the QCP, are crossover boundaries between the phases shown.

Crossover boundary I marks the transition from the TLL phase, in which quantum excitations play a dominant role, to the QC phase, in which thermal excitations become increasingly important as T is raised. As a guide, the crossover occurs for $k_B T = \mu = g\mu_B (B_C - B)$ using μ given by Equation 17.40. A numerical slope factor of order unity is introduced by taking as the crossover criterion the predicted maximum in the magnetic susceptibility, which is obtained in terms of T and B using the Fermi function and the magnon density of states for an antiferromagnetic linear chain. For $B > B_C$ an energy gap $g\mu_B (B - B_C)$ for magnon, excitation develops. The temperatures for crossover II are given by the relationship $k_B T = g\mu_B (B - B_C)$.

The phases are distinguished by the spin excitations which determine the observed magnetic properties. In the quantum liquid TLL phase, spinon excitations prevent a long-range spin order from occurring at low temperatures as noted previously. The QC phase involves both quantum and thermal excitations. In the gapped phase, thermal excitation of magnons in states above the gap occurs.

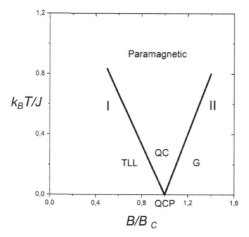

FIGURE 17.11 Schematic phase diagram for a spin ½ Heisenberg antiferromagnetic chain plotted using scaled variables $k_B T/J$ versus B/B_C with J as the exchange interaction between the nearest-neighbor spins. A QCP occurs at the critical field B_C. The three low-temperature phases are labeled Tomonaga–Luttinger liquid (TLL), quantum critical (QC), and gapped (G). A transition to paramagnetic behavior occurs for $k_B T/J \sim 1$.

Quantum phase transitions can be induced in a number of condensed matter systems by high magnetic fields, high pressures, or chemical composition. These transitions require competing interaction terms in the Hamiltonian whose relative importance can be varied in some way. Spin ½ Heisenberg antiferromagnetic chains are a good example of an applied field-driven quantum phase transition. Changes in the ratio of the J coupling term to the B-field term in the Heisenberg–Hamiltonian, due to changes in field, result in the continuous phase change to a long range ordered state at the QCP. Ising spin chains provide another example of a field-induced quantum phase transition as mentioned in Section 17.7.

17.7 INTRODUCTION TO THE ISING MODEL

The Ising model uses a simplified Hamiltonian to treat spin systems in which interactions are important. The approach may be extended to other order–disorder phenomena. Instead of the Heisenberg–Hamiltonian given by Equation 17.19, involving the scalar product of 3D operators S_i and S_j and the resultant terms such as S_{ix} and S_{jx}, the Ising model introduces 1D operators S_i and S_j, with eigenvalues ±1. The Ising Hamiltonian is written as

$$\mathcal{H} = \sum_{\langle i,j \rangle} J_{ij} S_i S_j, \tag{17.41}$$

where the summation is over all pairs i and j. For simplicity, it is usual to consider only nearest-neighbor interactions with a single coupling constant J. The sign of J determines whether parallel alignment ($J < 0$) or antiparallel alignment ($J > 0$) of neighbor spins is favored, and we shall consider the parallel alignment case. If only nearest-neighbor interactions are considered, Equation 17.41 becomes $\mathcal{H} = J \sum_{\langle ij \rangle} S_i S_j$. In spite of the simplified nature of the Ising model Hamiltonian, it has not been possible to obtain an exact (analytical) expression for the partition function in 3D. Expressions for Z have, however, been obtained in 1D and 2D. Numerical methods that permit exponents to be calculated to high precision have provided information for the 3D Ising model.

With an applied magnetic field present, the Ising Hamiltonian takes the form

$$\mathcal{H} = J \sum_{\langle ij \rangle} S_i S_j + g \mu_B B \sum_i S_i, \tag{17.42}$$

with g typically taken as the free electron g factor. For simplicity, we shall deal with the zero field case. The partition function for the 1D Ising model, with $J < 0$ and $B = 0$, is

$$Z = \sum_{\{S_i\}} e^{\beta |J| \sum_{i=1}^{N} S_i\, S_{i+1}}, \tag{17.43}$$

where the summation over $\{S_i\}$ covers the 2^N spin configurations. Using the properties of the exponential function, it follows that Equation 17.43 may be written as $Z = \prod_i \left(\sum_{(S_i)} e^{\beta |J| S_i\, S_{i+1}} \right)$. It is convenient to introduce a new variable x_i defined as follows: $x_i = 1$ for $S_i = S_{i+1}$ (parallel) and $x_i = -1$ for $S_i = -S_{i+1}$ (antiparallel). In terms of x_i, the partition function is

$$Z = \left[\sum_{x_1 = \pm 1} e^{\beta |J| x_1} \right]\left[\sum_{x_2 = \pm 1} e^{\beta |J| x_2} \right]\dots = \left[2\cosh \beta |J| \right]^N. \tag{17.44}$$

(a)

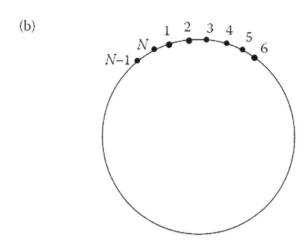

(b)

FIGURE 17.12 Representation of a 1D Ising chain of N spins in (a) open topology and (b) closed ring topology.

The number of spins N may be considered to be so large that the Ising chain end effects may be ignored. Alternatively, the chain can be chosen to form a closed loop with spin 1 and spin N as neighbors. This is illustrated in Figure 17.12.

Exercise 17. 6

Use the partition function for a 1D Ising spin system to obtain the Helmholtz potential F and hence an expression for the entropy S.
  The results are obtained immediately from Equation 17.44:

$$F = -k_\mathrm{B}T \ \ln Z = -Nk_\mathrm{B}T \ \ln\left[2 \ \cosh \beta|J|\right] \tag{17.45}$$

and

$$S = -\frac{\partial F}{\partial T} = -Nk_\mathrm{B}\left[\ln 2 \ \cosh \beta|J| - \beta J \tanh \beta|J|\right]. \tag{17.46}$$

The behavior of the entropy with T is shown in Figure 17.13, plotted in terms of dimensionless quantities.
  We see that the entropy increases smoothly from zero at low $k_\mathrm{B}T/|J|$ to a plateau value at high $k_\mathrm{B}T/ |J|$ that corresponds to a disordered chain, with $(1/2)N$ spins up and $(1/2)N$ down, and no long-range correlation of spin orientations. Complete order exists only at $T = 0$ K, and there is no phase transition at a finite temperature.

The temperature entropy diagram in Figure 17.13 shows that in the zero applied field, the entropy decreases smoothly from an upper limit $S = Nk_\mathrm{B} \ln 2$, to zero as the temperature decreases to 0 K. Application of a longitudinal magnetic field would shift the curve in Figure 17.13 to higher temperatures, but the continuous approach to zero at 0 K would still occur. However, the application of a *transverse* field at very low temperatures is predicted to induce a quantum phase transition

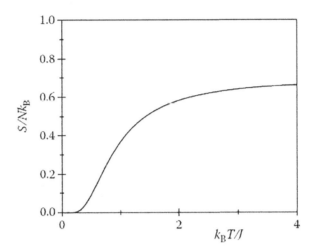

FIGURE 17.13 Predicted behavior of the scaled entropy S/Nk_B plotted versus the reduced temperature $k_B T/|J|$ for the N spin 1D Ising model. At high temperatures, the chain is completely disordered with half the spins up and the other half down.

with a QCP occurring at $B_C = 2J/g\mu_B$, where the spin system transforms from a ferromagnet into a *paramagnet*. This theoretical prediction is in agreement with experimental results for Ising chains.

In higher dimensions (2D and 3D), the Ising model predicts a phase transition from an ordered state to a disordered state at a finite temperature. Values for the critical exponents may be obtained using analytical solutions (2D) or numerical solutions (3D) for the Ising model. The modern theory of critical phenomena, involving scaling ideas and critical exponents, developed to a large extent from considerations of model systems such as the Ising model.

17.8 FERMI LIQUIDS

Chapter 13 deals with ideal Fermi gas systems, in which interactions between fermions are considered negligible. The states of such systems are single particle in a box state, and spin degeneracy allows two particles per state. Predicted properties of quantities such as the heat capacity are in reasonable agreement with the experiment for a number of Fermi systems, as discussed in Chapter 13. For situations in which interactions between fermions are not negligible, the approach must be modified to allow each fermion to move in the field of the other fermions. An important approach known as the Fermi liquid theory provides considerable physical insight into the nonideal Fermi fluid. The spirit of the theory is due to Landau. We briefly outline the basic ideas involved.

In the Fermi liquid theory, the fermions are replaced by quasi-particles with an effective mass m^*. It is assumed that the single-particle states are shifted by interaction effects but that the perturbed states can be labeled using the wave vector $\boldsymbol{k}$, just as for the noninteracting case. In the calculation of properties of interest, such as the specific heat or the magnetic susceptibility, it is states with $\boldsymbol{k}$ near $\boldsymbol{k}_F$ at the Fermi surface that are important. Quasi-particles may be thermally excited from occupied to unoccupied states within a range $k_B T$ of the Fermi energy. In general, the dependence of ε on $\boldsymbol{k}$ for a Fermi liquid is not known. Formally, the shift in energy of a state labeled by $\boldsymbol{k}$ may be written as

$$\delta\varepsilon(\boldsymbol{k}) = \varepsilon(\boldsymbol{k}) - \varepsilon_0(\boldsymbol{k}) = \sum_{\boldsymbol{k}'} f(\boldsymbol{k}, \boldsymbol{k}')\delta n(\boldsymbol{k}') \tag{17.47}$$

where $\varepsilon_0(\boldsymbol{k})$ is the energy at $T = 0$ K and $f(\boldsymbol{k}, \boldsymbol{k}')$ is a function which gives the change in energy of the system caused by a change in the quasi-particle distribution function $\delta n(\boldsymbol{k}')$. At temperatures $T \ll T_F$, only partially unoccupied states near the Fermi level need be considered in scattering processes and $k \simeq k' \simeq k_F$. Landau reasoned that the function $f(\boldsymbol{k}, \boldsymbol{k}')$ must depend only on the angle θ between $\boldsymbol{k}$ and $\boldsymbol{k}'$ so that $f(\boldsymbol{k}, \boldsymbol{k}') = f(\theta)$. If we define a new function $F(\theta) = N(\varepsilon_F) f(\theta)$ and expand the function in terms of the set of Legendre polynomials, with the retention of terms up to second order, we obtain

$$F(\theta) = \sum_n F_n P_n(\cos\theta) = F_0 + F_1[\cos\theta] + F_2\left[\frac{1}{2}(3\cos^2\theta - 1)\right]. \tag{17.48}$$

The coefficients F_n are called Landau parameters, and for many purposes, only a few are needed. If spin interactions are allowed for in the consideration of magnetic properties, an analogous function $G(\theta)$ is needed with Landau parameters G_n.

An important Fermi liquid is helium-3 in the temperature range of 3–70 mK. Many of the properties of this system can be accounted for using three Landau parameters, F_0, F_1, and G_0.

Detailed calculations for the effective mass give

$$m^* = m\left(1 + \frac{1}{3}F_1\right). \tag{17.49}$$

The specific heat of a Fermi liquid becomes

$$c^* = \left(\frac{m^*}{m}\right)c, \tag{17.50}$$

where m is the free particle mass and $c = \left(N_A \pi^2 k_B^2 / 2\varepsilon_F\right)$ is the Fermi gas specific heat.

The magnetic susceptibility of a Fermi liquid is given by

$$\chi^* = \left(\frac{m^*}{m}\right)\left[1 + \frac{1}{4}G_0\right]^{-1}\chi_{Pauli}. \tag{17.51}$$

For liquid helium-3, the Landau parameters are determined empirically using selected experimental results. A consistent description of various properties is obtained in this way. Figure 17.14 shows the specific heat as a function of temperature at pressures close to zero.

The effective mass for liquid helium-3 is given by $m^*/m = 3.01$ at fairly low pressures close to atmospheric. The Landau parameters are found to be pressure sensitive, as might be expected.

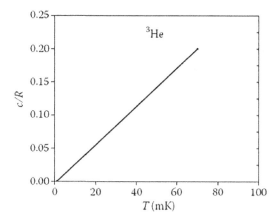

FIGURE 17.14 The plot shows the linear T dependence of the molar-specific heat of liquid helium-3 in the range below 70 mK. Transitions to superfluid phases occur below 2 mK.

The molar entropy of a Fermi liquid, such as helium-3, is given by $S = \int_0^T \left(c^*/T\right)dT = \gamma^* T$, using $c^* = \gamma^* T$, with $\gamma^* = \left(N_A \pi^2 k_B^2/2\varepsilon_F^*\right)$. We have treated the Landau parameters as phenomenological quantities but many body calculations that give estimates of these parameters for Fermi fluids have been carried out. Details may be found in advanced texts on the subject.

Figure 17.15 shows the phase diagram for helium-3, which, as shown by the specific heat plot in Figure 17.14, is a good example of a Fermi liquid over the temperature range of 3–70 mK.

It is of interest to consider the unusual thermodynamic features of the phase diagram, specifically the minimum in the melting curve. Below 320 mK, the melting curve exhibits a region of negative slope. Use of the Clausius–Clapeyron equation, $(dP/dT)_m = \Delta S_m/\Delta V_m$, with $(dP/dT)_m < 0$ and $\Delta V_m > 0$, shows that $\Delta S_m < 0$ in the negative slope region. The entropy *decreases* on melting, which means that, in the region of negative slope, the liquid is more ordered than the solid. This may be understood as follows. In the solid phase, the atoms are fairly well localized on lattice sites. The entropy is determined by the nuclear spin entropy, which, in low applied fields, is to a good approximation, given by $S = N_A k_B \ln 2$. For $T < 320$mK, this entropy value is greater than that of the delocalized liquid phase. Figure 17.16 shows a plot of the entropy of the coexisting solid and liquid phases for helium-3 as a function of temperature on the melting curve.

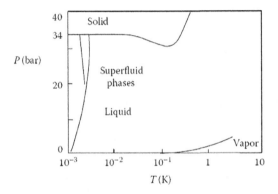

FIGURE 17.15 Phase diagram for helium-3 showing the melting curve. At the lowest temperatures below 2 mK, superfluid liquid phases are found. The melting curve shows a minimum with a region of negative slope below 320 mK.

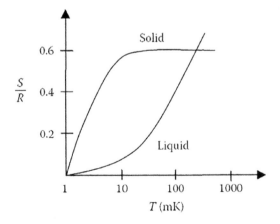

FIGURE 17.16 Molar entropy S divided by the gas constant R as a function of temperature for solid and liquid helium-3 on the melting curve. Note that below 300 mK, the liquid has lower entropy than the solid.

The roughly linear T dependence of the entropy of liquid helium-3 over the range of 3–70 mK is consistent with the predictions of Fermi liquid theory. In low-temperature research, the interesting and unusual properties of helium-3, which result in the negative slope of the melting curve below 320 mK, have been exploited to achieve cooling by means of adiabatic (isentropic) compression in Pomeranchuk cells. The rapid decrease in the entropy of solid helium-3 at the lowest temperatures shown is due to nuclear antiferromagnetic ordering, with a $T_C \sim 1$ mK. The ordering is due to exchange interactions between helium-3 atoms mediated by a coupled exchange motion of several neighboring atoms in the solid.

17.9 NONIDEAL BOSE SYSTEMS: BOSE LIQUIDS

The most extensively studied system of interacting bosons is liquid helium-4. Figure 14.4 shows the specific heat for liquid helium-4 in the vicinity of the superfluid transition $T_\lambda = 2.17$ K, together with the predicted specific heat curve for an ideal Bose gas of the same particle mass and particle density. The difference in the shapes of the curves is marked. Interatomic forces in helium-4 are of the van der Waals type. Figure 17.17 gives the phase diagram for helium-4, which shows that there are two liquid phases He I and He II separated by what is termed the λ line.

For pressures less than ~ 30 atm, helium-4 remains a liquid down to $T = 0$ K. This is a result of the large zero point vibrational motion of the atoms. Below 0.6 K, the heat capacity of liquid helium-4 (designated He II in the temperature range below T_λ) follows the T^3 law, which is found for the low-temperature heat capacity of solids. This suggests that phonons are important excitations in liquid helium-4. Neutron scattering methods have been used to investigate the excitation spectrum for liquid helium, and the results are shown in Figure 17.18.

At low energies, the dispersion relation has the Debye form, as discussed in Section 15.9,

$$\varepsilon(q) = v\hbar q, \tag{17.52}$$

which is consistent with longitudinal phonons with wavevector q. (Transverse phonons do not exist in a liquid.)

At higher energies, a new type of excitation becomes important. These excitations are called rotons. Experiment shows that vortices are formed in superfluid helium when the container of the fluid rotates about a vertical axis. Rotons are the corresponding excitation quanta. A minimum in the dispersion curve occurs for $q = q_0$. In the vicinity of the minimum, the spectrum may be represented, approximately, by the shifted parabolic form

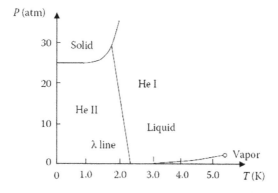

FIGURE 17.17 Phase diagram for helium-4. He I is the normal liquid phase and He II the superfluid phase with extremely interesting transport and other properties as described in the text. The λ line separates the two liquid phases. A solid phae exists at pressures above 25 atm. The liquid–vapor coexistence curve ends in a critical point as shown.

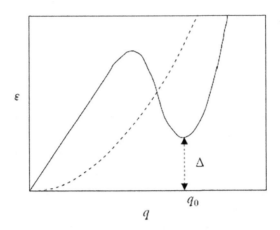

FIGURE 17.18 Dispersion curve for excitations in superfluid liquid helium-4 determined by neutron scattering. The dashed line shows a quadratic dispersion curve for comparison. Phonon and roton excitations are important in this system as discussed in the text.

$$\varepsilon(q) = \Delta + \hbar^2 \frac{(q - q_0)^2}{2m^*}, \tag{17.53}$$

with Δ as the energy at the minimum in the dispersion curve and m^* as the effective mass of a helium atom in the superfluid. Experiment yields the following values for the various quantities in Equation 17.53: $\Delta/k_B = 8.65$ K, $q_0 = 0.192$nm^{-1}, and $m^*/m = 0.16$.

Exercise 17.7

Use the approach developed in the discussion of the specific heats of solids in Chapter 15 with the Debye-type dispersion relation given in Equation 17.52 to obtain an expression for the specific heat of helium-4 at low T.

The low-temperature specific heat is given by

$$c_v = \left(\frac{2\pi^2 k_B V}{15(V\hbar)^3} \right) (k_B T)^3. \tag{17.54}$$

This gives the observed T^3 dependence for c_v.

Exercise 17.8

Use the Planck distribution to estimate the number of phonons and rotons present in liquid He II at a given temperature below T_λ.

For phonons, we have from Chapter 15 $N_p = (V/2\pi^2) \int (q^2 dq/(e^{\beta\hbar vq} - 1))$. Assume that the temperature is sufficiently low that the upper limit in the integral may be extended to infinity. With the variable $x = \beta\hbar vq$, we get

$$N_p = \left[\frac{V}{2\pi^2 (\beta\hbar v)^3} \right] \int_0^\infty \frac{x^2 dx}{e^x - 1} = \left[\frac{V}{2\pi^2 (\beta\hbar v)^3} \right] (k_B T)^3 \Gamma(3)\zeta(3). \tag{17.55}$$

$\Gamma(3)$ is a gamma function and $\zeta(3)$ is a Riemann zeta function with $\Gamma(3) \zeta(3) = 2.4$. We see that N_p follows the T^3 law.

The roton number is given by

$$N_r = \left(\frac{V}{2\pi^2}\right)\int_0^\infty \frac{q^2 dq}{e^{\beta\left[\Delta + \hbar^2(q-q_0)^2/2m^*\right]} - 1} \approx \left(\frac{V}{2\pi^2}\right)e^{-\beta\Delta}\int_0^\infty e^{-\frac{\beta\hbar^2(q-q_0)^2}{2m^*}} q^2 dq.$$

For $x = (\beta/2m^*)^{1/2}\hbar(q - q_0)$ and, with recognition that the region near $x = 0$ is of dominant importance, this results in

$$N_r = \left(\frac{V}{2\pi^2\hbar}\right)\left(\frac{4\pi m^*}{\beta}\right)^{1/2} q_0^2 e^{-\beta\Delta}, \tag{17.56}$$

where the use has been made of the definite integral $\int_0^\infty e^{-x^2}dx = \sqrt{2\pi}$, and as an approximation, we put $q^2 = q_0^2\left[\left(2m^*/\beta\right)^{1/2}\left(x/\hbar q_0\right) + 1\right] \approx q_0^2$. The number of rotons decreases almost exponentially as the temperature is lowered, consistent with the decrease in the roton contribution to the excitation spectrum as shown in Figure 17.18.

The properties of liquid He II have proved to be of great interest and the reader is referred to specialized texts on the subject for further details.

The present chapter has provided an introduction to some of the approaches that are used in classical and quantum systems made up of large numbers of interacting particles. Many areas in this field are the subject of continuing research activity. The final three chapters of this book deal with special topics in statistical and thermal physics.

PROBLEMS CHAPTER 17

17.1 Make use of the approximate expression for the partition function for a real gas of N particles in a volume V at temperature T obtained in Chapter 17 to show that the molar entropy of a monatomic classical real gas may be written as $S = S_{ideal} + S_{correction}$. Obtain an expression for $S_{correction}$ in terms of a definite integral involving the intermolecular potential and N, V, and T. Show that the correction term is negative and give a physical explanation for this result.

17.2 Consider an adsorbed layer of N molecules on a surface of area A. The layer may be regarded as a 2D nonideal gas with an intermolecular potential $u(r)$ dependent on the molecular separation r. Following similar procedures to those used for a 3D gas, derive the virial expansion for the film pressure, or force per unit length, at the boundary of this system.

17.3 Use the expression for the real gas partition function given in Section 17.2 to obtain an expression for the energy of a real gas as a function of temperature. Give your result in terms of the intermolecular potential $u(r)$ and the particle density. Compare your expression with that for the energy of an ideal gas.

17.4 A nonideal gas has an intermolecular potential that is approximated by a square well with a hard core of the form $u(r) = \infty(0 \le r \le a)$, $u(r) = -\varepsilon_0$ $(a < r \le b)$, and $u(r) = 0(r > b)$. Obtain an expression for the second virial coefficient in terms of the square well potential parameters. Compare your result with that for the pure hard-core potential discussed in Section 17.2.

17.5 Obtain the equation of state for a nonideal gas in which the intermolecular potential has the form $u(r) = u_0 e^{-ar^2}$. Take $u_0 \ll k_B T$, which permits expansion of the exponential function $e^{-\beta u(r)}$ in the configurational partition function expression.

17.6 Obtain expressions for the mean energy of a Heisenberg spin $\frac{1}{2}$ system in the mean field approximation, in the zero applied field, for the following temperature cases: $T \ll T_C$, $T \approx T_C$, and $T \gg T_C$, where T_C is the critical temperature. Hence, obtain the specific heat for the system at the same temperatures. Sketch the form of the specific heat curve as a function of T.

17.7 Obtain a mean field expression for the way in which the magnetization of the ferromagnet described in Question 17.6 tends toward its maximum value in the zero applied field for $T \gg T_c$. Give your result in terms of the T_c/T ratio.

17.8 Use the partition function expression for the N spin 1D Ising model to obtain the heat capacity C. Plot C/Nk_B versus J/k_BT, where J is the coupling constant. Qualitatively account for the form obtained.

17.9 The nearest-neighbor spin–spin correlation function for the 1D Ising model is given by the general expression $G(r = 1) = \langle S_i S_{i+1} \rangle = \partial \ln Z/\partial J_i$, where J_i is the coupling between spins i and $i + 1$. Use the nearest-neighbor correlation function expression to show that the n spin correlation function is given by $G(r = n) = \prod_k [\tanh \beta J_k] = [\tanh \beta J]^n$ assuming a uniform interaction J between spins.

Section IID

The Density Matrix, Reactions and Related Processes, and Introduction to Irreversible Thermodynamics

18 The Density Matrix

18.1 INTRODUCTION

In dealing with systems whose eigenstates are known, such as ideal spin systems in a magnetic field, it is sometimes useful to introduce a formalism involving the density matrix. Knowledge of the density matrix operator permits the expectation values of other operators to be determined in a way that is straightforward in principle. We show below that the ensemble average expectation value for an operator A of a system is given by $\langle \overline{A} \rangle = \text{Tr}(\rho A)$, where ρ is the density matrix and Tr is the trace or diagonal sum of the product matrix. The density matrix contains statistical information about the system.

The density matrix is the quantum mechanical analogue of the classical phase space density of representative points description of an ensemble of systems introduced in Chapter 4. For systems consisting of large numbers of particles, the density matrix will, in general, be very large. To simplify matters, systems for which the density matrix can be written in compact form are used to introduce the subject. Generalization to other systems follows directly. The basic ideas are developed with application to an ideal spin system. The form of the density matrix for the ideal gas case is briefly considered. The density matrix approach is extremely powerful and can, for example, provide insight into how systems that are perturbed in some controlled way tend toward equilibrium. Simple examples involving particle beams are used to illustrate the method.

18.2 THE DENSITY MATRIX FORMALISM

The density matrix formalism emerges from the basic quantum mechanical ideas that are briefly reviewed here. Consider a large system with energy in the range from E to $E+dE$. Let a particular eigenstate of the system be $|\phi\rangle$. This state may be written in terms of the complete set of eigenstates $|i\rangle$ of the Hamiltonian of the system as $|\phi\rangle = \sum_i C_i |i\rangle$, where the coefficients C_i are, in general, time dependent and complex. The expectation value of some operator A is given by $\langle A \rangle = \int \phi^* A \phi \, d\tau$ or, in Dirac notation,

$$\langle A \rangle = \sum_{i,j} C_j^* C_i \langle j|A|i \rangle. \tag{18.1}$$

It is convenient to consider the products $C_j^* C_i$ as forming a matrix representation of an operator P, with matrix elements $\langle i|P|j \rangle C_j^* C_i$. Substitution for $C_j^* C_i$ in Equation 18.1 gives

$$\langle A \rangle = \sum_{i,j} \langle i|P|j \rangle \langle j|A|i \rangle = \sum_i \langle i|PA|i \rangle = \text{Tr}(PA) = \text{Tr}(AP). \tag{18.2}$$

Tr denotes the diagonal sum of the matrix elements. Equations 18.1 and 18.2 serve as our starting point for the introduction of the density matrix operator.

In statistical physics, we are generally interested in ensemble averages, which for the operators we consider are denoted as $\langle \overline{A} \rangle$. From Equation 18.1, we obtain

$$\langle \overline{A} \rangle = \sum_{i,j} \overline{C_j^* C_i} \langle j|A|i \rangle. \tag{18.3}$$

The ensemble average of the products $\overline{C_j^* C_i}$ gives the matrix representation of the density matrix operator ρ, with elements $\langle i|\rho|j\rangle = \overline{C_i C_j^*}$. ρ is the ensemble average of the operator P that is introduced in Equation 18.2. Examples of the form of the density matrix in the various statistical ensembles are given in Section 18.3. Equation 18.3 becomes

$$\langle \overline{A} \rangle = \sum_{i,j} \langle i|\rho|j\rangle \langle j|A|i\rangle = \sum_i \langle i|\rho A|i\rangle = \mathrm{Tr}(\rho A). \tag{18.4}$$

Note that ρ is an Hermitian operator with $\langle i\,|\rho|j\rangle = \langle j\,|\rho|i\rangle^*$. If $|\phi\rangle$ is normalized so that $\langle \phi|\phi\rangle = 1$, then $\mathrm{Tr}(\rho) = \sum_i \langle i|\rho|i\rangle = 1$, which shows that for this case the diagonal elements of ρ sum to unity. Equation 18.4 provides the basis for density matrix calculations.

For systems that are not in equilibrium, the density matrix is time dependent, and it is therefore useful to allow for this possibility in our discussion. The time-dependent Schrödinger equation is $i\hbar \partial/\partial t|\phi\rangle \mathcal{H}(t)|\phi\rangle, =$ with $\mathcal{H}(t)$ as the Hamiltonian of the system. For the present, we allow the Hamiltonian to be time dependent. Substitution for $|\phi\rangle$ results in $i\hbar \sum_j \partial/\partial t\left(C_j|j\rangle\right) = \sum_j C_j \mathcal{H}(t)|j\rangle$. Multiplying by $\langle i\,|$ and making use of the orthonormality property give

$$i\hbar \frac{\partial}{\partial t} C_i \sum_j C_j \langle i|\mathcal{H}(t)|j\rangle. \tag{18.5}$$

Now,

$$\frac{\partial}{\partial t}\langle i|\rho|i\rangle = \frac{\partial}{\partial t} C_i C_j^* = \frac{\partial}{\partial t} C_i^* + \frac{\partial C_i}{\partial t} C_j^*, \tag{18.6}$$

and with Equation 18.5 substituted into Equation 18.6, we obtain

$$\frac{\partial}{\partial t}\langle i|\rho|j\rangle = \frac{i}{\hbar}\sum_k \left[\langle i|\rho|k\rangle\langle k|\mathcal{H}|j\rangle - \langle i|\mathcal{H}|k\rangle\langle k|\rho|j\rangle\right] = \frac{i}{\hbar}\langle i|\rho\mathcal{H} - \mathcal{H}\rho|j\rangle.$$

In operator form, the evolution of the density matrix with time is described by the equation

$$\frac{\partial}{\partial t}\rho = \frac{i}{\hbar}[\rho, \mathcal{H}], \tag{18.7}$$

where $[\rho, \mathcal{H}]$ is the commutator of the two operators.

When the Hamiltonian H is time independent, the solution to Equation 18.7 is

$$\rho(t) = e^{-(i/\hbar)\mathcal{H}t}\rho(0)e^{(i/\hbar)\mathcal{H}t}, \tag{18.8}$$

where $\rho(0)$ is the density matrix at time $t = 0$.

Exercise 18.1

Show by expanding the exponential operators that Equation 18.8 is a solution to Equation 18.7.
The exponential operators are expanded as follows:
$e^{\pm(i/\hbar)\mathcal{H}t} = 1 \pm (i/\hbar)\mathcal{H}t + (1/2!)(i/\hbar)^2 \mathcal{H}^2 t^2 \pm \dots$ Differentiation of $\rho(t)$ with respect to time gives

$$\frac{\partial}{\partial t}\rho(t)=\frac{\partial}{\partial t}\left[\left(1-\left(\frac{i}{\hbar}\right)\mathcal{H}t+\cdots\right)\rho(0)\left(1+\left(\frac{i}{\hbar}\right)\mathcal{H}t+\cdots\right)\right.$$

$$=-\frac{i}{\hbar}\mathcal{H}\left(1-\left(\frac{i}{\hbar}\right)\mathcal{H}t+\cdots\right)\rho(0)\left(1+\left(\frac{i}{\hbar}\right)\mathcal{H}t+\cdots\right)$$

$$+\left(1-\left(\frac{i}{\hbar}\right)\mathcal{H}t+\cdots\right)\rho(0)\left(1+\left(\frac{i}{\hbar}\right)\mathcal{H}t+\cdots\right)\left(\frac{i}{\hbar}\mathcal{H}\right).$$

Returning to exponential form, we obtain $\partial/\partial t\,\rho\,(t)=i/\hbar\,[\rho,\mathcal{H}]$, as required.

From Equation 18.8, the matrix elements of ρ are $\langle i|\rho(t)|j\rangle=\langle i|e^{(i/\hbar)\mathcal{H}t}\rho(0)e^{(i/\hbar)\mathcal{H}t}|j\rangle$, or in terms of the wave functions, we have $\int u_i^*\rho(t)u_j\mathrm{d}\tau=\int\left(e^{(i/\hbar)\mathcal{H}t}u_i\right)^*\rho(0)e^{(i/\hbar)\mathcal{H}t}u_j\mathrm{d}\tau$. With the use of the series expansion of the exponential operator, it follows that

$$\langle i|\rho(t)|j\rangle=e^{(i/\hbar)(E_j-E_i)t}\langle i|\rho(0)|j\rangle. \tag{18.9}$$

Equation 18.9 shows that the diagonal elements of the density matrix are time independent, provided $\mathcal{H}$ is time independent, whereas the off-diagonal elements oscillate with frequencies $\omega_{ij}=(E_j-E_i)/\hbar$.

For a system in equilibrium, no time dependence of observable properties is expected. This implies that for a system in equilibrium, all off-diagonal elements of $\rho(0)$ are zero, that is, $\langle i|\rho(0)|j\rangle=0$ for all $i\neq j$ and consequently $\langle i|\rho(t)|j\rangle=0$ for all times t. By separating the real and imaginary parts, the coefficients C_i in Equation 18.3 may be written as $|C_i|e^{i\alpha_i}$, where α_i is a phase factor. The ensemble average of the product of coefficients is $\overline{C_iC_j^*}=|C_i||C_j|\overline{e^{i(\alpha_j-\alpha_i)}}$. If the phases are assumed to be statistically independent, $\overline{C_iC_j^*}=\langle i|\rho|j\rangle=0$. Vanishing of the off-diagonal elements of the density matrix is a consequence of what is called the random phase hypothesis, according to which there is no phase correlation between different members of the large ensemble.

Following the introduction to the density matrix given in this section, which is based on fundamental quantum mechanical concepts and relationships, we are in a position to apply the formalism to various systems such as the ideal gas and the ideal spin system. Before we do this, we consider the form that the density matrix takes for the three statistical ensembles introduced in Chapter 10.

18.3 FORM OF THE DENSITY MATRIX IN THE THREE STATISTICAL ENSEMBLES

For a system in equilibrium, the diagonal elements of the density matrix clearly correspond to the probability of finding the system in a particular eigenstate because

$$\langle i|\rho|i\rangle=\overline{C_iC_i^*}=\overline{|C_i|^2}. \tag{18.10}$$

Furthermore, in equilibrium, the off-diagonal elements are all zero, as shown in Section 18.2. We now consider the three ensembles in turn.

a. *Microcanonical Ensemble.* In Chapter 4, the number of accessible microstates $\Omega(E)$ for a system with energy in the range from E to $E+\delta E$ is evaluated for ideal systems of spins and particles. According to the fundamental postulate of statistical physics, all accessible microstates are equally probable. It follows directly that in the microcanonical ensemble, the density matrix elements are given by

$$\langle i|\rho|j\rangle = \left(\frac{1}{\Omega(E)}\right)\delta_{ij},$$ (18.11)

where the set of states $|j\rangle$ corresponds to the accessible microstates and δ_{ij} is the Kronecker delta. In this ensemble, the density matrix is constant with all diagonal elements equal to each other. For a system of N spins in a magnetic field B, with n spins up and $(N-n)$ spins down, the number of accessible microstates is given by Equation 4.23 as $\Omega(E) = \left(\begin{array}{c} N \\ n \end{array}\right)\delta E/2\mu\beta$. The form of the density matrix given by Equation 18.11, although simple, is not particularly useful, and we now consider the canonical ensemble form.

b. *Canonical Ensemble.* Most calculations make use of the density matrix in the canonical ensemble case with a temperature parameter β determined by the heat baths with which members of the ensemble are in contact. The matrix elements are given by

$$\langle i|\rho|j\rangle = \left(\frac{1}{Z}\right)e^{-\beta E_i}\delta_{ij},$$ (18.12)

where Z is the partition function for the system considered and δ_{ij} is the Kronecker delta. The density matrix operator may be written as

$$\rho = \frac{e^{-\beta\mathcal{H}}}{\mathrm{Tr}\, e^{-\beta\mathcal{H}}} = \frac{e^{-\beta\mathcal{H}}}{Z},$$ (18.13)

with $\mathcal{H}$ as the Hamiltonian of the system. The following expression for the expectation value of a quantity of interest that corresponds to a particular operator A is obtained from Equation 18.4:

$$\langle\overline{A}\rangle = \frac{1}{Z}\mathrm{Tr}\, e^{-\beta\mathcal{H}} A.$$ (18.14)

This is a useful relationship, particularly for systems such as spin systems.

c. *Grand Canonical Ensemble.* Finally, and for completeness, the density matrix operator in the grand canonical ensemble has the form

$$\rho = \frac{1}{\mathbb{Z}}e^{-\beta(\mathcal{H}-\mu n)},$$ (18.15)

where μ is the chemical potential and n the particle number operator. $\mathbb{Z}$ is the grand partition function. We shall not make use of this form of the density matrix in the present discussion.

18.4   DENSITY MATRIX CALCULATIONS

a. *Spin Systems.* Spin systems provide instructive examples of applications of the density matrix formalism. Because of the upper and lower energy bounds for spin systems, the density matrix has a finite number of elements. For ideal systems of identical spins, the density matrix can be broken up into identical submatrices. Consider an ideal noninteracting system of N electron spins with operator **J** in a magnetic field **B**. Each spin has a magnetic dipole moment $\mu = -g\,\mu_B J$. The energy eigenvalues for representative spin i are $\varepsilon_i = g\mu_B B m_{ij}$, where m_{ij} takes values in the range from J to $-J$. The energy $\langle E \rangle$ of the system is given by the sum of the energies of the individual spins. Each spin has $(2J+1)$ energy levels, and the density matrix for the whole system is made up of $(2J+1)N \times (2J+1)N$ elements, most of which are zero for a system in equilibrium. Along the diagonal, there are groups of $(2J+1)$ elements that are repeated N times. The density matrix for each spin has the form

$$\rho = \frac{1}{Z}\begin{pmatrix} e^{-\beta \varepsilon^J} & & 0 & \\ & e^{-\beta \varepsilon^{J-1}} & & \\ 0 & & e^{-\beta \varepsilon^{J-2}} & \\ & & & \ddots \end{pmatrix}. \tag{18.16}$$

Note that for an ideal system, we need to consider only one of the submatrices in calculating the mean value of some quantity for a single spin, for example, the mean energy $\langle \varepsilon \rangle$. We then multiply this value by N to obtain the mean value $\langle E \rangle = N\langle \varepsilon \rangle$ for the entire system. The ensemble average energy $\langle \varepsilon \rangle$ of a single spin is $\langle \varepsilon \rangle = \mathrm{Tr}\rho\mathcal{H}$, with

$$\rho = \left(\frac{1}{z}\right)e^{-\beta H} = \frac{1}{z}\left(1 - \beta\mathcal{H} + \frac{1}{2}(\beta\mathcal{H})^2 + \cdots\right). \tag{18.17}$$

We have again made use of the expansion of the exponential operator. For paramagnetic systems, it is often permissible to work in the high-temperature approximation, where only the first-order term in the expansion of the operator need be retained. In this approximation, where $k_B T$ is much larger than energy-level spacing $2\mu B$, we obtain the simplified expressions

$$\langle \varepsilon \rangle \approx \frac{1}{z}\mathrm{Tr}\left(\mathcal{H} - \beta\mathcal{H}^2\right) \tag{18.18}$$

and

$$z \approx \mathrm{Tr}\left(1 - \beta\mathcal{H}\right). \tag{18.19}$$

Exercise 18.2

Use Equation 18.18 to obtain $\langle \varepsilon \rangle$ for a system of N spins J at temperature T in a magnetic field B. For each spin, $\mathrm{Tr}H$ is readily evaluated as follows:

$$\mathrm{Tr}\,\mathcal{H} = g\mu_B B\,\mathrm{Tr}Jz = \left[g\mu_B B \sum_{m_J = J, J-1, \cdots, -J} m_J \right] = 0.$$

We have chosen the eigenstate basis so that the matrix corresponding to the operator J_z is diagonal with elements $J, J-1, \ldots, -J$ and trace zero. From Equation 18.19, it follows immediately that, in the high-temperature approximation,

$$z \approx (2J+1), \tag{18.20}$$

because there are $(2J+1)$ identical terms, each with value close to unity, along the diagonal of the density matrix. To obtain $\langle E \rangle$, it is necessary to evaluate $\mathrm{Tr}\mathcal{H}^2$. Now $\mathrm{Tr}\mathcal{H}^2 = \left(g\mu_B B\right)^2 \mathrm{Tr}J_z^2$ and

$$\text{Tr}J_z^2 = \sum_{mJ=J,J-1,\dots,-J} m_J^2. \tag{18.21}$$

The sum in Equation 18.21 may be evaluated with the identity

$$\text{Tr}J_z^2 = \text{Tr}J_x^2 = \text{Tr}J_y^2 = \frac{1}{2}\text{Tr}J^2, \tag{18.22}$$

where the use is made of the theorem that the trace of a matrix is independent of the choice of basis states. We obtain

$$\text{Tr}J^2 = (2J+1)J(J+1), \tag{18.23}$$

because there are $(2J+1)$ diagonal terms in the matrix representation, each equal to $J(J+1)$. It follows from Equations 18.22 and 18.23 that

$$\text{Tr}J_z^2 = \frac{1}{3}J(J+1)(2J+1). \tag{18.24}$$

With Equations 18.19, 18.20, 18.21, and 18.24, we obtain, in the high-temperature approximation,

$$\langle\varepsilon\rangle = -\frac{1}{3}(g\mu_B B)^2 \beta J(J+1). \tag{18.25}$$

This is the mean energy of a single spin. For the ideal system of N spins, the total mean energy in the high-temperature approximation is finally

$$\langle E\rangle = -\frac{Ng^2\mu_B^2 B^2 J(J+1)}{3k_B T}. \tag{18.26}$$

For the special case $J = 1/2$ and $g = 2$, Equation 18.26 gives the mean energy as $\langle E\rangle = N\mu_B^2 B^2/k_B T$, which agrees with the result in Equation 10.37 obtained with direct use of the canonical distribution provided we make the approximation $\tanh \mu_B B/k_B T \approx \mu_B B/k_B T$.

Although the results for the ideal spin system may be obtained in a straightforward way with the use of the canonical distribution, the density matrix approach provides a more powerful means for carrying out a number of calculations. In particular, the off-diagonal elements that we have seen to be zero for systems in equilibrium will in general not be zero for systems that are prepared in a particular way. Situations of this kind arise in magnetic resonance and in laser optics, for example. The instantaneous properties and the evolution of such systems with time are elegantly treated within the density matrix formalism.

Exercise 18.3

Use the result in Equation 18.26 to obtain an expression for the magnetization and magnetic susceptibility of the N spin system at temperature T in a magnetic field B.

Because $\langle E\rangle = -\langle M_z\rangle B$, it follows from Equation 18.26 that

$$\langle M_z\rangle = \frac{Ng^2\mu_B^2 BJ(J+1)}{3k_B T}. \tag{18.27}$$

The magnetic susceptibility is given by

$$\chi = \frac{\langle M_z\rangle}{VH} = \frac{Ng^2\mu_0\mu_B^2 J(J+1)}{3Vk_B T} \tag{18.28}$$

This is Curie's law, as expected.

Exercise 18.4

Obtain an expression for the entropy S of an N spin system, and plot the behavior of S as a function of B/T.

The entropy of the spin system may be obtained with the use of the expression

$$S = k_B \left(\ln Z + \beta \langle E \rangle \right), \tag{18.29}$$

which follows directly from $F = E - TS = -k_B T \ln Z$. For N spins, Equation 18.20 gives the partition function as

$$Z = (2J + 1)^N . \tag{18.30}$$

On substitution of Equations 18.26 and 18.30 into Equation 18.29, we find the entropy is given by

$$S = Nk_B \left[\ln(2J+1) - \frac{1}{3} g^2 \mu_B^2 \beta^2 B^2 J(J+1) \right]. \tag{18.31}$$

For $J = 1/2$, this expression reduces to $S = Nk_B \left[\ln 2 - (g\mu_B B/2k_B T)^2 \right]$. Figure 18.1 shows the entropy for this case in reduced form as S/Nk_B plotted versus B/T in the high-temperature approximation. In the low-field–high-temperature limit, S/Nk_B tends to $\ln 2$. As B is raised and/or T is lowered, the entropy decreases as shown. The high-temperature approximation will break down when $k_B T$ becomes comparable with the spacing of the energy levels.

For an isentropic process that corresponds to a reversible adiabatic change of the magnetic field, the ratio $(B/T)^2$ remains constant so that $B_f/T_f = B_i/T_i$, as given previously in Chapters 2 and 6 in the discussion of magnetic cooling processes. In Figure 18.1, an isentropic process is represented by a fixed point on the curve.

b. *The Ideal Gas and Other Systems.* Although spin systems have been considered in carrying out density matrix calculations, the formalism is quite general. We have $\langle E \rangle = \mathrm{Tr}\rho\mathcal{M}$, with $\rho = 1/Z\, e^{-\beta\mathcal{H}}$; in matrix representation, the mean energy is given by forming the product

$$\langle E \rangle = \frac{1}{Z} \begin{pmatrix} e^{-\beta E_0} & & & \\ & e^{-\beta E_1} & & \\ & & \ddots & \end{pmatrix} \begin{pmatrix} E_0 & & & \\ & E_1 & & \\ & & \ddots & \end{pmatrix} = \frac{1}{Z} = \sum_n E_n e^{-\beta E_n}, \tag{18.32}$$

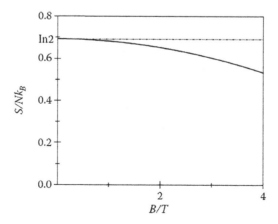

FIGURE 18.1 Entropy divided by Nk_B for a system of N noninteracting spins ($J = 1/2$) in a magnetic field B, plotted versus B/T in the high-temperature approximation. For $B/T = 0$, the scaled entropy is given by $\ln 2$.

with

$$z = \sum_n e^{-\beta E_n}. \tag{18.33}$$

These are familiar results for the canonical ensemble. Knowledge of the energy eigenvalues is clearly necessary to perform density matrix calculations. The ratio of successive diagonal elements in the density matrix is simply the Boltzmann factor for adjacent energy states.

For an ideal gas system, the energy eigenvalues are the particle in a box eigenvalue and, in the classical limit, the eigenstates form a quasi-continuum as described in Chapter 16. The summations in Equations 18.32 and 18.33 may again be converted to integrals. For the classical ideal gas system, we see that the density matrix approach reduces to the integration procedure used in Chapter 16.

Note that

$$-\frac{\partial \rho}{\partial \beta} = \mathcal{H}\rho = \rho\mathcal{H}, \tag{18.34}$$

and knowledge of the density matrix permits the mean energy to be obtained by differentiation of ρ with respect to β followed by the evaluation of the trace of the resulting matrix. Generalization of this approach to other mean values can clearly be made.

18.5   POLARIZED PARTICLE BEAMS

The density matrix formalism can be applied to a wide variety of systems. It is instructive to consider beams of particles. Consider electrons emitted by a hot filament in an evacuated enclosure. The energy of the electrons is determined by factors such as the temperature of the filament, the work function of the metal from which the filament is made, and any accelerating potential that is present. The beam will normally be unpolarized. We consider the electron's spin degrees of freedom with the spin operators represented by the Pauli spin matrices, $\sigma_x = \begin{pmatrix} 0 & 1 \\ 1 & 0 \end{pmatrix}, \sigma_y = i\begin{pmatrix} 0 & 1 \\ 1 & 0 \end{pmatrix},$

and $\sigma_x = \begin{pmatrix} 1 & 0 \\ 0 & -1 \end{pmatrix}$ giving

$$S_x = \frac{\hbar}{2}\sigma_x, \quad S_y = \frac{\hbar}{2}\sigma_y, \text{ and } S_z = \frac{\hbar}{2}\sigma_z, \tag{18.35}$$

with the axes x, y, and z chosen in some convenient way.

If no applied magnetic field is present, the energy levels that correspond to different spin orientations $S_z = \pm 1/2$ are degenerate, and the energy of the degenerate spin states may be taken to be zero. The density matrix for a single particle takes the form $\rho = 1/2\begin{pmatrix} 1 & 0 \\ 0 & 1 \end{pmatrix}$, and the expectation value of the spin for particles in the beam is given by

$$\langle S_z \rangle = \text{Tr}\rho S_z = \frac{\hbar}{4}\text{Tr}\begin{pmatrix} 1 & 0 \\ 0 & 1 \end{pmatrix}\begin{pmatrix} 1 & 0 \\ 0 & -1 \end{pmatrix} = 0.$$

Identical results are obtained for $\langle S_x \rangle$ and $\langle S_y \rangle$. In the presence of a magnetic field or some other polarizing mechanism, the form of the density matrix will change, and the expectation value $\langle S_z \rangle$, for example, may be nonzero. Although the situation that has been considered involves an electron beam, a related but somewhat more complicated approach is used for polarized light beams.

Exercise 18.5

Find the direction of spin polarization for a beam of electrons described by a density matrix of the form

$$\rho = \frac{1}{2} \begin{pmatrix} 1 & 1 \\ 1 & 1 \end{pmatrix}.$$

Consider the mean values of the spin components along orthogonal directions with the use of the Pauli spin matrices.

Along the x-direction, we obtain

$$\langle S_z \rangle = \frac{\hbar}{4} \mathrm{Tr} \begin{pmatrix} 1 & 1 \\ 1 & 1 \end{pmatrix} \begin{pmatrix} 1 & 0 \\ 0 & -1 \end{pmatrix} = \frac{\hbar}{4} \mathrm{Tr} \begin{pmatrix} 1 & 1 \\ 1 & -1 \end{pmatrix} = 0.$$

Similarly along y, $\langle S_y \rangle = 0$, and finally along x,

$$\langle S_x \rangle = \frac{\hbar}{4} \mathrm{Tr} \begin{pmatrix} 1 & 1 \\ 1 & 1 \end{pmatrix} \begin{pmatrix} 0 & 1 \\ 1 & 0 \end{pmatrix} = \frac{\hbar}{4} \mathrm{Tr} \begin{pmatrix} 1 & 1 \\ 1 & 1 \end{pmatrix} = \frac{\hbar}{2}.$$

The electron beam is clearly polarized along the x-direction.

18.6 CONNECTION OF THE DENSITY MATRIX TO THE CLASSICAL PHASE SPACE REPRESENTATION

Classical phase space concepts are introduced and discussed in Chapter 4. It is shown there that the state of a system of N particles in 3D is represented by a point in $6N$ dimensions phase space with $3N$ independent momentum coordinates ($p_1 \ldots p_{3N}$) and $3N$ independent position coordinates ($q_1 \ldots q_{3N}$). The coordinates will change with time, and the representative point will traverse the accessible regions of phase space.

An ensemble of identical systems may be represented by a set of representative points that swarm through phase space. For a sufficiently large ensemble, the representative points are regarded as constituting as *incompressible fluid* with density ρ. Using classical methods developed for fluid flow, it is possible to derive what is called Liouville's theorem for the rate of change of the phase space density with time:

$$\frac{d\rho}{dt} = \frac{\partial \rho}{\partial t} + [\rho, \mathcal{H}] = 0, \tag{18.36}$$

where H is the classical Hamiltonian and $[\rho, H]$ is a Poisson bracket given by

$$[\rho, \mathcal{H}] = \sum_{i=1}^{3N} \left(\frac{\partial \rho}{\partial q_i} \frac{\partial \mathcal{H}}{\partial p_i} - \frac{\partial \rho}{\partial p_i} \frac{\partial \mathcal{H}}{\partial q_i} \right). \tag{18.37}$$

We simply state Liouville's theorem without derivation. Our purpose here is to show the connection between the classical phase space approach and the approach given in this chapter. Note that the total time derivative $d\rho/dt$ corresponds to the change in density with time as seen by an observer moving with the fluid, whereas the partial derivative $\partial \rho/\partial t$ is the change in density at a point whose coordinates are fixed in phase space.

For a stationary ensemble for which $d\rho/dt = 0$, it follows from Equation 18.36 that $[\rho, H] = 0$.

From Equation 18.37, we obtain

$$\left(\frac{\partial \rho}{\partial q_i} \frac{\partial \mathcal{H}}{\partial p_i} - \frac{\partial \rho}{\partial p_i} \frac{\partial \mathcal{H}}{\partial q_i} \right) = 0 \tag{18.38}$$

because each term in the summation must be zero. It should be recalled that Hamilton's equations in analytical mechanics give $\dot{q}_i = \partial \mathcal{H} / \partial p_i$ and $\dot{p}_i = -\partial \mathcal{H} / \partial q_i$.

Equation 18.38 has a number of solutions. First, ρ may be constant over the *accessible* regions of phase space, that is,

$$\rho(q, p) = \text{constant}. \tag{18.39}$$

This solution corresponds to the microcanonical ensemble case, with all accessible microstates equally probable. An alternative solution corresponds to the density being some function of the Hamiltonian,

$$\rho = \rho(\mathcal{H}). \tag{18.40}$$

It follows immediately that $[\rho,] = 0$, because in this case

$$[\rho, \mathcal{H}] = \sum_{i-1}^{3N} \left[\left(\frac{\partial \rho}{\partial \mathcal{H}} \right) \left(\frac{\partial \mathcal{H}}{\partial q_i} \right) \left(\frac{\partial \mathcal{H}}{\partial p_i} \right) - \left(\frac{\partial \rho}{\partial \mathcal{H}} \right) \left(\frac{\partial \mathcal{H}}{\partial p_i} \right) \left(\frac{\partial \mathcal{H}}{\partial q_i} \right) \right]$$

and each term in the summation vanishes.

The canonical ensemble corresponds to a particular choice of function in Equation 18.40,

$$\rho = (q, p) \propto e^{-\beta \rho \mathcal{H}(q,p)}. \tag{18.41}$$

Equation 18.39 is the classical analogue of Equation 18.11 involving the density matrix, whereas Equation 18.41 corresponds to Equation 18.13. The connection between the classical phase density of representative points approach and the quantum mechanical ensemble probability density for various quantum states, as expressed in the density matrix, is therefore apparent.

PROBLEMS CHAPTER 18

18.1   The entropy of a system is given in terms of the density matrix ρ by $S = -k_B \text{Tr} \rho \ln \rho$. Show that this expression is equivalent to the entropy expression involving the partition function $S = k_B [\ln Z + \beta \langle E \rangle]$.

18.2   An electron beam has an isotropic spin distribution $\langle S_x \rangle = \langle S_y \rangle = \langle S_z \rangle = 0$. Make use of the Pauli spin matrices to obtain the form of the density matrix that describes the beam in a representation in which S_z is diagonal. What is the spin polarization of a beam of electrons described by the density matrix $\rho = 1/2 \begin{bmatrix} 1 & 1 \\ 1 & 1 \end{bmatrix}$?

18.3   Obtain the density matrix for a partially polarized electron beam in which a fraction f of electrons are polarized along the beam direction and $(1-f)$ are polarized in the opposite direction.

18.4   Write down the form of the density operator for a 1D ideal gas with temperature parameter β. Use this operator to obtain the density matrix for the gas. How would the expressions be modified in the case of a 2D ideal gas?

18.5   Use the density matrix approach to obtain expressions for the temperature dependence of the mean energy $\langle E \rangle$ and the entropy S of an ideal system of N spins, with $I = 1$ situated in a magnetic field B. Sketch the behavior of S as a function of B/T. Show that the magnetic susceptibility is given by Curie's law, and obtain the Curie constant for this system.

18.6   Show that the entropy of a system may be written as $S = k_B [\ln Z - (\beta^2/Z) \text{Tr} \mathcal{H}^2]$, and use this expression to discuss adiabatic demagnetization of an ideal paramagnetic spin system. Mention any approximations that are made.

18.7   For a nonideal spin system in which dipolar interactions are important, the Hamiltonian has the form $\mathcal{H} = \mathcal{H}_Z + \mathcal{H}_D$, where $\mathcal{H}_Z$ is the Zeeman Hamiltonian and $\mathcal{H}_D$ is the dipolar Hamiltonian. Show that in an adiabatic demagnetization process in which the applied field is reduced from H_i to H_f, the final temperature T_f is related to the initial temperature T_i by the expression $T_f = T_i \left[\left(H_f^2 + H_L^2 \right)^{1/2} / \left(H_i^2 + H_L^2 \right)^{1/2} \right]$, where $H_L^2 = \text{Tr} \mathcal{H}_D^2 / \text{Tr} M_Z^2$ defines a local field in the system. The magnetization is denoted by M_Z, and you should assume that $\text{Tr} (M_Z \mathcal{H}_D) = 0$. Comment on the role of the local field in demagnetization experiments.

18.8   Consider the demagnetization of a paramagnetic spin system in a process in which the applied magnetic field is suddenly reduced from an initial value H_i to a final value H_f. What is the form of the density matrix for the system immediately following the sudden reduction in the field? Show that the system proceeds to a new equilibrium condition characterized by an inverse temperature $\beta_f = \beta_i \, H_i H_f / H_f^2$.

19 Reactions and Related Processes

19.1 INTRODUCTION

This chapter deals with various chemical and physical processes that at first sight appear unrelated but which have features in common. In particular, the chemical potential plays a key role in the discussion of all the phenomena that are considered. In Section 7.6, the topic of chemical equilibrium in gaseous systems containing molecular constituents that react with one another is introduced and the law of mass action is stated. The chemical potential μ of the molecular species undergoing a reaction is of great importance in treating processes of this kind. For reaction processes taking place at constant volume, it is appropriate to make use of the Helmholtz potential F, which, in the classical high-temperature–low-density limit, may be written down using expressions for the ideal gas given in Chapter 16. It is then straightforward to obtain μ for each molecular species, and useful results that include the law of mass action follow.

The adsorption of gas molecules on a surface has certain similarities to chemical reaction processes, with μ again playing an important role in describing processes of this kind. It is important to distinguish between what are termed chemical and physical adsorption processes, and both of these processes are discussed in terms of simple models. Another example of a process in which dynamic equilibrium is reached at a particular temperature is the excitation of carriers into the conduction or valence bands of a semiconductor. It is shown that the law of mass action applies in situations of this kind.

19.2 THE PARTITION FUNCTION FOR A GASEOUS MIXTURE OF DIFFERENT MOLECULAR SPECIES

Consider a gaseous mixture of molecular species that interact to form a reaction product. As discussed in Chapter 7, chemical reactions are conveniently written in the form $\sum_{i} x_i X_i = 0$, where X_i represents a molecular species i and x_i the number of molecules of this species involved in a single reaction process. We are interested in the concentrations of the various molecules once equilibrium has been reached at a given temperature and pressure. In the calculation of the partition function for a particular constituent, we assume that the classical ideal gas approximation holds. The fact that the molecules undergo a chemical reaction shows that interactions between molecules are not weak. However, provided the ideal gas equation of state applies to the system, the intermolecular potential is important only for brief periods when molecules collide. The molecules possess, on average, a negligible potential energy.

Following the notation used in Section 7.6, the gaseous mixture consists of molecular species i with equilibrium numbers N_i. For each component i, the partition function is $Z_i = z_i^N / N!$, where z_i is the single-particle partition function. As a result of the properties of the exponential function, the partition function for the whole gaseous mixture is simply $Z = \prod_i Z_i$. Taking logarithms and applying Stirling's formula, given in Appendix A, we obtain

$$\ln Z = \sum_i N_i \left[\ln z_i - \ln N_i + 1 \right]. \tag{19.1}$$

For a monatomic species, the single-particle partition function is given by Equation 16.4 as $z = V(mk_BT/2\pi\hbar^2)^{3/2} \approx (V/V_Q)$, with V_Q as the quantum volume. For composite molecules such as diatomics and triatomics, it is necessary to consider the internal degrees of freedom which include electronic, vibrational, rotational, and nuclear spin contributions as discussed in Section 16.5. The partition function for polyatomic molecules is written as $z_i^{total} = (V/V_Q)z_i^{int}$, where z_i^{int}, the internal partition function for species i, may be calculated from available information. The form for Z given in Equation 19.1 together with z_i^{total} for each molecular type gives the partition function for the gas mixture.

19.3   THE LAW OF MASS ACTION

As noted above, chemical reactions are conveniently written in the compact form as $\sum_{ix_i} x_i \, X_i = 0$, with x_i as the number of molecules that participate in an individual reaction and X_i as the chemical symbol. In equilibrium, we minimize the Gibbs potential (constant P) or the Helmholtz potential (constant V) and obtain Equation 7.33, $\sum_{ix_i} x_i \, \mu_i = 0$. For constant volume conditions, the chemical potential is given by $\mu_i = (\partial F/\partial N)_{T,V,N_j}$. With the use of Equation 19.1, the Helmholtz potential follows immediately

$$F = -k_B T \ln Z = -k_B T \sum_i N_i [\ln z_i - \ln N_i + 1] \tag{19.2}$$

and the chemical potential is obtained as

$$\mu_i = \left(\frac{\partial F}{\partial N_i}\right)_{T,V,N_j} = -k_B T \ln\left(\frac{z_i}{N_i}\right). \tag{19.3}$$

Combining Equations 7.33 and 19.3, we find that the chemical equilibrium is governed by the equation $\sum_i x_i \mu_i = -k_B T \sum_i x_i \ln(z_i/N_i) = 0$, and it follows that

$$\sum_i (x_i \ln z_i - x_i \ln N_i) = 0. \tag{19.4}$$

From Equation 19.4, we obtain the law of mass action in the form

$$\prod_i N_i^{x_i} = \prod_i z_i^{x_i}. \tag{19.5}$$

The equilibrium constant for a given reaction is defined as $K_N(T,V) = \prod_i z_i^{x_i}$. By obtaining an explicit expression for the equilibrium constant in terms of the partition functions, we have significantly extended the treatment of chemical reactions given in Section 7.6. Knowledge of the molecular partition functions for each species permits $K_N(T, V)$ to be calculated and inserted in the law of mass action. This procedure establishes a relationship between the numbers of reactant and product molecules, for a specific reaction, under given conditions of volume and temperature.

As an illustrative example, consider the chemical reaction introduced in Section 7.6 in the form of Equation 7.32, $-1H_2 - 1Cl_2 + 2HCl = 0$. From the law of mass action, we have $N_{HCl}^2 N_{H_2}^{-1} N_{Cl_2}^{-1} = K_N$, and this relates the equilibrium numbers of molecules of each species, under given conditions, in terms of the equilibrium constant. To obtain K_N, the molecular partition functions z_i for the diatomic molecules must be evaluated using expressions given in Chapter 16 together with information on the moment of inertia and the vibrational frequencies for each type of molecule. If we choose the numbers of reactant H_2 and Cl_2 molecules to be the same, that is, $N_{Cl_2} = N_{H_2}$, it follows that $N_{HCl} = \sqrt{K_N N_{H_2}^2}$. Measured equilibrium constants, obtained by the analysis of equilibrium molecular compositions, are found to be in good agreement with the calculated values that make use of molecular partition functions. The approach outlined above provides a powerful theoretical method for dealing with chemical reactions and may be adapted to deal with other situations, such as ionization processes in gases at very high temperatures. Evaluation of the partition function for monatomic gases and for systems of particles such as electrons in a vapor is simple with the use of Exercise 16.1, $z_i = \left(V/V_Q^i \right)$, where V_Q^i is the quantum volume for species i.

An alternative form of the law of mass action is obtained using the particle concentration $c_i = N_i/V$ and has the form

$$\prod_i c_i^{x_i} = \prod_i \left(\frac{z_i}{V} \right)^{x_i} = K_C\left(T\right), \qquad (19.6)$$

with $K_C\left(T\right)$ as the equilibrium constant in terms of concentrations. When reactions are carried out at constant pressure instead of at constant volume, it is necessary to minimize the Gibbs potential rather than the Helmholtz potential. Similar expressions to those given above are obtained.

Exercise 19.1

Dissociation of hydrogen molecules is induced at high temperatures. Give the reaction equation for this process, and obtain an expression for the equilibrium constant under constant volume conditions.

The dissociation reaction is given by $H_2 - 2H = 0$.

From Equation 19.5 and the definition of the equilibrium constant, we obtain

$$K_N\left(T,V\right) = \frac{[N_H]^2}{N_{H_2}} = \frac{[z_H]^2}{z_{H_2}}.$$

For molecular hydrogen, the partition function is $z_{H_2} = z^{trans} z^{rot} z^{vib} z^{el}$, as discussed in Chapter 16.

Expressions for the various partition function contributions are given in Section 16.5, and with the use of these expressions for homonuclear hydrogen, we obtain

$$z_{H_2} = g_{H_2} \left(\frac{V}{\beta^{3/2}} \right) \left(\frac{2m_H}{2\pi\hbar^2} \right)^{3/2} \left(\frac{I_{H_2}}{\beta\hbar^2} \right) \left(\frac{e^{-\frac{1}{2}\beta\hbar\omega}}{1 - e^{-\beta\hbar\omega}} \right) e^{\beta\varepsilon_0}.$$

The factor g_{H_2} allows for spin degeneracy, and we put $g_{H_2} = 4$ for the ground state of the hydrogen molecule to allow for the four nuclear spin states. The spin state of the electrons is fixed in the molecular ground state. Note that a factor 1/2 has been introduced in the expression for z^{rot} to allow for the indistinguishability of the H atoms in the H_2 molecule, as pointed out for homonuclear diatomics in Section 16.5. The partition function for the hydrogen atom is $z_H = g_H(V/\beta^{3/2})$ $[m_H/2\pi\hbar^2]^{3/2}$, where the atomic degeneracy factor is $g_H = 4$, with allowance for the electron and nuclear spin states. We choose the zero of energy of the system to be the energy when the two H atoms are infinitely far apart. This simplifies the expression for z_H because, with this choice, the energy of a single H atom in its ground state is zero. Care must be taken to allow for zero-point vibrational energy in the value of the ground-state energy $-\varepsilon_0$ for the hydrogen molecule.

The equilibrium constant $K_N(T, V)$ can be calculated, for given T and V, with the use of molecular constants for hydrogen in the above partition function expressions. The numbers of reactant and product molecules under given temperature and volume conditions can then be calculated.

19.4 ADSORPTION ON SURFACES

The adsorption of molecules on a surface in contact with a gas can take place either via a process called chemical adsorption or through a different process termed physical adsorption. Chemical adsorption involves the formation of chemical bonds at the surface giving rise to a relatively immobile layer that cannot be removed simply by lowering the gas pressure. In contrast, physical adsorption involves weaker surface interactions, and the number of physically adsorbed molecules, at a given temperature, does depend on the gas pressure. Physically adsorbed molecules may be relatively mobile on the surface. In some systems, both types of adsorption may be important with the formation of physically adsorbed layers above a chemically adsorbed layer.

We consider two adsorption models. Model 1 allows for translational mobility of molecules on a surface and corresponds to a 2D gas, whereas model 2 assumes that the molecules stick or are bonded to particular sites on the surface and possess no translational mobility.

Model 1

Let the surface have area A and be in contact with a gas at pressure P and temperature T in a container of volume V. In equilibrium, the chemical potentials of the surface layer and the gas will be equal, with $\mu_g = \mu_s$. For the gas phase, we have from Equation 19.3, with omission of the subscript i for our single-component system, $\mu_g = -k_B T \ln(z/N_g)$, where N_g is the number of molecules in the gas phase. To simplify the discussion, we shall consider a monatomic gas for which, from Section 16.2, we have $z \approx (V/V_Q)$ and

$$\mu_g = -k_B T \ln\left(\frac{V}{N_g V_Q}\right). \tag{19.7}$$

Molecules adsorbed on the surface have a binding energy $-\varepsilon_0$. If the molecules are highly mobile on the surface, the surface layer may be treated as a 2D gas and the single-particle partition function, obtained by adaptation of the 3D expression (Equation 16.4), is

$$z = A\left(\frac{mk_B T}{2\pi\hbar^2}\right)e^{\beta\varepsilon_0} = \left(\frac{A}{A_Q}\right)e^{\beta\varepsilon_0}. \tag{19.8}$$

$A_Q = (2\pi\hbar^2/mk_B T)$ is the quantum area for a 2D system (analogous to the quantum volume for a 3D system), and the exponential factor allows for the potential energy because of binding of molecules to the surface. With the inclusion of the binding energy, the chemical potential of the *surface* gas of N_s molecules is

$$\mu_s = -k_B T \ln\left(\frac{A}{N_g A_Q}\right) - \varepsilon_0. \tag{19.9}$$

In equilibrium, we can equate the expressions for μ_g (Equation 19.7) and μ_s (Equation 19.9). With antilogs, we obtain

$$\frac{N_s}{N_g} = \left(\frac{V_Q A}{V A_Q}\right)e^{\beta\varepsilon_0} = \left(\frac{mk_B T}{2\pi\hbar^2}\right)^{1/2} = \left(\frac{A}{V}\right)e^{\beta\varepsilon_0}. \tag{19.10}$$

From the ideal gas equation of state for the gas phase, we rewrite Equation 19.10 in terms of the pressure P as

$$n_s = \frac{N_s}{A} = P\left(\frac{m}{2\pi\hbar^2 k_B T}\right)^{\frac{1}{2}} e^{\beta\varepsilon_0}. \tag{19.11}$$

The number of adsorbed molecules per unit area at constant T is proportional to the pressure. At high pressures, the adsorption process will saturate as all the available sites become occupied. Layers of adsorbed molecules may form above the primary layer, but we neglect such processes. As the temperature is increased, n_s decreases rapidly at a given pressure because of the exponential dependence on reciprocal temperature.

Model 2

In this model, it is assumed that there are specific surface sites at which adsorbed molecules are localized, as shown in Figure 19.1. In contrast to Model 1, the adsorbed molecules do not form a mobile surface layer.

We again let the binding energy be $-\varepsilon_0$ and equate the chemical potentials of the adsorbed surface layer and the gas phase $\mu_s = \mu_g$ with μ_g given by Equation 19.7. If we ignore vibrational degrees of freedom, the partition function for a single molecule bound to a surface site is given, to a good approximation, by the simple expression

$$z_s = e^{\beta\varepsilon_0}. \tag{19.12}$$

If there are N_s surface sites, and n_s of these are occupied, the partition function for the adsorbed layer is

$$z_s = \binom{N_s}{n_s}\left(e^{\beta\varepsilon_0}\right)^{n_s}. \tag{19.13}$$

The binomial coefficient allows for the adsorbed molecules to be distributed in various ways among the surface sites. With $\mu_s = -(1/\beta)\,\partial\ln Z_s/\partial n_s$ and the use of Stirling's formula from Appendix A, we obtain

$$\mu_s = \left(\frac{1}{\beta}\right)\ln\left(\frac{n_s}{N_s - n_s}\right) - \varepsilon_0. \tag{19.14}$$

In equilibrium, we equate μ_s from Equation 19.14 and μ_g from Equation 19.7 to give

$$\frac{N_A - V_Q}{V} = \left(\frac{n_s}{N_s - n_s}\right)e^{-\beta\varepsilon_0}. \tag{19.15}$$

From the ideal gas equation of state for the gas phase, we have $N_g/V = \beta P$ and, on substitution in Equation 19.15 with rearrangement, we get

$$\frac{n_s}{N_s} = \frac{P}{P + e^{-\beta\varepsilon_0}/(\beta V_Q)}. \tag{19.16}$$

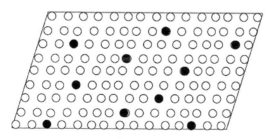

FIGURE 19.1 Adsorption sites on a solid surface showing a partial occupation of sites by adsorbed molecules (black circles). The adsorbed layer is in equilibrium with the gas phase.

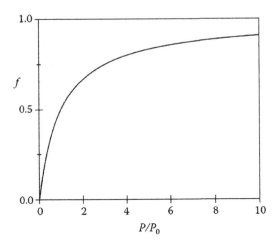

FIGURE 19.2 The Langmuir adsorption isotherm showing the change in the fraction of occupied sites as a function of the reduced gas pressure P/P_0 with P_0 defined following Equation 19.16.

Equation 19.16 gives the fraction, f, of surface sites that are occupied for a given P and T.

If, at a given temperature, we define $P_0 = e^{-\beta\varepsilon_0}/\beta V_Q$, then Equation 19.16 may be rewritten as

$$f = \frac{P}{P + P_0}, \tag{19.17}$$

which is the Langmuir adsorption isotherm expression. Equation 19.17 is in agreement with experiment for many adsorption processes. The form of the isotherm is shown in Figure 19.2.

For $P \leq P_0$, Figure 19.2 shows that f increases roughly linearly with pressure, whereas for $P > P_0$, saturation effects occur.

In the derivation of the Langmuir adsorption isotherm expression, we have treated the adsorbed molecules as a system in equilibrium with the gas reservoir in volume V. A somewhat different approach treats a single site as a system in equilibrium with the gas reservoir. The grand partition function for the single-site system is

$$Z = \sum_{n=0,1} e^{\beta n(\mu+\varepsilon_0)}. \tag{19.18}$$

The mean occupancy of the site $\langle n \rangle$ is given by

$$n = \frac{\displaystyle\sum_{n=0,1} n e^{\beta n(\mu+\varepsilon_0)}}{\displaystyle\sum_{n=0,1} e^{\beta n(\mu+\varepsilon_0)}} = \frac{e^{\beta(\mu+\varepsilon_0)}}{1 + e^{\beta(\mu+\varepsilon_0)}} = \frac{1}{e^{-\beta(\mu+\varepsilon_0)} + 1}. \tag{19.19}$$

With μ_g from Equation 19.7 together with the ideal gas equation of state $N_g/V = \beta P$, it follows that $n/N = f = \left(P/P + e^{-\beta\varepsilon_0}/\beta\right)$, in agreement with Equation 19.16.

19.5 CHARGE CARRIERS IN SEMICONDUCTORS

Although it may seem that the thermal excitation of carriers in semiconductors has little to do with chemical reactions discussed in Section 19.3, there are some common features. Consider an intrinsic semiconductor (e.g., Ge, Si, or GaAs) with a very low concentration of donor or acceptor impurities. There is a band gap E_g between the valence and conduction bands, as depicted for an intrinsic semiconductor in Figure 19.3.

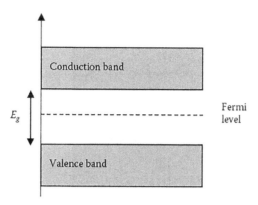

FIGURE 19.3 Schematic representation of the valence and conduction bands for a direct or indirect gap semiconductor. The band gap is E_g, and zero energy is chosen to coincide with the top of the valence band.

At low temperatures, states in the valence band are filled, while states in the conduction band are empty. As the temperature is raised, thermal excitation of carriers takes place, producing electrons in the conduction band and holes in the valence band. Semiconductors such as Ge and Si have band gaps of the order of 1 eV, which is much greater than the thermal energy $k_B T \sim 25$ meV at 300 K. An important question that arises for semiconductors concerns the position of the chemical potential μ on the energy scale. For a metal in which the conduction band is not filled, μ at low temperatures coincides with the Fermi level for the conduction band carriers. In semiconductors at low T, there are very few carriers in the conduction band, and it may be expected that μ will lie somewhere in the band gap. This is confirmed by calculation, as shown below.

The number of electrons in the conduction band at some temperature T is given by

$$N_e = \int_{E_g}^{\infty} f(\varepsilon)\rho(\varepsilon)\,d\varepsilon, \tag{19.20}$$

where $f(\varepsilon)$ is the Fermi function and $\rho(\varepsilon)$ is the density of states in the conduction band. If μ does not lie close to the conduction band edge but is somewhat lower in energy, it follows that the Fermi function may be approximated by $f(\varepsilon) \simeq e^{\beta(\mu-\varepsilon)}$ if we assume $\beta(\mu-\varepsilon) \gg 1$. Effectively, we are approximating the tail of the Fermi function by the Maxwell–Boltzmann distribution. The density of states for electrons with energies slightly greater than the band gap may be approximated by the familiar particle in a box expression, and with allowance for spin degeneracy, we have $\rho(k)d^3k = 2(V/(2\pi)^3)4\pi k^2 dk$. For a parabolic band, the $\varepsilon(k)$ dispersion relation has the form $\varepsilon = E_g + \hbar^2 k^2/2m_e^*$, with m_e^* as the effective mass of an electron near the bottom of the conduction band. We obtain a modified expression for the density of states,

$$\rho(\varepsilon)\,d\varepsilon = \left(\frac{V}{2\pi^2}\right)\left(\frac{2m_e^*}{\hbar^2}\right)^{3/2}\left(\varepsilon - E_g\right)^{1/2}\,d\varepsilon. \tag{19.21}$$

Equation 19.21 is similar to the particle in a box density of states with $\varepsilon^{1/2}$ replaced by $(\varepsilon - E_g)^{1/2}$ and with the electron mass replaced by the effective mass. Equation 19.21, together with the use of the modified Fermi function in Equation 19.20, gives

$$n_e = \frac{N_e}{V} = \left(\frac{1}{2\pi^2}\right)\left(\frac{2m_e^*}{\hbar^2}\right)^{3/2} e^{\beta\mu} \int_{E_g}^{\infty} e^{-\beta\varepsilon}\left(\varepsilon - E_g\right)^{1/2}\,d\varepsilon. \tag{19.22}$$

If the variable is changed to $x = \beta(\varepsilon - E_g)$, the number of electrons per unit volume in the conduction band is

$$n_e = \left(\frac{1}{2\pi^2}\right)\left(\frac{2m_e^*}{\beta\hbar^2}\right)^{3/2} e^{\beta(\mu - E_g)} \int_0^\infty e^{-x} x^{1/2}\, \mathrm{d}x = \frac{1}{\sqrt{2}}\left(\frac{m_e^*}{\pi\beta\hbar^2}\right)^{3/2} e^{\beta(\mu - E_g)}. \tag{19.23}$$

The integral is evaluated using $\Gamma(3/2)$ given in Table 14.1.

It follows that the number of holes in the valence band is

$$n_p = \int_{-\infty}^0 \left[1 - f(\varepsilon)\right]\rho(\varepsilon)\,\mathrm{d}\varepsilon.$$

With similar procedures to those used for electrons in the conduction band, the hole density in the valence band is

$$n_p = \frac{1}{\sqrt{2}}\left(\frac{m_h^*}{\pi\beta\hbar^2}\right)^{3/2} e^{-\beta\mu}. \tag{19.24}$$

where m_h^* is the effective hole mass in the valence band. The product of the electron and hole densities, obtained with the use of Equations 19.23 and 19.24, is given by

$$n_e n_p = \frac{1}{2}\left(\frac{1}{\pi\beta\hbar^2}\right)^3 \left(m_e^* m_h^*\right)^{3/2} e^{-\beta E_g} = K_N(T). \tag{19.25}$$

$K_N(T)$ is a constant at a given temperature for a particular semiconductor and from Equation 19.25 may be written in the alternative form

$$K_N(T) \propto \left(\frac{1}{V_{Qe}}\right)\left(\frac{1}{V_{Qh}}\right) e^{-\beta E_g}, \tag{19.26}$$

with V_{Qe} and V_{Qh} as the quantum volumes for electrons and holes, respectively, in the semiconductor. Equation 19.25 is simply the law of mass action used for chemical reactions in Chapter 7 and in Section 19.3. The thermally induced production of conduction band electrons and valence band holes may be viewed as an electron transfer reaction process with an activation energy E_g. For an intrinsic semiconductor with equal numbers of electrons and holes, we put $n_e = n_h$ in Equation 19.25 and obtain

$$n_e = n_p = \frac{1}{\sqrt{2}}\left(\frac{1}{\pi\beta\hbar^2}\right)^{3/2} \left(m_e^* m_h^*\right)^{3/4} e^{-\beta E_g/2}. \tag{19.27}$$

If we equate the expressions for n_e in Equations 19.24 and 19.27 and assume $m_e^* \approx m_h^*$, then $\mu = E_g/2$. The chemical potential is seen to lie in the middle of the band gap when the electron and hole effective masses are equal. This is an important result in semiconductor science.

For semiconductors doped with donor or acceptor centers with energy levels that lie close to the band edges, the chemical potential will shift from the middle of the band. Further details are given in books on solid-state physics.

PROBLEMS CHAPTER 19

19.1   The dissociation of iodine molecules into two iodine atoms occurs at high temperatures and is described by the chemical equation $I_2 \rightleftharpoons 2I$. Give the form of the equilibrium constant for this reaction, and find the partial pressure of the iodine atom component at temperature T in a container of volume V.

19.2   At very high temperatures, atomic hydrogen dissociates into a proton and an electron in a process represented by the reaction $H \rightleftharpoons p + e$. Derive an expression for the equilibrium constant for this reaction assuming that the states of all the particles involved are effectively ideal gases. Allow for spin degeneracy and assume as an approximation that only the ground state of the hydrogen atom with energy $-\varepsilon_0$ need be considered in calculating the internal partition function for this particle.

19.3   A mixture of hydrogen and deuterium undergoes the following reaction in the gas phase $H_2 + D_2 \rightleftharpoons 2HD$. Obtain the equilibrium constant for this reaction at some temperature T by assuming that the rotational motion can be described in the classical limit, whereas only the ground states are important for both the electronic and vibrational degrees of freedom. Give your result in terms of the vibrational frequency $\omega_{H_2} = (k/\mu_R)$ for the hydrogen molecule, where k is the effective vibrational spring constant and $\mu_R = (m_H)^2 / 2m_H$ is the reduced mass for H_2. Take k and the electronic ground-state energy $-\varepsilon_0$ to be the same for all three molecular species and the deuterium molecular mass to be twice that of hydrogen. The deuteron has spin 1, whereas the proton has spin $\frac{1}{2}$.

19.4   The Langmuir adsorption isotherm holds for large myoglobin molecules in solution in water. One oxygen molecule can be bound or adsorbed on each myoglobin molecule in a process described by $Mb + O_2 \rightleftharpoons MbO_2$. Show that the Langmuir adsorption isotherm form can be obtained for this system with the use of the law of mass action and replacing pressures by concentrations C_{Mb}, C_{O_2}, and C_{MbO_2}.

19.5   In doped semiconductors, the electron and hole densities in the conduction and valence bands, respectively, are not equal. For an n-type semiconductor containing donors, the chemical potential moves toward the conduction band. Show that the law of mass action holds for doped semiconductors in which transitions occur between the donor level and states both at the bottom of the conduction band and at the top of the valence band.

20 Introduction to Irreversible Thermodynamics

20.1 INTRODUCTION

Situations dealt with in the previous chapters in this book have involved systems in equilibrium or very close to equilibrium. We now consider phenomena that occur in systems that are not in equilibrium. The second law of thermodynamics states that in any process, the entropy of the local universe either remains constant or increases. Many processes in nature are irreversible and are accompanied by a net increase in entropy. There is a corresponding asymmetry with respect to time in the equations governing irreversible processes. In nature, living organisms that maintain a high degree of order are far from equilibrium. Life on earth relies on thermal energy from the sun that is captured and stored as chemical energy in plants through photosynthesis. The stored chemical energy is used by other organisms to sustain life in processes that lead to entropy production. Devising a detailed description of living systems that are far from equilibrium is an extremely difficult and challenging task. New approaches to the study of complex systems, including living systems, are currently being developed.

In this chapter, we consider irreversible processes in systems not too far from equilibrium. Examples of situations of this kind are phenomena associated with thermoelectric effects and thermo-osmosis. Thermoelectric effects that involve a coupled electrical transport and heat transport were discovered in the nineteenth century. The effects that we discuss bear the names of the discoverers. In particular, we analyze the Seebeck and Peltier effects that involve contacts between two different electrical conductors, which may be metals or semiconductors, which are maintained at different temperatures. In practical applications, the Seebeck effect is used for thermometry, in the form of thermocouple devices, whereas the Peltier effect provides the basis for heat pumps in small refrigerator systems. Thermo-osmosis with the related thermomechanical effect involves the flow of energy and particles between reservoirs at different temperatures and pressures. One of the most dramatic examples of the thermomechanical effect is the fountain effect that is found in superfluid helium.

20.2 ENTROPY PRODUCTION IN HEAT FLOW PROCESSES

As a starting point for the discussions of irreversible thermodynamics, consider heat energy transmission along a thermally conducting bar connecting two heat baths, as shown in Figure 20.1.

If the heat flow is constant, entropy is produced in this composite system at a rate

$$\frac{dS}{dt} = I_Q \left(\frac{1}{T_2} - \frac{1}{T_1} \right), \tag{20.1}$$

where $I_q = dQ/dt$ is the heat current with the unit joules per second. As an approximation, if $T_1 \approx T_2$, we introduce an average temperature $T = (T_1 + T_2)/2$, and with the system not far from equilibrium, Equation 20.1 becomes

$$\frac{dS}{dt} = I_Q \frac{\Delta T}{T^2}, \tag{20.2}$$

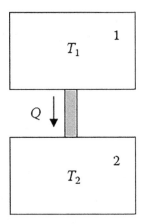

FIGURE 20.1 A conducting bar connects two heat baths at temperatures T_1 and T_2, with $T_1 > T_2$. Heat flow along the bar results in entropy production at a steady rate if the temperatures of the baths are held constant.

where $\Delta T = T_1 - T_2$. For convenience, an *entropy current* $I_s = I_Q/T$ may be defined in terms of the heat current, giving

$$\frac{\mathrm{d}S}{\mathrm{d}t} = I_S \frac{\Delta T}{T}. \tag{20.3}$$

Provided the heat baths have sufficiently large heat capacities so that ΔT remains approximately constant, entropy will continuously be produced at the rate given by Equation 20.3. The form of Equation 20.3, which applies to thermal energy flow, may be generalized to allow for coupled flows that involve, for example, energy and particles. This development is dealt with in the next section, first for the cases of discrete systems that consist of distinct parts and second for continuous systems such as a bar through which both heat and charge flow.

Exercise 20.1

A copper rod of diameter 5 mm and length 20 cm is connected to two heat baths at slightly different temperatures of 300 and 290 K, respectively. Find the rate of entropy production due to heat flow along the rod.

From Equation 20.2, $\mathrm{d}S/\mathrm{d}t = (I_Q \, \Delta T/T^2)$, and we calculate I_q from the given dimensions of the copper rod and the thermal conductivity coefficient of copper $\kappa = 4.0 \times 10^2$ Wm^{-1} K^{-1}. We obtain $I_Q = 0.4$ J sec^{-1}, and with $T = 295$ K, this gives $\mathrm{d}S/\mathrm{d}t = 46 \ \mu$WK^{-1}.

20.3   ENTROPY PRODUCTION IN COUPLED FLOW PROCESSES

a. *Discrete Systems.* Consider two reservoirs that are connected to allow energy and particles to flow between them, as shown in Figure 20.2.

The total energy and the total number of particles are held fixed so that

$$E = E_1 + E_2 \tag{20.4}$$

and

$$N = N_1 + N_2. \tag{20.5}$$

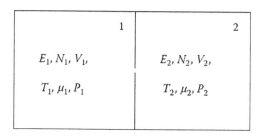

FIGURE 20.2 Two connected reservoirs 1 and 2 with gradients in pressure, temperature, and chemical potential across the connecting aperture. Coupled energy and particle transport processes take place provided the gradients are maintained.

To simplify matters, the volumes V_1 and V_2 are kept fixed. If at some instant the two reservoirs are not in equilibrium with each other, then energy and particles are transferred from one reservoir to the other because of differences in the temperature and the chemical potential. The total entropy of the two reservoirs is $S(E, N, V) = S_1(E_1, N_1, V_1) + S_2(E_2, N_2, V_2)$. If energy and particles flow from reservoir 1 to reservoir 2, the rate of change of entropy for the composite system is

$$\frac{dS}{dt} = \left(\frac{\partial S_1}{\partial N_1} - \frac{\partial S_2}{\partial N_2} \right) \frac{\partial N_1}{\partial_t} + \left(\frac{\partial S_1}{\partial E_1} - \frac{\partial S_2}{\partial E_2} \right) \frac{\partial E_1}{\partial_t}, \tag{20.6}$$

where the use has been made of Equations 20.4 and 20.5. With the use of the general form of the fundamental relation in Chapter 3, we have $\partial S/\partial N = -\mu/T$ and $\partial S/\partial E = 1/T$. In considering Equation 20.6, it is helpful to introduce the thermodynamic forces X_N and X_E that drive particle flow and energy flow, respectively. For particle flow, we define

$$X_N = -\left(\frac{\mu_1}{T_1} - \frac{\mu_1}{T_2} \right) = -\Delta \left(\frac{\mu}{T} \right), \tag{20.7}$$

and for energy flow,

$$X_E = -\left(\frac{1}{T_1} - \frac{1}{T_2} \right) = -\Delta \left(\frac{1}{T} \right). \tag{20.8}$$

The corresponding currents are defined as $I_N = \partial N_1/\partial t$ and $I_E = \partial E_1/\partial t$. Equation 20.6 is rewritten in terms of these forces and currents as

$$\frac{dS}{dt} \equiv X_N I_N + X_E I_E. \tag{20.9}$$

For coupled flow processes, the currents are functions of the thermodynamic forces: $I_N = I_n(XN, XE)$ and $I_E = I_e(X_N, X_E)$. Expanding both I_N and I_E in Taylor series about $X_N = 0$, $X_E = 0$ and retaining first-order terms as an approximation give

$$I_N = \left(\frac{\partial I_N}{\partial X_N} \right) X_N + \left(\frac{\partial I_N}{\partial X_E} \right) X_E = L_{NN} X_N + L_{NE} X_E \tag{20.10}$$

and

$$I_E = \left(\frac{\partial I_E}{\partial X_N} \right) X_N + \left(\frac{\partial I_E}{\partial X_E} \right) X_E = L_{EN} X_N + L_{EE} X_E. \tag{20.11}$$

The linear forms imply that conditions are not far from equilibrium with fairly small differences in T and μ between the reservoirs.

For arbitrary coupled flows I_1 and I_2, we write in general

$$I_1 = L_{11}X_1 + L_{12}X_2 \tag{20.12}$$

and

$$I_2 = L_{21}X_1 + L_{22}X_2 \tag{20.13}$$

In matrix form, Equations 20.12 and 20.13 are combined as follows:

$$\begin{pmatrix} I_1 \\ I_2 \end{pmatrix} = \begin{pmatrix} L_{11} & L_{12} \\ L_{21} & L_{22} \end{pmatrix} \begin{pmatrix} X_1 \\ X_2 \end{pmatrix}.$$

Importantly, the off-diagonal coefficients $L12$ and $L21$ are equal, as was shown by Onsager from microscopic considerations that involve the *reversibility* of dynamical processes on the *microscopic* scale. The general equality $L_{12}=L_{21}$, which is known as the Onsager reciprocal relation, is extremely helpful in simplifying the treatment of coupled flows.

b. *Continuous Systems.* Many situations of interest, particularly in thermoelectricity, involve continuous rather than discrete systems. For systems of this kind in situations involving the transport of energy and/or particles, there is a gradient in temperature and/or chemical potential. The results we obtain are similar to those given above for discrete systems. For continuous systems, it is convenient to introduce current densities per unit area J rather than currents I. In most cases we consider, transport occurs along a chosen direction in a material. For flow processes that involve just a single current density J, we can write $J = LX$, where X is a thermodynamic force expressed as a gradient function and L is a phenomenological coefficient. Examples are as follows:

i. Thermal conductivity (Fourier's equation),

$$J_Q = \kappa \nabla T, \tag{20.14}$$

with κ as the thermal conductivity coefficient;

ii. Electrical conductivity (Ohm's law),

$$J_e = -\sigma \nabla \phi, \tag{20.15}$$

with σ as the electrical conduction and ϕ as the electric potential; and

iii. Diffusion (Fick's law),

$$J_N = -D \nabla \mu, \tag{20.16}$$

with D as the diffusion coefficient.

In Cartesian coordinates, the del operator used in the above equations has the form $\nabla = i\partial/\partial x + j\partial/\partial y + k\partial/\partial z$, where $i, j,$ and k are the unit vectors. For many situations that we consider, the flows are parallel to a particular axis that may be chosen as the z-axis.

From the general fundamental relation introduced in Chapter 3, with the volume held constant, we have in terms of the entropy density s, energy density e, and particle density n,

$$ds = \left(\frac{1}{T}\right)de - \left(\frac{\mu}{T}\right)dn = \sum_i X_i \, dx_i. \qquad (20.17)$$

Equation 20.17 defines $\sum_i X_i dx_i$, and it follows that the thermodynamic forces are given by $X_i = \partial s/\partial x_i$.

In the presence of gradients in the temperature or the chemical potential, we introduce J_s as the entropy current associated with a volume element dV, and ∇J_s gives the net flux of entropy from the volume element. For each variable x_i, there is an equation of continuity on the basis of conservation considerations. It follows that the rate of increase of the entropy density because of gradients in the X_i has the form

$$\frac{ds}{dt} = \sum_i \nabla X_i \cdot J_i. \qquad (20.18)$$

Equation 20.18 is an important relation in irreversible thermodynamics and gives the rate of production of entropy per unit volume in a region of interest in terms of thermodynamic forces and currents. If we consider coupled energy and particle flow situations, Equation 20.18 becomes

$$\frac{ds}{dt} = \nabla\left(\frac{1}{T}\right) \cdot J_E - \nabla\left(\frac{\mu}{T}\right) \cdot J_N. \qquad (20.19)$$

Equation 20.19 is similar in form to Equation 20.9. If we again expand J_E and J_N and retain the first-order terms, the coupled flows are similar to those given in Equations 20.12 and 20.13:

$$J_1 = J_E = L_{11}X_1 + L_{12}X_2 \qquad (20.20)$$

and

$$J_2 = J_N = L_{21}X_1 + L_{22}X_2 \qquad (20.21)$$

with $X_1 = \nabla(1/T)$ and $X_2 = -\nabla(\mu/T)$. Equations 20.20 and 20.21 can again be combined in the matrix form. In the following sections, the formalism developed for both discrete and continuous systems is applied to a number of different flow situations.

20.4 THERMO-OSMOSIS, THERMOMOLECULAR PRESSURE DIFFERENCE, AND THERMOMECHANICAL EFFECT

Consider the flow of a fluid through a membrane, across which there is both a temperature and a pressure gradient. Using the equations derived in Section 20.3, we obtain a relationship between the temperature difference and the pressure difference. The arrangement for thermo-osmosis is shown in Figure 20.3.

The two chambers on either side of the membrane are each maintained at constant temperatures by means of large heat baths with which the reservoirs are in thermal contact. To relate the coefficients in Equations 20.20 and 20.21, we consider two special cases.

i. If the pistons are stationary, there is energy transport but no net particle transport. Equation 20.21 becomes

$$0 = L_{21}\nabla\left(\frac{1}{T}\right) - L_{22}\nabla\left(\frac{\mu}{T}\right), \qquad (20.22)$$

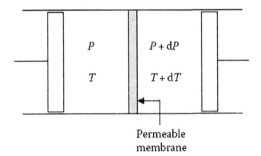

FIGURE 20.3 Thermo-osmosis through a permeable membrane. Pressure and temperature gradients are maintained across the membrane. Both energy and particle flow can occur.

and we obtain

$$\frac{L_{21}}{L_{22}} = \frac{\nabla(\mu/T)}{\nabla(1/T)} = \mu - T\left(\frac{\partial\mu}{\partial T}\right), \tag{20.23}$$

with the use made of the partial derivative identities given in Section 7.7. The expression Equation 20.23 establishes a useful relationship between L_{21} and L_{22}.

ii. If there is no temperature gradient, $\nabla(1/T)=0$. Equations 20.20 and 20.21 become, under isothermal conditions, $J_E = -L_{12}\nabla(\mu/T)$ and $J_N = -L_{22}\nabla(\mu/T)$, and the ratio gives

$$\frac{L_{12}}{L_{22}} = \left(\frac{J_E}{J_N}\right)_{\Delta T=0}. \tag{20.24}$$

The ratios of the coefficients in Equations 20.23 and 20.24 together with the Onsager reciprocal relation $L_{12}=L_{21}$ lead to the important relation for coupled flows under isothermal conditions

$$\left(\frac{J_E}{J_N}\right)_{\Delta T=0} = \mu - T\left(\frac{\partial\mu}{\partial T}\right). \tag{20.25}$$

In general for a single-component system, the Gibbs potential is $G=\mu N=E-TS+PV$. The differential of G, combined with the fundamental relation, gives the useful thermodynamic relationship $SdT-V\,dP+N\,d\mu=0$, which we rearrange with $v=V/N$ and $s=S/N$ as

$$\frac{d\mu}{dT} = \left(v\frac{dP}{dT} - s\right). \tag{20.26}$$

If $d\mu/dT\approx0$, Equation 20.26 takes the simple form

$$\frac{dP}{dT} = \frac{s}{v}. \tag{20.27}$$

If we write the entropy density as $s=q/T$ per particle, Equation 20.27 gives the ratio of the pressure difference to the temperature difference across the membrane in terms of the heat transfer per particle, q, and the volume change per particle, v.

It is necessary to consider microscopic effects to understand the origins of the heat transfer q. We assume that the energies of particles transported across the membrane are, on average, greater than the mean energy of the particles on the side from which they come. This effect will tend to lower the temperature on one side of the membrane and to raise the temperature on the other side.

Heat transfer occurs to the heat baths, which maintain the temperatures at T and $T + dT$, respectively, to compensate for this effect. The heat transfers to and from the heat baths that occur to maintain the temperatures constant result in entropy transfer between the two reservoirs. The nature of the membrane determines the sign of dP for a given dT. Note that the ratio of energy flow to particle flow under *isothermal* conditions in Equation 20.25 can be written as

$$\left(\frac{J_E}{J_N}\right)_{\Delta T = 0} = \frac{L_{12}}{L_{22}} = (e + Pv + q), \tag{20.28}$$

where the terms inside the brackets on the right-hand side represent the various contributions to the energy that are involved in the transfer of a particle through the membrane. For each particle transferred, e is the internal energy component, Pv is the work done by the piston, and q is the heat that is effectively transferred between the heat baths, which maintain the temperatures of the two chambers.

By considering various special cases, such as thermal transport, an expression for all of the coefficients L_{ij} may be obtained in terms of measurable quantities. With this information, the dynamical equations may be used to describe any coupled flow situation. In particular, we shall adapt the results obtained in this section to thermoelectric effects in Section 20.6.

Exercise 20.2

The liquid helium-II phase below the λ transition in helium-4 exhibits fascinating transport effects, of which the most dramatic is the fountain effect discovered in the 1930s. Assume that the membrane in Figure 20.3 contains very narrow channels that, in terms of the two-fluid model introduced in Chapter 14, allow the passage of only the superfluid helium component, but not the normal component. The superfluid helium-4 atoms transport no entropy. Show that the fountain effect, which is a special case of thermo-osmosis, can be described by Equation 20.27. Explain the origin of the entropy changes in this system.

As a simplification, we approximate the behavior of the chemical potential of liquid helium at low temperatures as similar to that of a Bose gas below the Bose–Einstein condensation temperature. We have from Equation 14.15 $\mu \simeq -k_B T/N\left[1-(T/T_0)^{3/2}\right]^{-1}$. Clearly, $d\mu/dT$ is, to a good approximation, close to zero, and Equation 20.27 is valid for superfluid helium. A temperature difference across the membrane gives rise to a pressure difference with an increase in temperature accompanied by an increase in pressure.

If the piston on the left-hand side (reservoir 1) is moved in slowly, superfluid atoms are transferred to the vessel on the right side (reservoir 2). Although the superfluid atoms transport no entropy, a decrease in the fraction of superfluid atoms in reservoir 1 would lead to a rise in temperature and heat is given up to the heat bath to keep T_1 constant. For a decrease of n superfluid atoms in 1 heat $Q_1 = nT_1 s_1$ is reversibly transferred to the heat bath, and this corresponds to a loss in entropy for 1. Similarly, the increase in superfluid concentration in reservoir 2 leads to the reversible absorption of heat $Q_2 = nT_2 s_2$ from the heat bath and an increase in entropy of reservoir 2. If we apply the first law to the coupled reservoir system for a process that involves the transfer of n superfluid atoms, we get $n(e_2 - e_1) = n(T_2 s_2 - T_1 s_1) - n(P_2 v_2 - P_1 v_1)$, and it follows that $g_1 = g_2$, where g is the Gibbs potential per particle. This shows that the chemical potentials in the two reservoirs are the same, that is, $\mu_1 = \mu_2$. Because $g = e - Ts + Pv$, if we combine the differential dg with the fundamental relation, we find that $-sdT + vdP = 0$ or $dP/dT = s/v$ in agreement with Equation 20.27.

The thermomechanical effect in liquid helium is related to the fountain effect and arises when the two coupled reservoirs are kept at the same temperature, but a pressure difference is maintained between them. From Equation 20.28, we have $(J_E/J_N)_{\Delta T=0} = L_{12}/L_{22} = e + Pv + q = h + q$, where h is the enthalpy per particle and q is the heat transferred per particle. Use of Equation 20.27 to obtain q gives for the ratio of the flows $(J_E/J_N)\Delta t = 0 = h + vT(dP/dT)$.

20.5 THERMOELECTRICITY

The Seebeck and Peltier thermoelectric effects mentioned in Section 20.1 involve a coupled thermal energy and electric charge flow. We make use of the results obtained in Section 20.3 to discuss these effects. Consider a long thin conducting bar connecting two heat baths at temperatures T_1 and T_2, as shown in Figure 20.4.

The particles that transport charge are electrons or holes, depending on the nature of the conducting bar, which may be metallic or semiconducting. Particle flow gives rise to an electric current density $\boldsymbol{J}_e = \pm e\,\boldsymbol{J}_N$, where the sign depends on whether the particles are negative electrons or positive holes. For the present discussion, we shall assume metallic conductors with electrons as the charge carriers. For energy flow, there are two contributions, $J_E = J_Q + \mu J_N$, where J_Q is the thermal current. It follows that Equation 20.19 may be written as

$$\frac{ds}{dt} = \nabla\left(\frac{1}{T}\right)\cdot\left(J_Q - \frac{\mu J e}{e}\right) + \nabla\left(\frac{\mu}{T}\right)\cdot\left(\frac{J e}{e}\right) = J_Q \cdot \nabla\left(\frac{1}{T}\right) + \left(\frac{1}{eT}\right)Je\cdot\nabla\mu. \tag{20.29}$$

The coupled heat and charge flow equations are similar in form to Equations 20.20 and 20.21,

$$J_Q = L_{11}X_1 + L_{12}X_2 \tag{20.30}$$

and

$$J_e = L_{21}X_1 + L_{22}X_2 \tag{20.31}$$

with the thermodynamic forces in the present case given by $X_1 = \nabla(1/T)$ and $X_2 = (1/eT)\nabla\mu$. The chemical potential must be modified in the presence of an electric field and is written as $\mu = \mu_c + \mu_e$, where μ_c is the concentration-dependent contribution and $\mu_e = e\phi$ is the potential energy of a charged particle in the electric potential ϕ. Just as in the thermo-osmosis case described in Section 20.4, the coefficients L_{21} and L_{22} are determined by considering special flow situations and the use of the Onsager reciprocal relation, $L_{12} = L_{21}$.

  i. Choose $T_1 = T_2 = T$, giving $\nabla T = 0$. From Equation 20.31, we obtain

$$J_e = \left(\frac{1}{eT}\right)L_{22}\nabla\mu, \tag{20.32}$$

  where $\nabla\mu = \nabla\mu_c + \nabla\mu_e$. For a homogeneous system with no temperature gradient, we expect $\nabla\mu_c = 0$, and this gives $\nabla\mu_c = e\nabla\phi$. Equation 20.32, with the use of Ohm's law (Equation 20.15), becomes $J_e = -(1/eT)L_{22}\,(eJ_e/\sigma)$, which leads to

$$L_{22} = -\sigma T. \tag{20.33}$$

 ii. As a second special flow case, put $\boldsymbol{J}_e = 0$ in Equation 20.31, which corresponds to zero electric current, to obtain the following expression for $\nabla\mu$:

$$\nabla\mu = eT\left(\frac{L_{21}}{L_{22}}\right)\nabla\left(\frac{1}{T}\right) = \left(\frac{e}{T}\right) - \left(\frac{L_{21}}{L_{22}}\right)\nabla T. \tag{20.34}$$

Insertion of $\nabla\mu$ from Equation 20.34 into Equation 20.30 leads to

$$J_Q = -\left(\frac{1}{T^2}\right)\left(L_{11} - \frac{L_{12}L_{21}}{L_{22}}\right)\nabla T. \tag{20.35}$$

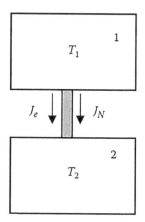

FIGURE 20.4 Coupled energy and particle flows along a conducting rod that connects heat baths at temperatures T_1 and T_2, respectively.

With the use of Equation 20.14, we obtain

$$\kappa = \frac{J_Q}{\nabla T} = -\frac{1}{T^2}\left(L_{11} - \frac{L_{12}L_{21}}{L_{22}}\right), \tag{20.36}$$

with κ as the thermal conductivity coefficient. Remembering that $L_{12}=L_{21}$, Equation 20.36 provides a relationship between L_{11}, L_{12}, and L_{22}.

Finally, for temperature gradients along the conducting rod that are not too large, we again take $\nabla\mu = e\nabla\phi$. With Equations 20.33 and 20.34, this gives

$$\frac{d\phi}{dT} = \frac{L_{21}}{\sigma T^2}. \tag{20.37}$$

The quantity $d\phi/dT$ is defined as the thermoelectric or Seebeck coefficient e and is of fundamental importance in our discussion. The value of ε is a characteristic of a particular conductor. Measured values of e for metals, semimetals, and semiconductors range from tens to hundreds of microvolts per kelvin. From Equation 20.37, we obtain

$$L_{21} = \sigma\varepsilon T^2. \tag{20.38}$$

All of the coefficients are now determined, and the coupled flow Equations 20.30 and 20.31 may be written as

$$J_Q = \left[\kappa + \sigma\varepsilon^2 T\right]\nabla T + \frac{\sigma\varepsilon T}{e}\nabla\mu \tag{20.39}$$

and

$$J_e = \sigma\varepsilon\nabla T - \frac{\sigma}{e}\nabla\mu. \tag{20.40}$$

Combining Equations 20.39 and 20.40 gives

$$J_Q = -\varepsilon T\, J_{e+}\kappa\nabla T. \tag{20.41}$$

This is a useful equation that relates the heat current density to the electric current density in a conductor along which there are both an electric potential gradient and a temperature gradient. For application to a particular material, it is necessary to insert values for the two coefficients: the thermoelectric coefficient ε and the thermal conductivity κ. We are now in a position to discuss the Seebeck and Peltier effects.

20.6 THE SEEBECK AND PELTIER EFFECTS

The discovery of thermoelectric effects was made by Seebeck in the nineteenth century, who found that a current was produced in an electric circuit made of two different metals joined to make a loop and whose junctions were kept at different temperatures. This effect bears his name. Roughly 10 years later, Peltier showed that when a current was passed through a junction of two metals, heat was either absorbed or given out dependent on the direction of the current. In the mid-nineteenth century, William Thomson (later Lord Kelvin) unified the description of thermoelectric effects. He predicted and then showed that heat should be either absorbed or emitted from a current-carrying conductor along which a temperature gradient is maintained. With the formalism that has been established in the preceding sections, we give a brief account of these effects.

Consider a composite electrical conductor consisting of two different materials a and b carrying an electric current I_e, as shown in Figure 20.5.

Because of the isothermal conditions at the junction of the conductors, the application of Equation 20.41 to the two materials with $\nabla T = 0$ gives

$$J_Q^a - J_Q^b = \left(\varepsilon^b - \varepsilon^a \right) T\, \boldsymbol{J}_e. \tag{20.42}$$

This shows that the heat current is not constant across the junction and that heat is exchanged with the reservoir at the junction. The Peltier coefficient π^{ab} (with SI unit joules per coulomb (JC^{-1}) or volts (V)) is defined as the heat current supplied to the junction per unit electric current through the junction:

$$\pi^{ab} = \left(\frac{J_Q^a - J_Q^b}{J_e} \right) = T\left[\varepsilon^b - \varepsilon^a \right] = T\Delta\varepsilon^{ab}. \tag{20.43}$$

We see that the heat absorbed or given out at the junction is proportional to the difference in the thermoelectric coefficients of the two materials making up the composite conductor and to the absolute temperature. With a suitable choice of materials, $\Delta\varepsilon^{ab}$ can be made sufficiently large and

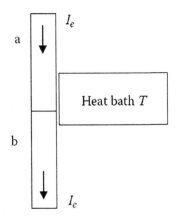

FIGURE 20.5 A long rod-shaped conductor, made of two materials a and b, which are joined as shown, and through which an electric current is passed. The junction is in contact with a heat bath at temperature T.

negative to provide useful refrigeration at the junction. The Peltier effect is used in compact special-ized commercial refrigeration units. These typically involve a series of p- and n-type semiconductor junctions, a and b, for which the thermoelectric coefficient difference $\Delta\varepsilon^{ab}$ is large. The junctions are arranged to have heat-absorbing junctions in a plane on one side and heat-rejecting junctions on the opposite side.

We turn to the Seebeck effect and consider two electrically conducting materials a and b connected, as shown in Figure 20.6, with junctions 1 and 2 at different temperatures T_1 and T_2, respectively.

For convenience in our discussion, the conductors at the upper junction are separated by an insu-lator that prevents charge flow in the circuit. The voltmeter that measures the potential difference across the upper junction has a very high electrical resistance. Application of Equation 20.40 to the Seebeck circuit, with $J_e = 0$, gives

$$0 = -\sigma\varepsilon\nabla T - \frac{\sigma}{e}\nabla\mu. \tag{20.44}$$

If we assume that the approximation $\nabla\mu = e\nabla\phi$ is valid, we get $\varepsilon\nabla T = -\nabla\phi$, and integration around the circuit leads to

$$\Delta\phi_{14} = -\int_{T_1}^{T_2}\left(\varepsilon^a - \varepsilon^b\right)dT \approx -\Delta\varepsilon^{ab}\,\Delta T. \tag{20.45}$$

This is the thermoelectric emf. Note that the connections to the voltmeter are at the same tempera-ture and any contact potential difference which arises at these points is cancelled out. If one of the two junctions is maintained at a fixed known temperature T_R, the Seebeck effect may be used for thermometry, with the temperature of the other junction given in terms of the measured emf. Calibration tables exist for various metal junctions. For selected metal pairs, $\Delta\varepsilon \sim 10^{-5}\,\mathrm{V\ K^{-1}}$, and the emfs are of the order of millivolts for temperature differences of 100 K between metal junctions. Note that both the Seebeck and the Peltier coefficients involve $\Delta\varepsilon^{ab}$ and do not give values for ε^a and ε^b separately. However, with the use of the Thomson effect and the relationships that link the various thermoelectric effects, it is possible to determine all of the coefficients for a particular conductor, which can then be used as a reference for other conductors.

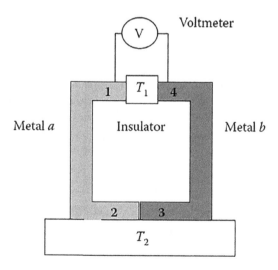

FIGURE 20.6 A thermocouple arrangement that consists of two conductors joined to form a loop with junc-tions at different temperatures T_1 and T_2. The electrical insulator at the upper junction is maintained at T_1 and a voltmeter is connected between points 1 and 4 as shown.

Exercise 20.3

Show that the thermoelectric coefficient ε^s of a superconductor is zero and that the thermoelectric coefficient of a normal metal can be measured with the use of a contact between the metal and a superconductor.

We first note that the Cooper pairs in a superconductor transport no entropy, which distinguishes superconductors from normal metals in an important way. From Equation 20.37 with the electrical conductivity $\sigma=\infty$, we obtain $\varepsilon^s=0$. For a thermocouple made of a normal metal in contact with a superconductor and with the junctions at different but sufficiently low temperatures, Equation 20.45 shows that the Peltier emf is $\Delta\phi_{14} = -\int_{T_1}^{T_2}\left(\varepsilon^a - \varepsilon^s\right)dT \approx -\varepsilon^a\Delta T$, which gives a direct measure of ε^a.

Microscopically, the origin of the Peltier emf may be understood by considering the Fermi levels in two conductors that are placed in contact. In general, the Fermi levels will be different in the two conductors, and transfer of charge carriers will occur to equalize the levels. As a result, a contact potential difference is established between the conductors that gives rise to the Seebeck effect. When charge is transported across the contact between two different metals, heat is either absorbed or emitted, dependent on the current direction, to conserve energy. This gives rise to the Peltier effect.

20.7   THE THOMSON EFFECT

As noted above, the Thomson effect involves heat evolution because of an electric current in a conductor along which there is a temperature gradient. Charge carriers from high-temperature regions of the conductor have a higher average energy than carriers in cooler regions. To maintain the temperature gradient, heat is rejected from the conductor surface as carriers from the high-temperature region move to cooler regions. The Thomson coefficient τ is defined as the heat transfer because of this process per unit current for unit temperature gradient. We do not consider this process in detail and simply quote the expression for the Thomson coefficient $\tau=-T(d\varepsilon/dT)$, where ε is again the thermoelectric coefficient of the conductor and $d\varepsilon/dT$ gives the variation of ε with T. If we consider a short length of the conductor with a temperature difference ΔT between the two ends, the Thomson heat per unit area is given by

$$J_Q = J_e\tau\int_{T+\Delta T}^{T} dT = -J_e\tau\Delta T. \tag{20.46}$$

We assume that T is approximately constant for the small temperature difference involved. The sign of J_Q depends on the direction of current flow with respect to the direction of the temperature gradient. In addition to the Thomson heat, there is of course the Joule heat that must be allowed for. The Joule heat depends on both the electrical resistance of the conductor and J_e^2. Note that measurements of the Thomson heat for a given material as a function of temperature provide the values of τ, which can be used in an integral to give values for the corresponding Seebeck coefficient.

PROBLEMS CHAPTER 20

20.1   Liquid helium-4 under reduced pressure at a temperature of 1.2 K (below the λ transition) is contained in two vessels connected by a superleak that allows the passage of only the superfluid helium component. By means of a heater, a small temperature difference of 1.5 mK is maintained between liquid in the two vessels. Calculate the pressure difference that is established between the two helium baths as a result of the temperature difference. The density of helium-4 below 2 K is $\rho=0.145$ g cm^{-3}, and the entropy per gram is $s=5\times10^{-2}$ Jg^{-1} K.

20.2 A conducting bar carries an electric current of density J_e and has a temperature gradient maintained along its length. Consider a short region of the bar with the upper and lower ends at $T+\Delta T$ and T, respectively. Derive the Thomson effect result for the heat per unit area absorbed from, or emitted to, the surroundings by this small region of the bar in terms of the Thomson coefficient τ defined in Section 20.7.

20.3 A thermocouple is made of two different metals with Seebeck coefficients of 35 μV K^{-1} and -6.5 μV K^{-1}. What is the emf for this thermocouple when the hot and cold junctions are at 300 and 77 K, respectively?

20.4 Design a small Peltier effect refrigerator that provides a cooling power of 10 W. Base your design on a stack of selected semiconducting materials that have large Seebeck coefficients of +320 and -280 μV K^{-1}, respectively. Give the design details. The refrigerator should operate with a current supply of a few amperes. Will cooling of some of the contacts be necessary?

Appendix A
Useful Mathematical Relationships

FINITE SERIES SUMMATIONS

1. Arithmetic progression $S_n = (a + 2a + 3a + \cdots + na) = \dfrac{1}{2}n(n+1)a$.

2. Geometric progression $S_n = (a + ra + r^2a + \cdots + r^{n-1}a) = \left(\dfrac{a(1-r^n)}{1-r}\right)$ for $r \neq 1$.

$$S_n \approx \frac{a}{1-r} \text{ for } 0 < r < 1.$$

3. Riemann zeta function

   This function is defined by $\zeta(p) = \displaystyle\sum_{n=1}^{\infty}\left(1/n^p\right)$ for $p > 1$. Values are $\zeta(3/2) = 2.612$, $\zeta(2) = \pi^2/6$, $\zeta(3) = 1.202$, $\zeta(4) = \pi^4/90$. A general expression may be given in terms of Bernoulli numbers, but the values quoted are sufficient for the material covered in this book.

STIRLING'S FORMULA FOR THE LOGARITHM OF *N*!

$$\ln N! = N\ln N - N + \ln(2\pi N)^{1/2} \approx N\ln N - N \text{ for large } N.$$

DEFINITE INTEGRALS INVOLVING EXPONENTIAL FUNCTIONS

$$\int_0^{\infty} xe^{-ax}\,dx = \frac{1}{a^2} \qquad \int_0^{\infty} e^{-ax^2}\,dx = \frac{1}{2}\sqrt{\frac{\pi}{a}},$$

$$\int_0^{\infty} xe^{-ax^2}\,dx = \frac{1}{2a} \qquad \int_0^{\infty} x^2 e^{-ax^2}\,dx = \frac{\sqrt{\pi}}{4a^{3/2}},$$

$$\int_0^{\infty} x^3 e^{-ax^2}\,dx = \frac{1}{2a^2} \qquad \int_0^{\infty} x^4 e^{-ax^2}\,dx = \frac{3\sqrt{\pi}}{8a^{5/2}}.$$

General expressions for these integrals may be given in terms of the gamma function

$$\Gamma(n) = \int_0^{\infty} x^{n-1}e^{-x}\,dx \text{ for } n > 0.$$

For integral n, the values of the gamma function are obtained using $\Gamma(n+1) = n!$ for $n = 0, 1, 2, \cdots$ and the recursion formula $\Gamma(n+1) = n\Gamma(n)$.

For fractional values of $n = m + \dfrac{1}{2}$, where m is an integer, we have $\Gamma\left(m + \dfrac{1}{2}\right) = \left[1 \cdot 3 \cdot 5 \cdots (2m-1)/2^m\right]\sqrt{\pi}$ for $m = 1, 2, \ldots$. This gives, for example, $\Gamma\left(\dfrac{3}{2}\right) = \dfrac{1}{2}\sqrt{\pi}$.

For the case $m = 0$, $\Gamma\left(\dfrac{1}{2}\right) = \sqrt{\pi}$.

In terms of the gamma function, the following expressions are obtained for integrals of interest:

$$\int_0^\infty x^n e^{-ax^2}\,\mathrm{d}x = \frac{\Gamma[(m+1)/2]}{2a^{[(m+1)/2]}} \qquad \int_0^\infty x^m e^{-ax}\,\mathrm{d}x = \frac{\Gamma(m+1)}{a^{m+1}}.$$

These expressions may be used to confirm the values for the integrals listed above.

In Chapter 15, the following integral is used in the derivation of the Stefan–Boltzmann law and in the discussion of the Debye model for the specific heat of solids,

$$\int_0^\infty \frac{x^3\,\mathrm{d}x}{e^x - 1} = \frac{\pi^4}{15}.$$

Appendix B
The Binomial Distribution

Consider a random process with two possible outcomes or events labeled X and Y governed by probabilities p and q, respectively, with $p + q = 1$. The probability that in N trials n outcomes X will be obtained is given by the binomial distribution

$$P(n) = \left(\begin{array}{c} N \\ n \end{array} \right) p^n q^{(N-n)}. \tag{B1}$$

The binomial coefficient $\left(\begin{array}{c} N \\ n \end{array} \right) = N! / \left(n!(N-n)! \right)$ occurs in the binomial expansion $\left(p + q \right)^N = \sum_{n=0}^{N} \left(\begin{array}{c} N \\ n \end{array} \right) p^n q^{N-n}$, and the probability distribution derives its name from this identification. The probability distribution is normalized for $p + q = 1$ because it follows immediately from the binomial expansion that in this case $\sum_{n} P(n) = 1$.

For large N, the mean number of events of type 1 is $\langle n \rangle = Np$. This result is obtained with the use of the binomial expansion as follows.

The mean value of n is given by

$$\langle n \rangle = \sum_{n=0}^{N} n \left(\begin{array}{c} N \\ n \end{array} \right) p^n q^{N-n} = \sum_{n=0}^{N} P \frac{\partial}{\partial p} \left(\begin{array}{c} N \\ n \end{array} \right) p^n q^{N-n}$$

$$= p \frac{\partial}{\partial p} \sum_{n=0}^{N} \left(\begin{array}{c} N \\ n \end{array} \right) p^n q^{N-n} = p \frac{\partial}{\partial p} (p+q)^n = Np. \tag{B2}$$

Summation and partial differentiation have been interchanged in the third step, and it is assumed that $p + q = 1$.

The dispersion $\sigma^2 = \langle \Delta n^2 \rangle = \langle n^2 \rangle - \langle \infty \rangle$ may be obtained using a similar approach by calculating $\langle n^2 \rangle$ and making use of the above expression for $\langle n \rangle$.

$$\langle \Delta n^2 \rangle = \left[\sum_{n=0}^{N} n^2 \left(\begin{array}{c} N \\ n \end{array} \right) p^n q^{N-n} \right] - N^2 p^2 = \left[\sum_{n=0}^{N} \left(p \frac{\partial}{\partial p} \right)^2 \left(\begin{array}{c} N \\ n \end{array} \right) p^n q^{N-n} \right] - N^2 p^2 . \tag{B3}$$

Interchanging the order of summation and integration leads to

$$\left\langle \Delta n^2 \right\rangle = \left[\left(p\frac{\partial}{\partial p} \right)^2 \sum_{n=0}^{N} \left(\begin{array}{c} N \\ n \end{array} \right) p^n q^{N-n} \right] - N^2 p^2$$

$$= \left[\left(p\frac{\partial}{\partial p} \right)^2 (p+q)^N \right] - N^2 p^2 = Npq. \tag{B4}$$

The ratio $\sigma/n = \sqrt{Npq}/Np = \left(1/\sqrt{N}\right)\left(\sqrt{q/p}\right)$ shows that for large N, the root mean square width of the distribution, expressed as a fraction of the mean value, decreases as $1\sqrt{N}$. Note that the factor $\sqrt{q/p}$ is often of order unity.

We expect the probability distribution given in Equation B1 to exhibit a maximum at $\langle n \rangle$. For very large N, it is permissible to replace the discrete variable n by a continuous variable. In this limit, the probability of an outcome in the range n to $n + dn$ is $P(n)dn$, with the interval dn spanning a range of values of n.

The probability density $P(n)$ is expected to peak at $n = \int nP(n)dn = Np$. We can confirm this peak value by differentiating $P(n)$ to obtain the value of n at the extremum, which we designate for the moment as $\bar{n}$. It is convenient to use the function $\ln P(n)$ in performing the differentiation because this avoids the factorials. Applying Stirling's formula, we obtain

$$\left(\frac{d \ln P(n)}{dn} \right)_{\bar{n}} = \left(\frac{d}{dn} \left(N \ln N - N - n\ln n + n - (N-n)\ln(N-n) + (N-n) + n\ln p + (N-n)\ln q \right) \right)_{\bar{n}}$$

$$= \left(-\ln n + \ln(N-n) + \ln p - \ln q \right)_{\bar{n}} = 0 \tag{B5}$$

Rearranging it follows that $\bar{n} = Np = n$ as expected.

GAUSSIAN APPROXIMATION TO THE BINOMIAL DISTRIBUTION

For large N, we now show that the binomial distribution can be well approximated by the Gaussian function for values of n not too far from the peak. We use a Taylor expansion to approximate the probability distribution and again find it convenient to consider $\ln P(n)$ because this is a more slowly varying function than $P(n)$ and the expansion may be expected to converge rapidly. Retaining terms up to second order, the expansion has the form

$$\ln P(n) = \ln P(\langle n \rangle) + \left(\frac{d \ln P(n)}{dn} \right)_n (n - \langle n \rangle) + \frac{1}{2} \left(\frac{d^2 \ln P(n)}{dn^2} \right)_n (n - \langle n \rangle)^2 + \dots . \tag{B6}$$

The first derivative of $\ln P(\langle n \rangle)$ evaluated at $\langle n \rangle$ is zero, and the second derivative is given by

$$\left(\frac{d^2 \ln P(n)}{dn^2} \right)_n = -\frac{1}{\langle n \rangle} - \frac{1}{(N - \langle n \rangle)} = -\frac{N}{\langle n \rangle(N - \langle n \rangle)} = -\frac{1}{Npq}. \tag{B7}$$

Substituting in the Taylor expansion and then taking antilogarithms give the Gaussian distribution form $P(n) = P\{\langle n \rangle\}e^{-(n-\langle n \rangle)^2/2Npq}$. The coefficient $P(\langle n \rangle)$ may be obtained from the normalization condition by integrating over the range of 0–N. For sufficiently large N, the upper limit may be extended to infinity because the integral will converge. With the standard integral form given in Appendix A, we obtain finally in this approximation,

$$P(n) = \frac{1}{\sqrt{2\pi Npq}} e^{-(n-Np)^2/2Npq}. \tag{B8}$$

This has the Gaussian distribution form $P(x) = 1/\sqrt{2\pi\sigma} \, e^{-(x-\bar{x})^2/2\sigma^2}$, and we identify the dispersion as $\sigma^2 = Npq$ in agreement with the value obtained directly from the binomial distribution.

For large N, the Gaussian approximation is very close in form to the exact binomial distribution. Figure B1.1 for $N = 30$ shows the exact binomial distribution $P(n)$ versus n as the plotted points and the Gaussian approximation as the curve. The agreement is seen to be very good, showing that the Gaussian approximation works well even for fairly small N.

It is clear that for very large values of N, comparable with Avogadro's number, the Gaussian approximation gives an excellent fit to the binomial distribution.

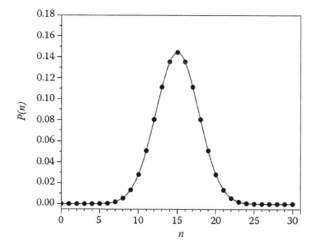

FIGURE B1.1   The binomial distribution for $N = 30$ and $p = q = 0.5$ shown as plotted points and the Gaussian approximation as the full curve.

Appendix C
Elements of Quantum Mechanics

The time-dependent Schrödinger equation for a particle of mass m moving in a fixed potential $V(r)$ is

$$-\frac{\hbar^2}{2m}\nabla^2\Psi(r,t)+V(r)\Psi(r,t)=i\hbar\frac{\partial\Psi(r,t)}{\partial t}. \tag{C1}$$

The wave function $\Psi(r, t)$ gives the probability amplitude of finding the particle at position r at time t. For the purposes of this book, we are generally interested in time-independent or stationary states. Writing the wave function as a product of a spatial part and a time-dependent part as $\Psi(r, t) = \psi(r)\,e^{-i\omega t}$ and substituting into Equation C1 lead to the time-independent Schrödinger equation

$$-\frac{\hbar^2}{2m}\nabla^2\psi(r)+V(r)\psi(r)=E\psi(r). \tag{C2}$$

In more compact form, we have $\mathcal{H}\psi(r)=E\psi(r)$, with the Hamiltonian operator defined as $\mathcal{H}=-\left(\hbar^2/2m\right)\nabla^2+V(r)$. The energy eigenvalues $E=\hbar\omega$ are obtained by solving this equation for a given potential function $V(r)$. Note that $|\Psi(r,t)|^2=\left|\psi(r)e^{-i\omega t}\right|^2=|\psi(r)|^2$, showing that the eigenstates are stationary states with a time-independent probability density for finding the particle at any given point.

PARTICLE IN A BOX EIGENSTATE AND EIGENVALUE

Consider a particle moving in one dimension in a potential well $V(x) = 0$ with infinite walls at $x = 0$ and $x = L$. This is the one-dimensional box situation, and the corresponding time-independent Schrödinger equation is

$$\frac{d^2\psi(x)}{dx^2}=\frac{2mE}{\hbar^2}\psi(x)=-\kappa^2\psi(x), \tag{C3}$$

with $\sqrt{2mE}\,/\,\hbar=|\kappa|$. The general solution to Equation C3 may be written as $\psi(x)=C_1e^{i\kappa x}+C_2e^{-i\kappa x}$.

Using the boundary conditions $\psi(0) = 0$ gives $C_1 = -C_2$, and it follows that $\psi(x) = C_1\sin\kappa x$.

At the other boundary of the well, we require $\psi(L) = 0$, which implies $\sin\kappa L = 0$, and this leads to $\kappa L = n\pi$ or $\kappa_n = n\pi/L$, with $n = 1, 2, 3,...$ The boundary conditions lead to the quantization of κ and E. The situation is similar to the case of vibrational standing waves on a string stretched between two fixed points or nodes. The energy eigenvalues for the particle are

$$E_n=\frac{\hbar^2\kappa_n^2}{2m}=\frac{n^2h^2}{8mL^2}. \tag{C4}$$

The probability of finding the particle in the box is given by $\int_0^L\left|\psi(x)\right|^2 dx=1$, and inserting $\psi(x) = C_1\sin\kappa x = C_1\sin(n\pi x/L)$ in the integral gives the constant $C_1=\sqrt{2/L}$.

For a particle in a three-dimensional box with infinite potential barriers at the sides, the energy eigenvalues may be obtained in a similar fashion to the one-dimensional box case, and we obtain

$$E_n = \frac{h^2}{8m}\left(\frac{n_x^2}{L_x^2}+\frac{n_y^2}{L_y^2}+\frac{n_z^2}{L_z^2}\right), \tag{C5}$$

with n_x, n_y, n_z = 1,2,3,.... For a cubical box $L_x = L_y = L_z = L$ and putting $V = L^3$, Equation C5 becomes

$$E_n = \frac{h^2}{8mV^{2/3}}\left(n_x^2 + n_y^2 + n_z^2\right). \tag{C6}$$

THE HARMONIC OSCILLATOR

The potential function for a one-dimensional harmonic oscillator is $V(x)=\frac{1}{2}kx^2$, with k as the effective spring constant for a particular system such as a diatomic molecule. The displacement x from the origin can take positive or negative values, and the potential has the form of a parabolic well with a minimum at $x = 0$. The energy levels for a particle of mass m moving in this static potential are given by the time-independent Schrödinger equation

$$-\frac{\hbar^2}{2m}\frac{d^2\psi(x)}{dx^2}+\left(\frac{1}{2}kx^2\right)\psi(x)=E\psi(x). \tag{C7}$$

Finding solutions to the harmonic oscillator Equation C7 is not straightforward, and we simply outline the procedure. Further details can be found in texts on quantum mechanics.

For a given energy, the wave function will fall off rapidly with increasing x because of the strong dependence of $V(x)$ on x. It is convenient to rearrange Equation C7 in the following form:

$$\left(\frac{\hbar^2}{2m}\frac{d^2}{dx^2}+E-\frac{1}{2}kx^2\right)\psi(x)=0. \tag{C8}$$

In the asymptotic large x limit, Equation C8 can be written to a good approximation as

$$\left(\frac{\hbar^2}{2m}\frac{d^2}{dx^2}-\frac{1}{2}kx^2\right)\psi(x)=0. \tag{C9}$$

This is similar in form to the differential equation $[(d^2/dx^2)-(x^2-1)]f(x)=0$, which has the following solution $f(x)= Ae^{-(1/2)x^2}$, where A is a constant. We expect the solution to Equation C9 to have the Gaussian form $\psi(x)=Ce^{-(1/2)\alpha x^2}$, with C and α as constants to be determined. Substituting $\psi(x)$ in Equation C1.9, and canceling the common exponential factor and the constant C lead to $(\alpha^2 - mk/\hbar^2)$ $x^2 - \alpha = 0$. In the large x limit, the first term is dominant, and we require that the coefficient of x^2 is zero, giving $\alpha = \sqrt{mk}/\hbar$.

For small x, a solution to the Schrödinger equation may be obtained by multiplying the Gaussian function by a polynomial in x of the form $H_n(x) = (a_0+a_1 x+a_2 x^2+...+a_n x^n)$, with the order n to be determined for each eigenstate. Note that successive terms in the polynomial have even and odd parity, and because the Gaussian function has even parity under a sign change of x, the parity of the wave function is determined by the parity of the nonzero terms in the polynomial. The polynomials $H_n(x)$ are known as Hermite polynomials. The simplest solution is obtained by retaining only the zeroth-order term in the polynomial, giving $\psi(x)=a_0Ce^{-(1/2)\alpha x^2}$. Substituting in Equation C8 and using $\alpha = \sqrt{mk}/\hbar$, we obtain the eigenvalue

$$E_0 = \frac{1}{2}\hbar\sqrt{k/m} = \frac{1}{2}\hbar\omega_0. \tag{C10}$$

We have put $\omega_0 = \sqrt{k/m}$, which is the angular frequency of the classical harmonic oscillator. Equation C10 gives the ground-state energy for the quantum mechanical harmonic oscillator. Following a similar procedure to the above, we obtain the next eigenvalue using the first-order term in the polynomial, corresponding to odd parity, so that $\psi(x) = a_1 x C e^{-(1/2)\alpha x^2}$, and find

$$E_1 = \frac{3}{2}\hbar\omega_0. \tag{C11}$$

This shows that the first excited state is at an energy $\hbar\omega_0$ above the ground state.

Proceeding in this fashion, we can obtain successive eigenvalues by choosing alternating even and odd parity terms in the polynomial function. The next eigenvalue is found using the zeroth- and second-order terms $(a_0 + a_2 x^2)$, and this leads to $E_2 = \frac{5}{2}\hbar\omega_0$. We find that the energy eigenvalues are given in terms of the quantum number n by the expression

$$E_n = \left(n + \frac{1}{2}\right)\hbar\omega_0, \text{ where } n = 0,1,2,3,\cdots \tag{C12}$$

The eigenfunctions are obtained using the normalization condition

$$\int_{-\infty}^{\infty} |\psi(x)|^2 \, dx = 1.$$

For the ground-state wave function, we obtain, for example, the Gaussian function

$$\psi_0(x) = \left(\frac{\alpha}{\sqrt{\pi}}\right)^{1/2} e^{-(1/2)\alpha x^2}. \tag{C13}$$

This form may be readily verified using the integral for the Gaussian function given in Appendix A. Wave functions for the excited states are usually expressed in terms of the appropriate Hermite polynomial and are given in texts on quantum mechanics.

STATE VECTORS AND DIRAC NOTATION

The quantum state of a system is specified by a state vector, which is independent of the basis states used to describe the system. Projections of the state vector onto the basis states give the components of the state vector in a particular basis or representation much like the components of classical vector with respect to a particular set of axes. The Dirac "bra-ket" notation provides a convenient and compact way for specifying the quantum state of a system. In general, we specify a state vector in this notation by means of the ket $|\phi\rangle$, where ϕ specifies the particular eigenstate for the system considered. For the one-dimensional particle in a box case, for example, we can specify a given state as $|n\rangle$ using the single quantum number n. For three-dimensional N particles, we require $3N$ quantum numbers to specify the state vector, which is written as $|n_{1x}, n_{1y}, n_{1z}; n_{2x} n_{2y} n_{2z}; \ldots; n_{Nx}, n_{Ny}, n_{nz}, \rangle$.

Appendix D
The Legendre Transform in Thermodynamics

INTRODUCTION TO THE LEGENDRE TRANSFORM

The three thermodynamic potentials H, F, and G that are introduced in Chapter 7 can be shown to take their particular forms by making use of the Legendre transform. For a function $F(x)$ of a single variable x, Legendre transforms allow us to represent the function $F(x)$ by another function $L(s)$, where s is a variable given by $s = dF(x)/dx$, which is the slope of the original function at a given point. The Legendre transform $L(s)$ of $F(x)$ is defined by the relation

$$L(s) = F\big(x(s)\big) - s(x)x(s), \tag{D1}$$

with $x(s)$ as the value of x for which the slope s is obtained. This unusual form can be understood by making use of the geometrical representation shown in Figure D1 in which $s(x)$ is the slope of the tangent to the curve at a point $x(s)$ and $L(s)$ is the intercept for this tangent.

Legendre transforms are applied to functions which are convex (with $\partial^2 F/\partial x^2 > 0$) for which the slope increases with an increase in x. Note that the Legendre transform may be defined with a change in sign for mathematical reasons and the definition given above is chosen for our applications in thermodynamics. Before considering the thermodynamic potentials, we examine the Legendre transform for a function of two variables. The procedure can of course be extended to any number of variables, but two variables are often sufficient for our purposes. Consider a function $F(x, y)$ of the variables x and y. The function can be represented as a surface in a three-dimensional plot. If one of the variables is held fixed, the situation is similar to that discussed above and as depicted in Figure D1. More generally, the function can be represented as the envelope of the set of tangent

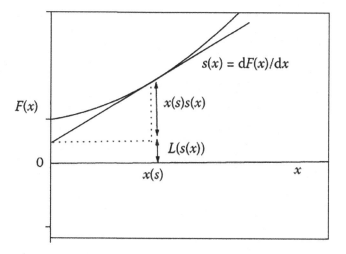

FIGURE D1 Graphical illustration of the Legendre transform of a function $F(x)$ of a single variable x. The relationship $L(s) = F(x(s)) - s(x)x(s)$, which is given in the text, is readily established from this plot. The set of slopes s and intercepts $L(s)$ specifies the function completely.

planes to the surface. For a given point (x, y), we have the slopes $s_x = \partial F/\partial x$ and $s_y = \partial F/\partial y$, and it follows that

$$L(s_x, s_y) = F(x, y) - s_x x - s_y y. \tag{D2}$$

THE LEGENDRE TRANSFORM AND THERMODYNAMIC POTENTIALS

In applying the Legendre transformation to thermodynamic relationships for a fluid system, we choose as our function the internal energy $E(S, V)$ expressed in terms of the entropy S and the volume V. As discussed in Section 3.12, this choice corresponds to the energy representation for a system with a fixed number of particles N. If S is kept fixed, then we obtain the partial Legendre transform as

$$L = E - \left(\frac{\partial E}{\partial V}\right)_S V = E + PV = H. \tag{D3}$$

The identity $P = -(\partial E/\partial V)_S$ follows from the fundamental relation (Equation 3.18) $TdS = dE + PdV$, and we have made use of the definition of the enthalpy given in Section 7.1. We see that H is simply the partial Legendre transform of E with S kept constant. In Equation D3, the extensive state function S, which is in general not readily controlled, has been replaced by the intensive variable P, which is readily controlled as an independent variable.

The Helmholtz potential $F = E - TS$ as defined in Equation 7.2 is obtained as the partial Legendre transform of E with V held constant. We find in this case

$$F = E - \left(\frac{\partial E}{\partial S}\right)_V S = E - TS. \tag{D4}$$

Processes in which E can change at constant V correspond to the canonical ensemble case in statistical physics as introduced in Chapter 10. The bridge relation between F and the partition function Z can be seen to follow in a natural way.

The Gibbs potential G defined in Equation 7.3 is the complete Legendre transform of E corresponding to both S and V being allowed to change

$$G = E - \left(\frac{\partial E}{\partial S}\right)_V S - \left(\frac{\partial E}{\partial V}\right)_S V = E - TS + PV. \tag{D5}$$

The three thermodynamic potentials given above are of great importance in the development of thermal physics as discussed in Chapters 3, 7, and elsewhere in this book.

Finally, to obtain the grand potential $\Omega_G = -PV$, which is used in Chapter 11 in connection with the grand canonical distribution, we allow the internal energy E to be a function not only of S and V but in addition of particle number N so that $E = E(S, V, N)$. The grand canonical ensemble corresponds to a set of systems each in thermal and diffusive contact with a reservoir at temperature T and with chemical potential μ. We therefore allow S and N to vary but keep V fixed. In this case, the Legendre transform of E is given by

$$\Omega_G = E - \left(\frac{\partial E}{\partial S}\right)_{V,N} S - \left(\frac{\partial E}{\partial N}\right)_{V,S} N = E - TS - \mu N. \tag{D6}$$

From the expressions for F and G, it follows that $\Omega_G = F - G = -PV$, as shown in Chapter 11.

Appendix E
Recommended Texts on Statistical and Thermal Physics

INTRODUCTORY LEVEL

Betts, D.S. and Turner, R.E., *Introductory Statistical Mechanics,* Addison Wesley, Wokingham, 1992.

Blundell, S. and Blundell, K.M., *Concepts in Thermal Physics,* Oxford University Press, New York, 2006.

Callen, H.B., *Thermodynamics and an Introduction to Thermostatistics,* second edition, Wiley, New York, 1985.

Huang, K., *Introduction to Statistical Physics,* second edition, Chapman and Hall/CRC Press, Boca Raton, FL, 2010.

Kittel, C. and Kroemer, H., *Thermal Physics,* second edition, W.H. Freeman, New York, 1980.

Mandl, F., *Statistical Physics*, second edition, Wiley, Chichester/New York, 1988.

Reif, F., *Fundamentals of Statistical and Thermal* Physics, McGraw-Hill, New York, 1965.

Schroeder, D.V., *An Introduction to Thermal Physics*, Addison Wesley Longman, New York, 2000.

Zemansky, M.W. and Dittman, R.H., *Heat and Thermodynamics,* seventh edition, McGraw-Hill, New York, 1997.

ADVANCED LEVEL

Landau, L.D. and Lifshitz, E.M., *Statistical Physics*, third edition (trans. Sykes, J.B. and Kearsley, M.J.), Pergamon Press, Oxford, 1980.

Pathria, R.K., *Statistical Mechanics,* second edition, Butterworth-Heinemann, Oxford, 1996.

Reichl, L., *A Modern Course in Statistical Physics,* second edition, Wiley, New York, 1998.

COMPUTER SIMULATIONS

Gould, H, Tobochnik, J. and Christian, W., *An Introduction to Computer Simulation Methods: Applications to Physical Systems*, third edition, Addison-Wesley, Reading MA, 2006.

Index

Printed in the United States
by Baker & Taylor Publisher Services